GWRC Report Series

Waterborne Pathogens

AF631572

GWRC Report Series

Waterborne Pathogens:
Review for the Drinking-water Industry

Prepared by
Emmanuelle Guillot, Jean-François Loret
Suez Environnement – CIRSEE

Publishing
London • New York

Global Water
Research Coalition

Published by **IWA Publishing**
Alliance House
12 Caxton Street
London SW1H 0QS, UK
Telephone: +44 (0)20 7654 5500
Fax: +44 (0)20 7654 5555
Email: publications@iwap.co.uk
Web: www.iwapublishing.com

First published 2010
© 2010 Global Water Research Coalition

Cover design by www.designforpublishing.co.uk
Typeset by Alden Prepress Services Private Limited.

Apart from any fair dealing for the purposes of research or private study, or criticism or review, as permitted under the UK Copyright, Designs and Patents Act (1998), no part of this publication may be reproduced, stored or transmitted in any form or by any means, without the prior permission in writing of the publisher, or, in the case of photographic reproduction in accordance with the terms of licences issued by the Copyright Licensing Agency in the UK, or in accordance with the terms of licenses issued by the appropriate reproduction rights organization outside the UK. Enquiries concerning reproduction outside the terms stated here should be sent to IWA Publishing at the address printed above.

The publisher makes no representation, express or implied, with regard to the accuracy of the information contained in this book and cannot accept any legal responsibility or liability for errors or omissions that may be made.

Disclaimer
The information provided and the opinions given in this publication are not necessarily those of IWA and should not be acted upon without independent consideration and professional advice. IWA and the Author will not accept responsibility for any loss or damage suffered by any person acting or refraining from acting upon any material contained in this publication

This study was jointly funded by GWRC members. GWRC and its members assume no responsibility for the content of the research study reported in this publication or for the opinion or statements of fact expressed in the report. The mention of trade names for commercial products does not represent or imply the approval or endorsement of GWRC and its members. This report is presented solely for informational purposes

British Library Cataloguing in Publication Data
A CIP catalogue record for this book is available from the British Library

Library of Congress Cataloging-in-Publication Data
A catalog record for this book is available from the Library of Congress

ISBN 10: 1843391791
ISBN 13: 9781843391791

Contents

© 2010 Global Water Research Coalition. Waterborne Pathogens: Review for the Drinking-water Industry. Prepared by Emmanuelle Guillot and Jean-François Loret. ISBN: 9781843391791. Published by IWA Publishing, London, UK.

Foreword

The Global Water Research Coalition (GWRC) is a non-profit organisation that serves as a collaborative mechanism for water research. The benefits that the GWRC offers its members are water research information and knowledge. The Coalition focuses on water supply and wastewater issues and renewable water resources: the urban water cycle.

GWRC was officially formed in April 2002 with the signing of a partnership agreement and a partnership agreement was signed with the U.S. Environmental Protection Agency in July 2003. GWRC is affiliated with the International Water Association (IWA).

The members of the GWRC are: Anjou Recherche – Water Operations Research Center of Veolia Water (France); EAWAG – Swiss Federal Institute for Aquatic Science and Technology; KWR – Watercycle Research Institute (Netherlands); PUB – National Water Agency of Singapore; SUEZ Environmental – CIRSEE – International Research Center on Water and Environment (France); Stowa – Foundation for Applied Water Management Research (Netherlands); TZW – Water Technology Center of the German Waterworks Association; UKWIR – UK Water Industry Research; Water Environment Research Foundation (USA); WQRA – Water Quality Research Australia; WRC – Water Research Commission (South Africa); Water Research Foundation (USA); WateReuse Foundation (USA); WSAA – Water Services Association of Australia.

These organisations have national research programs addressing different parts of the water cycle. They provide the impetus, credibility, and funding for the GWRC. Each member brings a unique set of skills and knowledge to the Coalition. Through its member organisations GWRC represents the interests and needs of 500 million consumers.

Microbial contamination is one of the major risks drinking water suppliers have to face, in developing and developed countries as well. Outbreaks of waterborne diseases affecting millions of people worldwide are reported from countries at all levels of development and from all types of water supply.
Because of the implication for public health, the members of the Global Water Research Coalition selected waterborne pathogens as one of the priority areas of the GWRC's research agenda.

In choosing waterborne pathogens as a priority area, the GWRC was interested in identifying the potential for collaborative research. The purpose of this book is to provide an updated, comprehensive review of current knowledge on a selection of pathogens of interest for the drinking water industry, and to identify the remaining gaps of knowledge and thus the necessary research to be conducted. This review is based on the literature published up to 2008. Emphasis has been placed on the information needed to assess and manage the risks related to each of these pathogens in drinking water production and distribution. It is hoped that this book will be useful to all who are active in research in the field of microbial water quality and public health.

© 2010 Global Water Research Coalition. Waterborne Pathogens: Review for the Drinking-water Industry. Prepared by Emmanuelle Guillot and Jean-François Loret. ISBN: 9781843391791. Published by IWA Publishing, London, UK.

Acknowledgements

The Global Water Research Coalition wishes to express its appreciation to Suez Environnement – CIRSEE for acting as the GWRC's lead organisation for this research area. The efforts by the authors of this book Emmanuelle Guillot and Jean-François Loret are gratefully acknowledged.

The authors are especially grateful to Li Chen and Alexia Borrel for their help in the collection and review of literature references, and also to the following persons for reviewing the document and supplying useful information:

Misha Hasan (AWWA Research Foundation, USA), Chris Saint (The Cooperative Research Centre for Water Quality and Treatment, Australia), Beate Hambsch (DVGW Technologiezentrum Wasser, Germany), Hans-Peter Füchslin (EAWAG, Switzerland), Gertjan Medema (KIWA Water Research, The Netherlands), Zdravka Do-Quang, Nadine Dumoutier, Jean-Michel Laîné, Olivier Schlosser (Suez Environnement, France), Albert Bosch (Universitat de Barcelona, Spain), Philippe Hartemann (Université de Nancy, France), Fanus Venter (University of Pretoria, on behalf of the Water Research Commission, South Africa), Liu Wen-Tso, Jiang Yong Hu (University of Singapore, on behalf of the Centre for Advanced Water Technology – Public Utility Board of Singapore), Gin Yew-Hoong Karina (Nanyang Technological University, on behalf of the Centre for Advanced Water Technology – Public Utility Board of Singapore), Ann Grimm, Eric Villegas, Elizabeth Hilborn, Sandhya Parshionikar, Jim Sinclair, Jennifer Best, Tom Carpenter, Jeet Saxena (US Environmental Protection Agency, USA), Karine Delabre, Jean-Claude Joret, Florence Ménard-Szczebara (Veolia, France), Robert Pitchers (Water Research Centre – NSF, on behalf of UKWIR, United Kingdom).

© 2010 Global Water Research Coalition. Waterborne Pathogens: Review for the Drinking-water Industry. Prepared by Emmanuelle Guillot and Jean-François Loret. ISBN: 9781843391791. Published by IWA Publishing, London, UK.

Introduction

Infectious diseases caused by pathogenic bacteria, viruses and protozoa are the most common and widespread health risk associated with drinking water. Most waterborne pathogens are introduced into drinking water supplies by human or animal faeces (enteric pathogens) but they can also exist naturally in water environments as indigenous aquatic micro-organisms.

Controlling the risks related to these pathogens is a permanent challenge for the water industry. In addition to the constantly evolving range of pathogens to consider, assessing and managing such risks requires the integration of information issued by a wide range of disciplines. The necessary knowledge is however still incomplete for most pathogens, and research efforts are necessary to fill the remaining gaps of knowledge.

The purpose of this book is to provide an updated, comprehensive review of current knowledge on a selection of pathogens of interest for the drinking water industry, and to identify the remaining gaps of knowledge and thus the necessary research to be conducted. This review is based on the literature published up to 2008. Emphasis has been placed on the information needed to assess and manage the risks related to each of these pathogens in drinking water production and distribution.

The pathogens considered in this review were selected on the basis of:

- Their recognized or highly suspected transmission by drinking water through ingestion, inhalation, or contact
- Their recognized pathogenic character for humans
- The severity of their health effects.

Due to the similarity between these selection criteria and those used by the USEPA for the establishment of the drinking water Contaminant Candidate List (CCL), the micro-organisms listed in the first two versions of the CCL (USEPA 1998, USEPA 2005) were automatically included in this review.

The micro-organisms identified as the etiologic agents in waterborne disease outbreaks in Europe, Canada, and the USA (Risebro 2006, Lee 2002, Blackburn 2004, and for *Legionella* in Europe: http://www.ewgli.org), were also considered in this study. *Burkholderia* was added to the list because of its implication in outbreaks in Australia and southeast Asia (Inglis 2000), and hepatitis E virus and *Cyclospora* because of their recent recognition as emerging water-related pathogens (WHO 2003).

As a result of this selection, the micro-organisms addressed in this study include enteric bacteria: *Campylobacter* spp., *Escherichia coli* O157H7, *Helicobacter pylori*, *Salmonella* spp., *Shigella* spp., *Yersinia* spp., aquatic bacteria : *Aeromonas* spp., *Burkholderia pseudomallei*, cyanobacteria, *Legionella* spp., non-tuberculosis mycobacteria, enteric viruses : adenovirus, calicivirus, enterovirus (coxsackievirus and echovirus), hepatitis A virus (HAV), hepatitis E virus (HEV), rotavirus, enteric protozoa: *Cryptosporidium*, *Giardia*, *Cyclospora*, microsporidia and the aquatic protozoan, *Acanthamoeba* spp.

This list is not exhaustive and it should evolve in the future, since the number of known pathogens for which water is a transmission route is increasing and new pathogens continue to be discovered.

Each micro-organism in this review is described in a summary fact sheet consisting of the following items: microbiology, human health effects, geographical distribution, epidemiology, ecology, inactivation/removal, surrogates, environmental detection and research needs.

Since much remains unknown about many aspects that are common to viruses, the document features a general section on «virus research needs».

Finally, the conclusion gives an overview of the main gaps in current knowledge on these waterborne pathogens, but does not propose any priority for research, since this would require to take into account financial and political considerations which are not in the scope of this review.

© 2010 Global Water Research Coalition. Waterborne Pathogens: Review for the Drinking-water Industry. Prepared by Emmanuelle Guillot and Jean-François Loret. ISBN: 9781843391791. Published by IWA Publishing, London, UK.

Bacteria

Campylobacter

Health significance	**High**
Persistence in water supplies	**Moderate**
Resistance to chlorine	**Low**
Relative infectivity	**High**
Important animal source	**Yes**

Adapted from table 7.1 Waterborne pathogens and their significance in water supplies (WHO 2004).

Microbiology

General description

Since the late 1970s, *Campylobacter* spp. have become one of the leading causes of acute human gastroenteritis worldwide (Hunter 1997).

Campylobacters are slender, spirally curved, motile, Gram-negative bacteria. They are strictly microaerophilic, non-saccharolytic and oxidase positive.

Classification

There are 16 species and six subspecies assigned to the genus *Campylobacter*, of which the most frequently reported in human disease are *C. jejuni* (formerly known as *Campylobacter fetus* subsp. *jejuni*) and *C. coli.* These species are also called thermophilic campylobacters in that they are able to grow at 43°C.

Human health effects

Clinical features

Campylobacteriosis is an infection of the gastrointestinal tract. Symptoms of the infection include diarrhoea (often including the presence of mucus and blood), abdominal pain, malaise, fever, nausea and vomiting. The illness usually lasts 2 to 5 days but may be prolonged by relapses, especially in adults. Many of those infected show no symptoms. In some individuals reactive arthritis (painful inflammation of the joints) can occur. Rare complications include seizures due to high fever or neurological disorders such as Guillain-Barre syndrome or meningitis. Death from campylobacteriosis is rare and is more likely in the very young, the very old, or those already suffering from a serious disease such as AIDS.

The pathogenic mechanisms of *C. jejuni* are still not completely understood, but it does produce a heat-labile toxin that may cause diarrhea. *C. jejuni* may also be an invasive organism.

Dose – response data

The infective dose is low: it has been estimated that approximately 500 cells of *C jejuni* can cause human illness (Black 1988). Different β-Poisson dose-response models based on experimentation on human volunteers have been described (Medema 1996). More recently, a model fitting the dose – response relationship observed from both the human feeding study and outbreak data has been described (Teunis 2005). This model more appropriately fits the results observed with low doses.

© 2010 Global Water Research Coalition. Waterborne Pathogens: Review for the Drinking-water Industry. Prepared by Emmanuelle Guillot and Jean-François Loret. ISBN: 9781843391791. Published by IWA Publishing, London, UK.

Target populations

Although anyone can suffer from *Campylobacter* infection, children under 5 years and young adults (15–29) are more frequently afflicted than individuals in other age groups.

Diagnosis

Diagnosis is performed by culture of faeces on a selective agar, usually a commercial *Campylobacter* medium containing special supplements. Incubation has to be carried out in a microaerophilic atmosphere (5% oxygen).

Treatment

As with all diarrhoeal illnesses, treatment involves rehydration therapy plus antibiotic therapy for those with severe infection.

Geographical distribution

Worldwide.

Epidemiology

Approximately 5%–14% of all diarrhoea worldwide is believed to be caused by *Campylobacter*. In almost all developed countries, the incidence of human campylobacter infections has been steadily increasing for several years. The reasons for this are unknown (WHO 2004).

In addition, the emergence of antibiotic-resistant strains of *Campylobacter* all over the world has been observed.

Transmission routes

Transmitted by the faecal-oral route.

Campylobacteriosis is considered to be a zoonosis, a disease transmitted to humans from animals (mainly poultry) or animal products. In animals, campylobacters seldom cause disease. People are exposed to the bacteria after consuming contaminated food such as undercooked meats, contaminated water, or raw milk and by direct or indirect contact with animals. Contact with raw poultry and ingestion of undercooked poultry are the major sources of human infections in the USA (Tauxe 1988). Person to person transmission is relatively uncommon (Hunter 1997).

Waterborne outbreaks

Although the majority of cases are sporadic, numerous outbreaks have been related to drinking untreated or inadequately chlorinated water, especially in Nordic-European countries (Rautelin 1990) (Melby 1991) (Koenraad 1997) (Miettinen 2001) (Nygard 2003a) (Kuusi 2004) (Martin 2006) (Jakopanec 2008), in the USA (Craun 1996), in England and Wales (Said 2003) and also in Canada (Alary 1990) and in New Zealand (Stehr-Green 1991). Recently, a waterborne outbreak occurred in Wales due to a lack of maintenance and inspection of the holding reservoir (Richardson 2007).

Two outbreaks involving both *Campylobacter* and *E. coli 0157* infection associated with contamination of a drinking water supply have been reported: the first occurred in 1996 in a Fife village affecting 711 residents (Jones 1996) and the second occurred in May 2000 in Walkerton, Ontario, affecting over 2,000 persons and killing seven (Holme 2003).

One study suggested that swimming in natural sources of water was a novel risk factor (Schonberg-Norio 2004).

Ecology

Regrowth in distribution systems/biofilms

Cannot multiply in water.

However, one study (Buswell 1998) suggested that integration into aquatic biofilms can enhance *Campylobacter* survival.

Occurrence

Campylobacters have frequently been isolated from surface water, such as streams, rivers, and lakes, due to discharges from wastewater treatment plants, runoff from pastures after rain, and direct contamination by wild birds (Brennhovd 1992) (Arvanitidou 1995) (Koenraad 1997) (Jones 2001) (Diergaardt 2004) (Hörman 2004) most frequently at a concentration of less than 10 MPN (Most Probable Number) per 100 mL. They have occasionally been isolated from groundwater (Stanley 1998) and drinking water (Moore 2001) (Arvanitidou 1994) but the numbers detected were low (maximum 0.3 MPN/ 100mL) (Savill 2001).

Studies have shown that sewage and sewage sludge contain campylobacters in concentrations of 10^2 to 10^5 CFU/100 mL and 10^1 to 10^3 CFU/100 mL, respectively (Jones 1990).

Reservoirs

The bacteria are widely distributed and found in most warm-blooded domestic and wild animals. They are common in food animals such as poultry, cattle, pigs, sheep, ostriches, and shellfish and in pets, including cats and dogs. Recently, Axelsson-Olsson *et al.* (Axelsson-Olsson 2005) suggested that amoebae such as *Acanthamoeba polyphaga* may serve as a nonvertebrate reservoir for *C. jejuni* in the environment.

Survival

Several studies reported the persistence of *Campylobacter* spp. in water in a viable but non-cultivable form (VBNC) which is thought to be capable of causing infection (Rollins 1986) (Jones 1991) (Talibart 2000). However the role of this form as a survival mechanism or simply a degenerative form remains controversial.

C. jejuni and *C. coli* survive in cold water at temperatures below 10°C much longer (for up to 4 months) (Rollins 1986) than they survive in water at temperatures higher than 18°C (Korhonen 1991a) (Korhonen 1991b) (Thomas C. 1999).

In water microcosm experiments, factors such as temperature, oxygenation, anaerobic conditions, nutrient and biofilms influenced their survival. Survival was enhanced by the presence of autochthonous microflora and nutrients, and decreasing temperature and anaerobic conditions (Thomas C 1999) (Leclerc 2004). Buswell *et al.* (Buswell 1998) and Lehtola *et al.* (Lehtola 2006a) suggested that the integration into aquatic biofilms can enhance *Campylobacter* survival, even if it is not able to multiply in biofilms.

Recently one study showed that *C. jejuni* cells survived longer when co-cultured with amoebae than when cultured alone (Axelsson-Olsson, 2005).

It has been shown that the origin of the strain is a determining factor for the survival of *C. jejuni* in drinking water at 4°C: poultry isolates showed a prolonged survival, which could be an indication that these strains could play an important role in the transmission of campylobacteriosis through water (Cools 2003).

Inactivation/removal

Published data on inactivation are summarized in annex 1. Blaser *et al.* (Blaser 1986) suggested that disinfection procedures (using chlorine and monochloramine) commonly used for treatment of drinking water to remove coliform bacteria are adequate to eliminate *C. jejuni.*

However, ingestion of *C. jejuni* by protozoans has been shown to protect against subsequent disinfection (King 1988).

Surrogates

As *Campylobacter* spp. are faecally borne pathogens and are not particularly resistant to disinfection, *E. coli* is an appropriate indicator for the presence/absence of *Campylobacter* spp. in drinking water supplies (WHO 2004).

Environmental detection

Traditional culture methods are available but they are time-consuming (4 to 5 days) and laborious, requiring prolonged incubation, selective enrichment and biochemical identification (Fricker 1987). Moreover, they fail to detect viable but not culturable states. A novel method for enrichment of Campylobacter species based on *Acanthamoeba-Campylobacter* co-culture has been reported (Axelsson-Olsson 2007).

Different molecular methods have been developed to detect *Campylobacter* in water.

Direct PCR assays have been developed for *Campylobacter* spp. (Moore 2001) (Moreno 2003a) or specifically for *C jejuni* and *C. coli* (Oyofo 1993) (Kirk 1994). A triplex-PCR assay was developed in order to detect and differentiate the thermophilic species of *Campylobacter* (Khan 2007).

These tests are rapid and sensitive but cannot distinguish viable and dead micro-organisms.

Several approaches have been described to detect only viable campylobacters in water:

- PCR or seminested PCR assays combined with enrichment culture (Hernandez 1995a) (Waage 1999b); a PCR enzyme-linked immunosorbent assay (ELISA) combined with 48 h of enrichment culture to detect viable *C. jejuni* and *E. coli* (Sails 2002); an in situ hybridization assay to detect viable thermotolerant *Campylobacter* species in river water and sewage (Moreno 2003a); a (PNA) probe for the specific detection by FISH of thermophilic species of *Campylobacter* (Lehtola 2005); a reverse transcriptase polymerase chain reaction assay targeting mRNA (Sails 1998).
- A real-time PCR assay for the quantitative detection of *Campylobacter jejuni* in food after 48 h of enrichment culture was developed (Sails 2003), but to date, no rapid quantitative method has been developed to detect viable *Campylobacter* in water.
- A wide range of genotyping methods have been used in molecular epidemiology and taxonomy studies: ribotypes (Hernandez 1996), RAPD fingerprinting (Hernandez 1995b) (Hernandez 1996) (Yakoob 2000) (Karita 2003), PCR and sequencing (Lu 2002), PCR-RFLP (restriction fragment length polymorphism) and AFLP (amplified fragment length polymorphism) (Ronner 2004) (Moreno 2002) or by pulsed field gel electrophoresis (Hanninen 1998) (Steele 1998).

Currently all these available alternative molecular methods are at the experimental stage and validated protocols do not yet exist. In the future, evaluation studies and standardization of these new methods will be necessary.

Research needs

Partly for lack of rapid and specific methods to detect this micro-organism in environmental water, much remains unknown about the epidemiology, ecology and resistance to treatment.

Analytical methods

It is necessary to develop and validate molecular methods for rapid, specific, and quantitative detection of viable *C. jejuni and C. coli* in environmental water. Evaluation studies including comparative studies with the reference culture method and standardization of these new methods will also be necessary.

Route of transmission

There is a lack of knowledge concerning the role of various environments in the transmission of *Campylobacter*. One of the major gaps to our knowledge at present is the relative contribution of each of the known and unknown sources to the overall burden of disease. Research needs should include the contribution from drinking water as opposed to other exposure routes associated with human campylobacteriosis infections.

Occurrence

Occurrence needs to be more documented with levels of contamination on various types of water, including groundwater and treated drinking water.

Reservoirs

Studies on potential new reservoirs for campylobacters need to continue.

Environmental survival

Further studies should investigate the ability of *Campylobacter* to enter the VBNC state, and the role of the VBNC state in this organism.

The impact of the majority of environmental variables on *Campylobacter* strains survival and the risk that this poses to human health remain unclear.

Biofilms: the survival of the pathogen in potable water biofilms and the role of biofilm as a reservoir of *Campylobacter* infections need to be further investigated. How are *Campylobacter*'s infectivity, exposure route and resistance to treatment altered by association with biofilms and by intracellular survival within protozoa?

Inactivation/removal

Specific data on *Campylobacter* removal and inactivation by conventional clarification and disinfection processes are lacking.

Escherichia coli O157:H7

Health significance	**High**
Persistence in water supplies	**Moderate**
Resistance to chlorine	**Low**
Relative infectivity	**High**
Important animal source	**Yes**

Adapted from table 7.1 Waterborne pathogens and their significance in water supplies (WHO 2004).

Microbiology

General description

Whereas *Escherichia coli* bacteria are usually symbiotic as part of the normal intestinal flora of animals and humans, some strains are capable of causing serious diarrhoeal infections in humans. Pathogenic *E. coli* are divided into six groups based on serological and virulence characteristics: enterohemorrhagic (EHEC), enterotoxigenic (ETEC), enteropathogenic (EPEC), enteroinvasive (EIEC), enteroaggregative (EAggEC), and diffuse adherence (DAEC).

The most prominent representative of the EHEC, previously called VTEC (Verotoxin-producing *E. coli*) or STEC (shiga toxin-producing *E. coli*) is *E. coli* O157:H7, first recognized as a human pathogen in 1982 (Riley 1983). Its pathogenicity has been attributed in part to the production of a shiga-like toxin that is cytotoxic for Vero cells.

Classification

E. coli O157:H7 is a specific serotype of *E. coli*. The combination of letters and numbers in the name of the bacterium refers to the specific markers found on its surface and distinguishes it from other types of *E. coli.*

Human health effects

Clinical features

E. coli O157:H7 causes acute bloody diarrhoea and abdominal cramps with little or no fever; sometimes the infection causes nonbloody diarrhoea or no symptoms. The incubation period is 3–4 days, and usually the illness resolves in 5 to 10 days.

In some persons, particularly children under 5 years of age and the elderly, the infection can also cause a complication called hemolytic uremic syndrome (HUS), in which the red blood cells are destroyed and the kidneys fail. About 2%–7% of infections lead to this complication (Mead 1998).

Except in the case of humans with a severe infection, the number of *E.coli* O157 in faecal material will be several orders of magnitude lower than the concentration of other *E. coli* species.

Dose – response data

The infectious dose is estimated at less than 100 bacteria (Griffin 1991) (Keene 1994). Due to its high virulence, no human dose-response study has been conducted with this micro-organism. However,

© 2010 Global Water Research Coalition. Waterborne Pathogens: Review for the Drinking-water Industry. Prepared by Emmanuelle Guillot and Jean-François Loret. ISBN: 9781843391791. Published by IWA Publishing, London, UK.

a β-Poisson model based on animal studies has been shown to be applicable to man (Haas 2000). Application of β-Poisson models developed for *Shigella dysenteriae* has also been proposed (Powell 2000) (Teunis 2004).

Target populations

All persons. Children under 5 years old and the elderly are more likely to develop serious complications.

Diagnosis

Diagnosis is based on clinical presentation and culture of faeces using a modified sorbitol MacConkey agar. A positive isolation is confirmed by biochemical tests and agglutination by specific O157 antiserum (Hunter 1997). The O157:H7 strain is distinguished microbiologically from other *E. coli* by its inability to ferment sorbitol and, most importantly, by its production of "shiga-like" toxins.

Treatment

Most persons recover in 5–10 days without antibiotics or any other specific treatment.

Hemolytic uremic syndrome is a life-threatening condition usually treated in an intensive care unit. Blood transfusions and kidney dialysis are often required. With intensive care, the death rate for hemolytic uremic syndrome is 3%–5%.

Geographical distribution

Worldwide.

Since the first description of this illness in 1982, infections have been reported from more than 30 countries.

Epidemiology

Transmission routes

E. coli O157:H7 is mainly transmitted by food. The major source is ground beef; other sources include unpasteurized milk and juice, sprouts, lettuce, and salami. Transmission via drinking water and recreational water, person-to-person contact as well as direct animal-to-man contact has also been documented (WHO 2004). Contamination of produce may occur after using contaminated water for irrigation or processing.

Waterborne outbreaks

In recent years, several outbreaks of *E. coli* O157:H7 have been associated with consumption of contaminated municipal water, well water, and contact with recreational waters (Dev 1991) (Swerdlow 1992) (Isaacson 1993) (Keene 1994) (Centers for Disease Control 2000) (Chalmers 2000) (Bopp 2003) (Bruce 2003) (Rangel 2005). Two outbreaks involving both *Campylobacter* spp. and *E. coli 0157* co-infection associated with contamination of a drinking water supply have been reported: the first occurred in 1996 in a Fife village affecting 711 residents (Jones 1996) and the second occurred in May 2000 in Walkerton, Ontario, affecting over 2,000 persons and killing seven (Holme 2003).

Despite the potential for large contamination of the environment, waterborne infection is relatively rare because *E. coli* O157 is as susceptible to chlorination as bacterial indicators. In most outbreaks associated with drinking water, the cause was inadequate treatment or contamination of the water supply after treatment and during distribution.

Ecology

Regrowth in distribution systems/biofilms

Not extensively documented.

The persistence of *E. coli* O157:H7 in drinking water biofilms has not yet been studied.

Occurrence

Sources of environmental contaminations are thought to be animal and human faeces or sewage.

E. coli O157 have been detected in a variety of water environments: in raw water (Gannon 2004) and surface water (Kogure 1997) (Johnson J. 2003). It is commonly present in animal and human wastewaters (Garcia-Aljaro 2005).

Levels of contamination and distribution are not well documented, particularly in developing countries.

Reservoirs

Man, livestock, such as healthy cattle and sheep (Chapman 1993) and, to a lesser extent, goats, pigs, chickens and wild birds are a major source of EHEC strains. These sources can be regarded as threats to the quality of water supplies, particularly private supplies in which adequate treatment is not guaranteed.

Survival

Few studies have evaluated the ability of *E. coli* O157:H7 to persist in environmental water.

Limited observations, however, reported that *E. coli* O157:H7 is particularly well adapted for surviving in the aquatic environment (Kogure 1997) (Maule 2000). It survives for long periods of time in water, especially at cold temperatures ($<10°C$) (Porter 1997) (Wang 1998). *E. coli* O157:H7 survived in inoculated water for up to >300 days, depending on the type of water (Warburton 1998).

Kerr *et al.* (Kerr 1999) confirmed that *E. coli* O157:H7 can survive for long periods in the nutrient-starved conditions offered by bottled mineral water.

Artz *et al.* (Artz 2002) reported that wells with higher water quality as assessed using the European Union Drinking Water Directive standards may allow survival of *E. coli* O157:H7 for long periods.

E. coli O157:H7 in soil persisted for over 5 months after application of contaminated compost or irrigation water (Islam 2004).

Several reports indicate that *E. coli* O157 becomes nonculturable during prolonged storage in water (Rigsbee 1997) (Wang 1998) (Mizunoe 1999) (Kolling 2001) but the fate of *E. coli* O157 as viable but not culturable bacteria (VBNC) has not been investigated.

As for other agents *(Legionella,* mycobacteria...) *Acanthamoeba* may serve as a host for *E. coli* serotype O157 survival and growth in water and may protect them (Barker 1999) against chlorination and other disinfection measures.

One study reported that *E. coli* O157:H7, *Salmonella* spp., and *Vibrio cholerae* can grow in nutrient-limited, reconditioned wastewater over the temperature range of 4 to 46°C when the biological oxygen demand of this water is <2 (Rajkowski 2001).

Inactivation/removal

Published data on removal and inactivation are summarized in annex 1. Coagulation, filtration and disinfection are effective in removing enteric bacteria. Recently, a newly developed UVA irradiation system was evaluated (Hamamoto 2007). The incidence of *E. coli* O157 in treated drinking water is likely to be a problem principally if there is a failure of the chlorination process at the treatment works or contamination during distribution.

Chlorine or monochloramine levels typically maintained in water systems are sufficient to inactivate these organisms, as well as UV doses applied for drinking water treatment (Rice 1999) (Sommer 2000), although significant variations in susceptibility to inactivation between environmental strains have been demonstrated (Wojcicka 2007). One study demonstrated that *E. coli* O157:H7 adapts to starvation conditions by developing a chlorine resistance phenotype (Lisle 1998). Further studies should be carried out on this topic.

Surrogates

There is no indication that the response of *E. coli* O157 to water treatment and disinfection procedures differs from that of other *E. coli*. Hence conventional testing for *E. coli* (or alternatively, thermotolerant coliform bacteria) provides an appropriate index for the presence of *E. coli* O157 in drinking water (WHO 2004).

Environmental detection

Concentration: the most widely used filtration method for recovering bacteria is membrane filtration using microporous membranes typically composed of cellulose esters. Following filtration, the cells recovered on the membrane filter can be directly assayed or cultured.

Traditional culture methods rely upon a preliminary enrichment step followed by plating on selective media, such as Sorbitol MacConkey agar and screening negative sorbitol-fermenting colonies followed by confirmation by biochemical and serological tests. These methods are laborious and time-consuming (at least 3 days) and cannot detect injured cells nor the viable but non culturable state.

A variety of immunological methods have been developed.

The use of enrichment in conjunction with immunological detection has been reported to detect viable *E. coli* O157:H7 in surface water (Shelton 2004).

An approach combining fluorescent antibody and tetrazolium dye reduction, which detect repiratory activity, has been used to enumerate viable *E. coli* O157:H7 in water (Pyle 1995).

Another approach has been described which combines immunomagnetic separation with cyanoditolyl tetrazolium chloride (CTC) incubation to determine respiratory activity and fluorescent antibodies for enumeration by either fluorescence microscopy or laser cytometry (Pyle 1999).

One study (Bukhari 2004) reported the rapid detection (8h) of *E. coli* O157:H7 using commercial immunological lateral diffusion assays.

The combined use of an immunomagnetic separation method and immunoblotting has been described for the enumeration and isolation of *E. coli O157* in wastewater (Garcia-Aljaro 2005).

A rapid (5 hours) two dot filter assay able to detect low levels of *E. coli O157* in water or in food products has been developed (Kamma 2008).

For several years, PCR-based detection has been increasingly used because it is highly sensitive and rapid and it allows the detection of bacteria in a VBNC state.

Campbell *et al.* (Campbell 2001) described a multiplex PCR assay in combination with an enrichment step for sensitive detection of *E. coli* O157:H7 in water. Another study (Tims 2003) reported the detection of viable *E. coli* O157:H7 by enrichment and PCR after rapid biosensor detection. Direct PCR has also been used for rapid detection (Jothikumar 2002) (Fode-Vaughan 2003). Recently, a reverse transcriptase PCR targeting mRNA from rfbE and flicC genes was developed for specific detection of viable *E. coli* O157:H7 cells including VBNC cells (Liu 2008a).

Some studies reported the simultaneous detection of several bacterial pathogens in addition to *E. coli* O157:H7, e.g. simultaneous detection of *E. coli* O157:H7, *Listeria monocytogenes* and *Salmonella* strains by real-time PCR (Bhagwat 2003). Morin *et al.* (Morin 2004) developed a reverse transcription-multiplex PCR assay for simultaneous detection of viable *E. coli* O157:H7, *Vibrio cholerae* O1, and *Salmonella Typhi* by targeting mRNA; Wolter *et al.* (Wolter 2008) developed a flow-through chemiluminescence microarray readout system to simultaneously detect *E. coli* O157:H7, *Salmonella typhimurium* and *Legionella pneumophila*.

Only two studies (Ibekwe 2002) (Ibekwe 2003) reported quantitative detection in environmental samples based on real-time PCR. This promising method needs to be optimized, validated and compared with the standard culture method.

Commercial PCR kits are available for screening food and environmental samples: the Probelia kit from Diagnostic Pasteur or the Bax® system from DupontQualicon, but these kits are not suitable for the detection of *E. coli* O157 in water samples with low contamination.

Genotyping *E. coli* O157 strains by pulsed-field gel electrophoresis (PFGE) has been commonly used to track sources of *E. coli* O157 outbreaks (Bender 1997) (Preston 2000) (Bopp 2003) as well as to study the genetic diversity of strains from range cattle environments (Renter 2003).

Research needs

To date, research has focused on agricultural sources of *E. coli* O157 and how it enters the food chain. Not much work has been done on *E. coli* O157 in water although it is an increasing concern to the water industry. Much remains unclear about the ecology of this micro-organism and its resistance to water treatment.

Analytical methods

Numerous PCR-based methods have been described for the detection of *E. coli* O157:H7 in water, but optimization and validation are still needed in order to use them as an alternative to the traditional culture method.

Ecology

There is a need to clarify the source, distribution, and fate of *E. coli* O157 in environmental water.

The different animal reservoirs need to be identified and their relative importance clarified.

Environmental survival

Insufficient data are available on the survival of *E. coli* O157 in water. Many environmental factors including temperature, nutrient availability, predation and sunlight should be evaluated.

Need to determine whether *E. coli* O157 exhibits similar decay rates in natural waters as faecal indicator organisms. This would demonstrate whether or not monitoring microbiological quality using standard parameters such as coliforms and *E. coli* could be an adequate indication of the potential presence of *E. coli* O157.

Long-term survival of non-culturable yet viable *E. coli* O157 may have implications for the transmission of the disease and should be studied further.

Does *E. coli* O157 survive long enough on land to contaminate surface or groundwater sources, via surface runoff ?

Treatment

Future research must be carried out to ensure that *E. coli* O157 is removed by normal water treatment processes as other *E. coli*. Furthermore, research must be carried out to confirm whether *E. coli* O157 in VBNC state may develop a chlorine resistance phenotype.

Helicobacter Pylori

Health significance	**High**
Persistence in water supplies	**Moderate**
Resistance to chlorine	**Low**
Relative infectivity	**Unknown**
Important animal source	**No**

Microbiology

General description

Helicobacter pylori was recognized as a human pathogen in 1983 (Marshall 1984) and was subsequently identified as a human carcinogen by the International Agency for Research on Cancers (IARC). It is included in the Contaminant Candidate List (CCL) (U.S. Environmental Protection Agency 2005) (U.S. Environmental Protection Agency 1998).

H. pylori is a Gram-negative microaerophilic, spiral-shaped, motile bacterium. Two morphologically distinct forms, a spiral shape and a coccoid form have been identified. The spiral shape is routinely cultured from clinical samples. To date, the coccoid form has never been cultured but it remains metabolically active (Hunter 1997). Transformation from the spiral-shaped bacterium to the coccoid form is thought to result from variations in the environment such as oxygen stress, temperature changes, the presence of antibiotics and other stress-inducing conditions (Engstrand 2001). It has been suggested that the coccoid form is responsible for transmission in the environment (Hulten 1998).

Classification

When it was first described in 1983, the species was placed into the genus *Campylobacter*, but it was given its own genus later in 1989. There are at least 14 species of *Helicobacter*, but only *H. pylori* has been identified as a human pathogen.

Human health effects

Clinical features

H. pylori is found in the stomach; although most infections are asymptomatic, the organism is the causative agent of gastritis and duodenal ulcers (as demonstrated by 2005 Nobel Prize winners Marshall B. and Warren R.) and plays a role in the development of gastric cancer. Infections are more prevalent in developing countries and are associated with overcrowded living conditions (WHO 2004). It has been estimated that over half of the world's population is infected with this organism (Lambert 1995).

Dose – response data

The infectious dose is unknown.

Target populations

All persons, but the majority of *H. pylori* infections are initiated in childhood and are chronic without treatment.

© 2010 Global Water Research Coalition. Waterborne Pathogens: Review for the Drinking-water Industry. Prepared by Emmanuelle Guillot and Jean-François Loret. ISBN: 9781843391791. Published by IWA Publishing, London, UK.

Diagnosis

Diagnosis can be confirmed by endoscopy followed by microscopy or culture of biopsy material onto appropriate media. A serological blood test and a urease breast test using radiolabelled carbon meals are also available.

Treatment

Therapy for *H. pylori* infection consists of 10 days to 2 weeks of one or two effective antibiotics, such as amoxicillin plus metronidazole.

Geographical distribution

H. pylori infection is widespread throughout the world (Goodwin 1997).

Epidemiology

The epidemiology of *H. Pylori* is still unclear. Studies on prevalence or seroprevalence suggest that drinking water might play some role in infection with *H. pylori* (McKeown 1999); this has not been confirmed by other work (Teh 1994) (Begue 1998).

Transmission routes

The mode of transmission remains unclear (Stone 1999) but four different pathways have been suggested (Leclerc 2002): (1) by the faecal-oral route, the bacterium excreted by faeces, might colonize water sources, becoming available to be transmitted to humans; (2) by oral-oral route, *H. pylori* which colonizes dental plaque and saliva, may be transmitted by saliva to other individuals (person to person transmission); (3) by gastric-oral route, the typical modality of transmission in childhood, by contaminated vomitus; (4) by gastric-gastric route, the bacterium being transmitted by endoscopic procedures (Weber 2001).

Recently, several studies suggested that water may be a vector of *H. pylori* in its faecal-oral route (Horiuchi 2001) (Karita 2003) (Rolle-Kampczyk 2004) (Watson 2004) (Queralt 2005) (Azevedo 2008).

Waterborne outbreaks

Not documented.

Ecology

Very little is known about the ecology of *H. pylori*.

Regrowth in distribution systems/biofilms

Cannot multiply in water. However, several studies have shown the presence of *H. pylori* in biofilms in water distribution systems (Park 2001) (Bunn 2002) (Azevedo 2003) (Braganra 2007).

Recently, several studies suggested that biofilms in water distribution systems may act either as sites for the passive accumulation of helicobacters or as potentially important reservoirs of infection (Watson 2004) (Azevedo 2008). Gião *et al.* (Gião 2008a) showed that *H. pylori* can persist and agglomerate in biofilms in a nonculturable state for at least 31 days.

Occurrence

H. pylori has not been isolated from environmental waters samples by traditional culture techniques except in one study (Lu 2002) which isolated *H. pylori* from untreated wastewater.

Using molecular techniques (PCR), *H. pylori* specific DNA was detected in sewage (Vincent 1995), surface water (Hegarty 1999) (Queralt 2005), and water supplies (Hulten 1996) (Hulten 1998) (Horiuchi 2001) (Watson 2004) including biofilms (Park 2001) (Bunn 2002). One study (Cellini 2004) reported the presence of free and plankton-associated *H. pylori* in seawater.

The health significance of finding *Helicobater* spp. DNA in drinking water still remains unknown.

Occurrence of *H. pylori* in surface, ground and finished waters needs to be further studied.

Levels of contamination are unknown.

Reservoirs

Not determined precisely.

Until recently the natural reservoir for *H. pylori* was thought to be the human gastrointestinal tract (Axon 1996).

However, the association of *Helicobacter* with nonhuman sources, such as livestock (Vaira 1992), domestic cats (Bode 1998) and vegetables (Hopkins 1993) suggested potential environmental reservoirs.

Survival

Data on the environmental survival of *H. pylori* are limited.

H. pylori in the environment transforms into a nonculturable, coccoid form (Adams 2003), which frequently results in the failure to detect this bacterium in environmental samples by conventional culture techniques. It is unclear whether the coccoid form, suggested to be similar to the viable but non culturable (VBNC) state of waterborne bacteria, could revert to the spiral form and therefore be able to infect humans (Engstrand 2001). Only one study has shown a successful reversion of coccoid *H. pylori* to the rod shaped culturable form. This study was performed in mice (Wang 1997) but contradictory results had been reported in pigs (Eaton 1995).

The non cultivable coccoid form may persist for one month or more in water or in biofilms as well as in food (Leclerc 2002) (Gião 2008a). It can survive for up to 10 days at 4°C in tap water (Fan 1998).

Azevedo *et al.* (Azevedo 2008) assessed the survival of multiple gastric and enterohepatic *Helicobacter* spp. in water for the first time: *H. pylori* was the species that survived the longest time in water (>96 hours in the dark at 25°C). Konishi *et al.* (Konishi 2007) reported that *H. pylori* has a longer survival in seawater and in deep ground water than in a nutrient-rich medium.

One study showed that free-living amoebae promote growth and survival of *H. pylori* under experimental conditions (Winiecka-Krusnell 2002). The putative dependence of *H. pylori* on free-living amoebae in nature could be important with respect to transmission and prevalence and needs to be studied.

Inactivation/removal

Published data on inactivation are summarized in annex 1. *H pylori* is readily inactivated by free chlorine (Johnson 1997). But Baker *et al.* (Baker K 2002) argued later that *H. pylori* could tolerate disinfectants better than the classical faecal indicator, *Escherichia coli,* and Moreno *et al.* (Moreno 2007) observed that *H. pylori* can survive for a few hours in chlorinated drinking water in a viable but non cultivable form. Hayes *et al.* (Hayes 2006) demonstrated that monodispersed cultures of *H. pylori* are inactivated by UV light to approximately the same degree as other Gram-negative organisms.

Surrogates

E. coli (or alternatively, thermotolerant coliforms) cannot be used as an index for the presence/absence of *H. pylori* (Hegarty 1999) (WHO 2004).

Environmental detection

To date, there is no standard culture method for detecting *H. pylori* in water samples.

Traditional media may be too nutrient rich to recover potentially stressed *H. pylori* from water (Azevedo 2004). One study reported the development of a plating medium for selection of *H. pylori* from water samples in seven days at 37°C (Degnan 2003), but the performance of this culture method has not yet been determined.

Immunological methods have been developed (Hegarty 1999), but they are limited because *H. pylori*-monoclonal antibodies may cross-react to closely related bacterial species or may not react with the non culturable coccoid form which may have a different antigenicity (Enroth 1995).

Different molecular methods have been developed to detect *H. pylori* in water:

A fluorescent in situ hybridization method (FISH) was developed to be applied in river and wastewater (Moreno 2003b). A fluorescently labeled peptide nucleic acid (PNA) probe was tested for the in situ detection of *H. pylori* in drinking water biofilms (Braganra 2007). A combination of direct viable count method and fluorescent in situ hybridization was proposed for the specific detection of viable cells of *H. pylori* (Piqueres 2006).

A direct PCR assay (Benson 2004), two PCR-based techniques for detecting helical and coccoid forms of *H. pylori* (Shahamat 2004) and an IMS-PCR (Enroth 1995) have been developed.

Several studies reported the development of quantitative PCR methods to determine *H. pylori* concentration in drinking water (McDaniels 2005) (Sen 2007) and in sewage (Nayak 2007).

Molecular typing methods have been used in epidemiological studies: RAPD fingerprinting (Karita 2003) (Yakoob 2000), PCR and sequencing (Lu 2002) and PCR-RAPD.

Currently all these available methods are at the experimental stage and validated protocols do not yet exist. As no reference method exists, it is difficult to compare the performance of existing methods.

Research needs

Much remains unknown about this newly described bacteria. There are significant research needs about the pathogenesis of human disease, therapeutic drugs and infective dose.

But research priorities for water utilities concern the epidemiology of *H. pylori,* in particular its possible waterborne transmission as well as the ecology which has not yet been extensively studied.

Analytical methods

There is a need to develop a reference culture method capable of detecting viable and culturable *H. pylori* in potable waters, and to optimize and validate molecular methods for rapid, specific, and quantitative detection of *H. pylori* in environmental water.

Route of transmission

The waterborne transmission route has been suggested by a large number of studies but there is still no evidence of transmission by this route. More epidemiological studies, including molecular methodologies for microbial source tracking, should be carried out, in order to assess the risk of waterborne transmission.

It would be important to estimate the relative importance of waterborne as opposed to other forms of transmission in both developed and developing countries.

Occurrence

Large studies on occurrence including treated, municipal drinking water and groundwater need to be carried out. The health significance of finding *Helicobater* spp. DNA in drinking water needs to be documented.

Reservoirs

Studies on potential reservoirs for *H. pylori* infections need to continue.

Environmental survival

Further studies should investigate the ability of *H. pylori* to enter the VBNC state, whether the coccoid form is viable and infectious and the role of the VBNC state in this organism.

The survival and development of the pathogen in potable water biofilms and the role of biofilm as a reservoir of *H. pylori* infections need to be further investigated.

Inactivation

Efficacy of different inactivation treatment processes (disinfectants, UV...) must be evaluated, with a special focus on the organism's ability to be present in aggregates.

Salmonella

	S. typhi	Other salmonellae
Health significance	High	High
Persistence in water supplies	Moderate	May multiply
Resistance to chlorine	Low	Low
Relative infectivity	Low	Low
Important animal source	No	Yes

Adapted from table 7.1 Waterborne pathogens and their significance in water supplies (WHO 2004).

Microbiology

General description

Salmonella spp. are considered one of the most important causal agents of foodborne illness in developed countries.

The genus *Salmonella* is mostly comprised of facultatively anaerobic, oxidase-negative, catalase-positive, Gram-negative, rod-shaped bacteria. Most strains are motile and ferment glucose with production of both acid and gas.

Classification

The genus *Salmonella* belongs to the family Enterobacteriaceae. *Salmonella* nomenclature is complex, and scientists use different systems to refer to and communicate about this genus. The nomenclature for the genus *Salmonella* has evolved from the initial one serotype-one species concept proposed by Kauffmann (Kauffmann 1966) on the basis of the serologic identification of O (somatic) and H (flagellar) antigens. In one nomenclature system (Le Minor 1982), there is only one species recognized, *Salmonella choleraesuis,* divided into seven subspecies. In 1987, Le Minor and Popoff (Le Minor 1987) requested that *Salmonella enterica* (type strain LT2) be the type and the only species of the genus *Salmonella*.

Recently, Tindall *et al.* provided a clear interpretation of the nomenclature and taxonomy of the genus *Salmonella* (Tindall 2005). According to the CDC system (essentially based on the recommendations established by the WHO Collaborating Centre), the genus *Salmonella* contains two species, each of which contains multiple serovars. The two species are *S. enterica* – the type species which is divided into six subspecies – and *S. bongori*. Currently, most of the journals all over the world have adopted this system.

The genus comprises more than 2,500 serovars, most of which are considered potential human pathogens, but only a reduced number of serovars have been associated with human infections. Both serovars, *S. typhi* and *S. paratyphi,* cause enteric fever only in humans and in higher primates. *Salmonella enteriditis* and *Salmonella typhimurium* have a broad host-spectrum and constitute the two most important serovars for salmonellosis transmitted from animals to humans.

© 2010 Global Water Research Coalition. Waterborne Pathogens: Review for the Drinking-water Industry. Prepared by Emmanuelle Guillot and Jean-François Loret. ISBN: 9781843391791. Published by IWA Publishing, London, UK.

Human health effects

Clinical features

Salmonella are the cause of two enteric diseases called salmonellosis: enteric fever (typhoid), resulting from bacterial invasion of the bloodstream, and acute gastroenteritis, resulting from a foodborne infection/intoxication.

Symptoms of gastrointestinal disease include diarrhoea, fever, and abdominal cramps that develop 12 to 72 hours after infection. The illness usually lasts 4 to 7 days, and most persons recover without treatment. A few may develop a chronic condition called Reiter's syndrome. This syndrome can last for months or years and can lead to arthritis.

The incubation period for typhoidal disease is usually 10–14 days (range 5–23 days). Typhoid fever is a more severe illness and can be fatal. Although typhoid is uncommon in areas with good sanitary systems, it is still prevalent elsewhere, and there are many millions of cases each year.

Dose – response data

The infectious dose is variable according to the pathogenic *Salmonella* strains. It is estimated typically at over 10^5 for non typhoidal salmonellosis (Kothary 2001). As few as 10^3 organisms can cause a typhoid infection (Hunter 1997). Different β-Poisson dose-response models based on experiments on humans have been described (Haas 1999) (Teunis 1999) (FAO/WHO 2002).

Target populations

Anyone can get salmonellosis, but it occurs more often in young children, the elderly, and the immunocompromised.

Diagnosis

Diagnosis of both gastrointestinal and typhoidal diseases is based on the culture of the responsible pathogen from faeces or blood culture. Confirmation of identity is based on biochemical and serological agglutination reactions (Hunter 1997).

Treatment

Persons with severe diarrhoea may require rehydration, often with intravenous fluids. Antibiotics are not usually necessary unless the infection spreads from the intestines. The antibiotic drugs of choice for typhoid are now ciprofloxacin and chloramphenicol. Several vaccines are effective against typhoid (Hunter 1997).

Geographical distribution

Worldwide, but according to the CDC, most cases of *Salmonella* illness have been reported in North America and Europe.

Epidemiology

Multidrug-resistant (MDR) strains of *Salmonella* are now encountered frequently and the rates of multidrug resistance have increased considerably in recent years, probably in response to antimicrobial usage in food animals. In the 1990s, an antibiotic-resistant strain of *S. typhimurium*, called Definitive Type

104 (DT104), was first found in the United Kingdom (Baggesen 2000), in Western Europe, and recently in the United-States. It is thc second most common strain (after *S. enteritidis*) of *Salmonella* found in humans. This strain poses a major threat because it is resistant to several antibiotics normally used to treat people with *Salmonella* infections.

Transmission routes

Salmonella is spread by the faecal-oral route.

Infections with non typhoidal serovars are primarily associated with person-to person contact, the consumption of a variety of contaminated foods (mainly contaminated raw meat, poultry and poultry-derived products) and exposure to animals, including domestic animals such as cats and dogs. Infection by typhoid serovars (*S. typhi* and *S. paratyphi*) is associated with ingestion of contaminated water or food, direct person-to-person transmission being uncommon (WHO 2004).

Waterborne outbreaks

In the early twentieth century, typhoid was the most common known cause of waterborne outbreaks in both Europe and the USA. Typhoid and paratyphoid are now uncommon in the western world but are still a major problem in many less affluent countries. In 2001, however, an explosive waterborne outbreak caused by multidrug-resistant (MDR) *S. Typhi* in a non-endemic community with otherwise good sanitation was reported (Swaddiwudhipong 2001). An outbreak caused by rainwater collection tanks contaminated by *S. Typhimurium 9* occurred at a school camp in Australia (Franklin 2008).

Despite their widespread distribution in the water environment, non-typhoidal salmonellae rarely cause waterborne outbreaks (Hunter 1997).

Ecology

Regrowth in distribution systems/biofilms

The ability of *Salmonella* strains to persist/multiply in the biofilms of distribution systems remains poorly documented.

Occurrence

Salmonellae are disseminated in the natural environment (water, soil, sometimes plants used as food) and through human or animal excretion. Humans and animals (either wild or domesticated) can excrete *Salmonella* either when clinically diseased or after having had salmonellosis, if they remain carriers.

Salmonella species are often detected in sewage, river and irrigation water, freshwater, marine coastal water and ground water (Morinigo 1986) (Fewtrell 1994) (Baudart 2000) (Catalao Dionisio 2000) (Ho 2000) (Gannon 2004). One study reported the isolation of *S. typhimurium* from drinking bottled water (Radhakrishna 2003).

S. typhi and *paratyphi* only colonize humans. They are disseminated by human sewage. *S. typhi* has occasionally been isolated from water and sewage but has never been detected in drinking water (Hunter 1997).

Reservoirs

The only reservoir for *S. typhi* and *S. paratyphi* is human.

A large number of serovars, including *S. typhimurium* and *S. enteritidis*, infect humans and also a wide range of animals, including poultry, cows, pigs, sheep, birds and even reptiles which constitute important reservoirs for *Salmonella*.

Survival

The persistence of *Salmonella* strains in water in a viable but non-cultivable form (VBNC) has been reported (Roszak 1984) but the health risk posed by such VBNC forms remains unclear (Santo Domingo 2000). Caro *et al.* (Caro 1999) demonstrated that *S. typhimurium* in the VBNC state was not pathogenic but further studies should investigate if VBNC cells can be resuscitated and regain pathogenicity and virulence.

Cho *et al.* (Cho 1999) reported that *S. typhi* survived longer in groundwater than in pond water in both a culturable and a VBNC state.

Environmental factors such as temperature, pH, and water activity affect pathogen survival. *Salmonella* grows in foods within the range of 7–54°C. The optimum temperature for growth is 37°C. Recently, one study showed that sunlight was the most important inactivating factor of *E. coli* and *S. typhimurium* in tropical estuarine water and that *E. coli* cells under all test conditions showed better survival capacity in comparison to *S. typhimurium* (Chandran 2005).

One study reported that *Escherichia coli* O157:H7, *Salmonella* spp., and *Vibrio cholerae* can grow in nutrient-limited, reconditioned wastewater over the temperature range of 4 to 46°C when biological oxygen demand of this water is <2 (Rajkowski 2001).

Inactivation/removal

Published data on inactivation are summarized in annex 1. Disinfection procedures commonly used for treatment of drinking water to remove coliform bacteria are adequate to eliminate *Salmonella* (WHO 2004). However, ingestion of *S. typhimurium* by predatory protozoa increases resistance to free chlorine residuals (King 1988).

Surrogates

Escherichia coli (or thermotolerant coliforms) is a generally reliable index for *Salmonella* spp. in drinking water supplies (WHO 2004).

Environmental detection

Standard culture methods followed by conventional biochemical methods are available for identification of *Salmonella* spp. in environmental water samples (Morinigo 1986).

Conventional typing methods such as serotyping and phage typing are used for the identification of *Salmonella* strains (serovars). But conventional identification is laborious, time-consuming (4–5 days) and cannot detect VBNC bacteria.

An immunological method based on a magnetoelastic biosensor was designed in order to detect *S. typhimurium* in a mixed microbial population (Guntupalli 2007).

In the past ten years, a large number of PCR-based detection methods have been developed mainly for food and clinical samples but also for environmental water samples (Aabo 1993) (Way 1993) (Kwang 1996) (Touron 2005) (Vazquez-Novelle 2005) (Kumar 2006). A combination of immunomagnetic separation and real-time PCR or PCR assays has also been reported (Fluit 1991) (Mercanoglu 2005) (Thompson 2006). Some PCR assays are specific of *Salmonella* serovars (Lin 2004), *S. enteritidis* (Lampel 1996) or *S. typhi* (Sharma 1995) (Kumar 2006). Recently, real-time PCR has been increasingly used for rapid screening of food samples (Malorny 2004) or quantification of RNA targets belonging to *Salmonella* spp. in water (Fey 2004).

Morin *et al.* (Morin 2004) described a rapid and reliable technique for the simultaneous detection of three viable pathogens *Escherichia coli* O157:H7, *Vibrio cholerae* O1, *and Salmonella Typhi* by targeting mRNA. Kong *et al.* (Kong 2002) described the rapid detection of six types of bacterial pathogens including

S. typhimurium in marine waters by multiplex PCR. A quantitative PCR assay was also designed for the detection of the four enteric bacteria strains, *Shigella dysenteriae*, *Vibrio cholerae*, *Salmonella typhimurium*, and *Escherichia coli* (Liu 2008b). A new approach based on a flow-through chemiluminescence microarray readout system was developed to simultaneously detect *Escherichia coli* O157:H7, *Salmonella typhimurium* and *Legionella pneumophila* (Wolter 2008).

At the present time, all these available alternative molecular methods are at the experimental stage and none has yet been optimized and validated on water samples.

Commercial PCR-based kits are available for screening food samples or environmental samples. In particular the Probelia kit from Sanofi Diagnostic Pasteur, or the Bax® system from DupontQualicon; both are AFNOR (French standardization agency) certified. But these kits are not suitable for water samples with low contamination levels.

Molecular typing of *Salmonella* strains is increasingly used with various techniques, such as pulsed-field gel electrophoresis (Baudart 2000), RAPD (Laconcha 1998), PCR ribotyping (Lagatolla 1996) and microarray technology (Alvarez 2003).

Research needs

Salmonella constitutes a major public health concern for the food industry, but waterborne transmission is scarce in developed countries. Waterborne outbreaks associated to *S. typhi* and *paratyphi* are now uncommon in the western world. Nevertheless the disease is still very common elsewhere, affecting an estimated 12.5 million people each year (Hunter 1997).

Because it is mainly a foodborne pathogen, the environmental ecology of *Salmonella* strains has not been investigated in depth, particularly as regards the diversity and survival of strains in natural ecosystems. The role of biofilms in the survival of pathogenic *Salmonella* strains remains unknown. Transfer of antibiotics resistance and virulence factors in drinking water biofilms could be an interesting area of future research.

Shigella

Health significance	High
Persistence in water supplies	Short
Resistance to chlorine	Low
Relative infectivity	Moderate
Important animal source	No

Adapted from table 7.1 Waterborne pathogens and their significance in water supplies (WHO 2004).

Microbiology

General description

Shigellosis is a global human health problem. Worldwide, *Shigella* infections probably affect over 2 million people and cause the death of approximately 600,000 each year, predominantly in developing countries (WHO 2004) (Niyogi 2005).

Shigella spp. are Gram-negative, non-spore-forming, non-motile bacilli, which grow in the presence or absence of oxygen.

Through plasmid transfer these bacteria can resist a large spectrum of antibiotics, which currently leads to a resurgence of shigellosis.

Classification

Shigella spp. are members of the family Enterobacteriaceae. The genus comprises four species: *Sh. dysenteriae*, *Sh. flexneri*, *Sh. boydii*, and *Sh. sonnei*. These species are subdivided into serotypes on the basis of their somatic O antigens, many of which are shared with other enteric bacilli, including *E. coli*.

Human health effects

Clinical features

Shigella spp. are known to cause disease only in humans and primates.

The symptoms of shigellosis include diarrhoea and/or dysentery with frequent mucoid bloody stools, abdominal cramps and tenesmus. The incubation period is usually 24–72 hours.

All species can produce severe disease, but illness due to *Sh. sonnei* is usually relatively mild and self-limiting. *Shigella dysenteriae* type 1 (previously known as *Sh. shigae*) is the causative agent of the most severe form of bacillary dysentery, at the origin of outbreaks in many developing countries.

Dose – response data

The infectious dose is low, ranging from less than 10 organisms (Kothary 2001) to 200 viable organisms (DuPont 1989). Several β-Poisson models have been developed for *Sh. flexneri* and *Sh. dysenteriae* (Haas 1983) (Crockett 1996).

© 2010 Global Water Research Coalition. Waterborne Pathogens: Review for the Drinking-water Industry. Prepared by Emmanuelle Guillot and Jean-François Loret. ISBN: 9781843391791. Published by IWA Publishing, London, UK.

Target populations

Most cases of *Shigella* infection occur in children under 10 years of age.

Diagnosis

Laboratory diagnosis is performed by culturing stool samples using selective/differential agar media. Enzyme-linked immuno-sorbent assays (ELISA) have been developed to screen stool samples for *Shigella* species. Several PCR protocols have also been developed for detection of *Shigella* spp. in faeces (Li 2004a); in particular, Islam *et al.* (Islam 1992) described an IMS-PCR to detect *Sh. dysenteriae* type 1 and *Sh. flexneri* in faeces.

Treatment

Antimicrobial agents are the mainstay of therapy of all cases of shigellosis. Due to the global emergence of drug resistance, the choice of antimicrobial agents for treating shigellosis is limited.

Currently, no vaccines against *Shigella* infection exist. Both live and subunit parenteral vaccine candidates are under development (Niyogi 2005).

Geographical distribution

Worldwide.

In industrialized countries, *Sh. sonnei* predominates. In tropical countries the various serotypes of *Sh. flexneri* and *Sh. dysenteriae* predominate, although *Sh. boydii* and *Sh. sonnei* are also found (Hunter 1997).

Epidemiology

Transmission routes

Shigella spp. are enteric pathogens predominantly transmitted by the faecal-oral route through person-to-person contact, contaminated food and water. Flies have also been identified as a transmission vector from contaminated faecal waste (WHO 2004). *Shigella* infections can also be venereally transmitted by oral-anal contact in male homosexuals (Hunter 1997).

Waterborne outbreaks

Numerous waterborne outbreaks associated with drinking water have been recorded worldwide mainly due to *Sh. sonnei* in developed countries (Egoz 1991) (Levy 1998) (Alamanos 2000) (CDC 2000) (CDC 2001) (Chen 2001) (Arias 2006). In 1998, a waterborne outbreak involving a small round structured virus, *Campylobacter* and *Shigella* co-infections occurred in Switzerland, 1998 (Maurer 2000). In most cases, the drinking water was not treated or was improperly treated prior to consumption.

Outbreaks associated with swimming in unchlorinated surface water were also reported (Blostein 1991) (Keene 1994) (Fleming 2000).

Ecology

Regrowth in distribution systems/biofilms

Not proved to replicate in water.

The role of biofilms in the survival of *Shigella* spp. remains unknown.

Occurrence

Data are limited. Available data on prevalence in water supplies may be underestimated, because detection techniques generally used can have relatively low sensitivity and reliability (WHO 2004).

The distribution of *Shigella* spp. in the environment reflects human faecal contamination. *Shigella* spp. have been isolated from some water supply systems in tropical countries (Khalil 1994). *Sh. dysenteriae* type 1 and *Sh. flexneri* strains were isolated from surface waters in Bangladesh (Faruque 2002). Faruque *et al.* (Faruque 2003) suggested that the rapid and simple detection of phages specific for *Shigella* strains may be a useful tool for predicting the presence of *Shigella* in environmental waters.

Reservoirs

Humans and other higher primates appear to be the only natural hosts for shigellae.

Survival

Studies on the survival of *Shigella* spp. have given differing results.

One study (Popovitch 1982) reported survival of *Sh. sonnei* and *Sh. flexneri* of up to about 50 days at 25°C in river water. By contrast another study found survival of *Sh. flexneri* for only 4–7 days at 25–35°C (El-sharkwai 1989). Ghosh *et al.* (Ghosh 2001) found that in laboratory conditions *Sh. dysenteriae* type 1 survived for a mean duration of 3.33 days and *Sh. flexneri* for a mean 11.167 days in field water samples. Concentration of bacteria and water temperatures were positively correlated with the duration of survival.

The microbial survival in seawater was studied: *Sh. typhi* and *Sh. sonnei* showed similar or greater survival than enteric viruses and *E. coli*. There was no clear association between microbial survival and water temperature in that case (Wait 2001).

In addition, *Shigella* spp. did not grow but survived for >28 days at 4 to 25°C in reconditioned wastewater (Rajkowski 2001).

The survival of *Shigella* strains (*Sh. sonnei* and *Sh. dysenteriae*) in water in a viable but non-cultivable form (VBNC) has been reported (Colwell 1985) (Islam 1993) but the health risk posed by such VBNC form needs to be further investigated.

One recent study reported the growth and survival of *Sh. dysenteriae* and *Sh. sonnei* in an amoeba (*Acanthamoeba castellanii*) for more than 3 weeks (Saeed 2009).

Inactivation/removal

Published data on inactivation are summarized in annex 1. Disinfection procedures commonly used for treatment of drinking water to remove coliform bacteria are adequate to eliminate *Shigella* spp. (WHO 2004).

Selma *et al.* (Selma 2007) suggested that ozone can constitute an alternative treatment to chlorine for disinfection of wash water used on fresh vegetables.

However, ingestion of *Sh. sonnei* by predatory protozoa has been shown to increase resistance to free chlorine (King 1988). The authors suggested that this is a mechanism allowing survival of fastidious bacteria in dilute and inhospitable aquatic environments.

Surrogates

Escherichia coli (or thermotolerant coliforms) is a generally reliable index for *Shigella* spp. in drinking water supplies (WHO 2004).

Environmental detection

The standard method is based on filtration of large volumes of water (2–5 L) through membranes, then isolation of *Shigella* by selective culture media followed by identification by biochemical tests and serotyping (June 1993). This process may take 48–72 h or even longer and cannot detect VBNC bacteria.

A colony blot immunoassay to detect enteroinvasive *Escherichia coli* and *Shigella* in water samples has been reported (Szakal 2001), but this approach is not very sensitive because of interaction with background bacteria.

Few PCR-based detection methods have been developed to specifically detect *Shigella* species in water (Bej 1991) (Islam 1993) (Theron 2001) (Hsu 2007), but no validated protocol is available.

Techniques for the multidetection of pathogens have emerged in the recent years:

Kong *et al.* (Kong 2002) described the rapid detection of six types of bacterial pathogens including *Sh. flexneri* in marine waters by multiplex PCR. A duplex real-time PCR assay for detection of 17 species of food- or waterborne pathogens including *Shigella* spp. in stools has been developed (Fukushima 2003).

Recently, Liu *et al.* (Liu 2008b) described a rapid quantitative PCR method to simultaneously detect four enteric bacteria: *Shigella dysenteriae*, *Vibrio cholerae*, *Salmonella typhimurium*, and *Escherichia coli*.

There are promising approaches which are at the experimental stage and need to be optimized and validated on water samples.

Molecular typing techniques such as RAPD (Yamada 1999) (Bando 1998), pulsed-field gel electrophoresis, enterobacterial repetitive intergenic consensus sequence-based PCR typing (Surdeanu 2003) and ribotypes (Faruque 2003), are often used in epidemiological studies.

Research needs

The resurgence of shigellosis in particular in the USA due to the emergence of drug resistant *Shigella* strains requires research on the virulence characteristics of environmental *Shigella* isolates in order to understand their pathogenic potential. The drug resistance patterns of the environmental and clinical *Shigella* strains should be evaluated.

The microbial ecology of *Shigella* species remains poorly known. In particular, very little is known about the presence and the survival of pathogenic strains in the aquatic environment, their interaction with biofilms and the role of protozoa. Transfer of antibiotics resistance and virulence factors in drinking water biofilms should be investigated in the future.

Several PCR-based methods are available for rapid and specific detection of *Shigella* species in water, but viability detection and quantitation are still lacking.

Yersinia

Health significance	**High**
Persistence in water supplies	**Long**
Resistance to chlorine	**Low**
Relative infectivity	**Low**
Important animal source	**yes**

Adapted from table 7.1 Waterborne pathogens and their significance in water supplies (WHO 2004).

Microbiology

General description

Yersinia spp. are Gram-negative, facultatively anaerobic, non-spore forming bacilli. They are motile at 25°C but not at 37°C.

Classification

The genus *Yersinia* is classified in the family Enterobacteriaceae and features seven species.

The species *Y. pestis*, *Y. pseudotuberculosis* and certain serotypes of *Y. enterolitica* are pathogens for humans.

Y. pestis is the causative agent of plague through contact with rodents and their fleas. Bubonic plague is the most widespread clinical form. Pulmonary plague may occur when the bacillus directly colonizes the lungs following aerosol inhalation (Carniel 2005). *Y. Pestis* is considered as a potential bioterrorism agent.

Both *Y. pseudotuberculosis* and *Y. enterocolitica* species are potential waterborne pathogens and will be addressed. Six serotypes and four sub-types of *Y. pseudotuberculosis* and more than 50 serotypes of *Y. enterocolitica* have been described.

Human health effects

Clinical features

Yersiniosis causes fever, diarrhoea and abdominal pain, which lasts for about one to three weeks. Other clinical manifestations include greatly enlarged painful lymph nodes referred to as "buboes".

Y. pseudotuberculosis most commonly causes a mesenteric adenitis, which can be indistinguishable from acute appendicitis. The incubation period for both species is between 3 and 7 days (Hunter 1997).

Dose – response data

The infectious dose is probably high, up to 10^9 organisms (Hunter 1997).

Target populations

Y. enterocolitica mainly affects children under five years old.

© 2010 Global Water Research Coalition. Waterborne Pathogens: Review for the Drinking-water Industry. Prepared by Emmanuelle Guillot and Jean-François Loret. ISBN: 9781843391791. Published by IWA Publishing, London, UK.

Diagnosis

The diagnosis may be made by culture from faeces, mesenteric lymph nodes, blood culture. Positive culture is confirmed by biochemical and serological investigations. Because it is uncommon, many laboratories do not routinely perform the identification of *Yersinia* spp. from faeces.

Treatment

Most cases recover on their own without treatment. Those with severe symptoms or bloodstream infections are generally treated with antibiotics.

Geographical distribution

Worldwide.

Epidemiology

Only *Y. enterocolitica* strains that possess virulent factors are pathogenic for humans.

Environmental strains of *Y. enterocolitica* of serotypes O:3, O:9, O:5, O:27 and O:8 are the most frequently associated with human infections in Europe, Japan, Canada and the USA (Schieman 1990).

Transmission routes

Transmitted by the faecal-oral route, with the major source of infection considered to be food, particularly meat and meat products, milk and dairy products. Ingestion of contaminated water is also a potential source of infection as well as contact with an infected person or animal (WHO 2004).

Waterborne outbreaks

Epidemiological data concerning waterborne yersiniosis are scarce (Schieman 1990), involving only a small number of individuals (Thompson 1986) (Inoue 1988a) (Han 2003). A large waterborne outbreak of yersiniosis was reported in Montana in 1974 (Eden 1977), but many of the strains isolated from water sources were later identified as non-pathogenic, leaving the cause of this outbreak uncertain (Hunter 1997).

Ecology

Regrowth in distribution systems/biofilms

Not extensively documented. Some strains may persist and/or multiply in water.

The role of biofilms in the persistence of *Yersinia* spp. has not yet been investigated.

Occurrence

Y. enterocolitica are ubiquitous, being isolated frequently from soil, water, animals, and a variety of foods. Y. *pseudotuberculosis* is less ubiquitous than *Y. enterocolitica*, and although frequently associated with animals, has only rarely been isolated from soil, water, and food.

Various authors have described the isolation of *Yersinia* spp. from surface water samples. *Y. pseudotuberculosis* were isolated from river water in Japan (Inoue 1988b) (Fukushima 1995).

Aleksic *et al.* (Aleksic 1988) reported the isolation of *Yersinia* spp. strains from well water and drinking water plants in Germany but none of the isolates was positive in virulence tests. The occurrence of non-pathogenic strains of *Yersinia* in surface water sources was also reported in Norway (Brennhovd 1992), in Argentina (Escudero 1994), in Italy (Massa 1988) and in Greece (Arvanitidou 1995).

Pathogenic strains of *Y. enterocolitica* were isolated in sewage and polluted surface waters (Sandery 1996) (Falcao 2004).

Y. enterocolitica strains detected in drinking water are more probably non-pathogenic strains of probable environmental origin (Arvanitidou 1994) (Romano 1997) (WHO 2004).

Reservoirs

Domestic and wild animals are the principal reservoir for *Yersinia* spp. Pigs are the major reservoir of pathogenic *Y. enterocolitica*; rodents are the major reservoir of *Y. pseudotuberculosis*.

Survival

Y. enterocolitica is a psychotropic organism, which can grow in refrigerated foods. Some species and strains of *Yersinia* may replicate in water environments under the conditions of minimum nutrition and low temperature (4°C) (Buzoleva 2000) (WHO 2004).

The lengthy survival of *Y. enterocolitica* in water was noted by (Karapinar 1991), who detected survivors in sterile spring water after 64 weeks at 4°C. Liao *et al.* (Liao 2003) showed the long-term survival of *Y. enterocolitica* in sterile water (several years at room temperature). Other works reported the long-term survival of *Y. pseudotuberculosis*, *Y. enterocolitica* and other *Yersinia* species at reduced temperatures (4–6°C) (Terzieva 1991) (Kuznetsov 1998). Chao *et al.* (Chao 1988) found that survival was greatly reduced with increasing temperatures. In addition, the longevity of *Y. enterocolitica* was chiefly regulated by predators and toxin producers.

The persistence of *Yersinia* strains in water in a viable but non-cultivable form (VBNC) has not yet been investigated.

Inactivation/removal

Published data on inactivation are summarized in annex 1. *Yersinia* spp. are sensitive to disinfection processes (Korol 1995) (WHO 2004). However, ingestion of *Y. enterocolitica* by protozoans has been shown to protect them against subsequent disinfection (King 1988). Control measures that can be used to minimize the presence of pathogenic *Yersinia* spp. in drinking water supplies include protection of raw water supplies from human and animal waste, adequate disinfection and protection of water during distribution (WHO 2004).

Surrogates

Due to the long survival and/or growth of some strains of *Yersinia* spp. in water, *E. coli* (or, alternatively, thermotolerant coliforms) is not a suitable index for the presence/absence of these organisms in drinking water.

Environmental detection

The currently recommended method for isolation of *Yersinia* spp. is based on membrane filtration followed by enrichment and isolation on selective media. Confirmation of typical colonies is by the demonstration of motility and biochemical reactions. This method is time-consuming (2 weeks).

For *Y. enterocolitica*, the subsequent differentiation between pathogenic and non-pathogenic strains can be difficult and unreliable. API 20E has been reported to be the system of choice for identifying pathogenic *Yersinia* isolates (Neubauer 1998). Immunomagnetic separation and agar layer methods have been reported to successfully isolate freeze-injured pathogenic *Y. enterocolitica* from water (Koujitani 2006).

Several PCR-based methods have been described for the detection of pathogenic *Y. enterocolitica* in environmental water samples (Kapperud 1995) (Sandery 1996) (Waage 1999a). Alexandrino *et al.* (Alexandrino 2004) reported a PCR protocol for rapid detection of *Y. enterocolitica* serovar 0:3 in wastewater samples.

Currently all these available alternative molecular methods are at the experimental stage for research purposes and validated protocols do not yet exist.

The multidetection of pathogens has emerged for several years:

Kong *et al.* (Kong 2002) described the rapid detection of six types of bacterial pathogens including *Y. enterocolitica* in marine waters by multiplex PCR.

A duplex real-time PCR assay for the detection of 17 species of food- or waterborne pathogens including *Yersinia* spp. in stools has been developed (Fukushima 2003).

There are promising approaches which are at the experimental stage and need to be optimized and validated on water samples.

Molecular typing of *Y. enterocolitica* in epidemiology investigations has been reported using restriction endonuclease analysis (Kapperud 1991) (Han 2003) and pulsed-field gel electrophoresis (PFGE) (Marranzano 2003).

Research needs

The epidemiology and the ecology of the micro-organism require further research.

Transmission route

More epidemiological studies, including molecular methodologies for microbial source tracking, should be carried out in order to document the role of waterborne transmission of *Y. enterocolitica* and *Y. pseudotuberculosis*.

Occurrence and distribution

The presence and distribution in water of these pathogens together with their survival capacity should be investigated in greater depth. In particular the impacts of microbial interactions (including biofilms/aggregates and protozoan) on pathogen survival require special attention.

The health significance of *Y. enterocolitica* in environmental water remains controversial and needs to be further studied.

Survival

The determination of the effects of various environmental factors (for example, ultraviolet irradiation) on the survival and growth of pathogenic *Yersinia* remains unknown.

Analytical methods

In order to carry out the research work described above there is a need to develop and validate additional PCR-based methods to detect and quantify 1) pathogenic strains of *Y. enterocolitica* and 2) *Y. pseudotuberculosis* in surface and treated waters.

Aeromonas

Health significance	Moderate
Persistence in water supplies	High
Resistance to chlorine	Low
Relative infectivity	Unknown
Important animal source	No

RMicrobiology

General description

Aeromonas species are Gram-negative, non-spore-forming, rod-shaped, facultatively anaerobic bacteria that are ubiquitous and autochthonous in aquatic environments. Morphologically, aeromonads are indistinguishable from members of the Enterobacteriaceae family from which they primarily differ by being oxidase-positive (WHO 2002).

Aeromonas hydrophila was included in the first two versions of the Contaminant Candidate List (CCL) (U.S. Environmental Protection Agency 1998) (U.S. Environmental Protection Agency 2005).

Classification

The taxonomy of the genus is still debated. Although historically in the family Vibrionaceae, they have been placed into their own family, the Aeromonadaceae (Bergey's Manual 2005).

The genus is divided into two groups. The group of psychrophilic non motile aeromonads consists of only one species, *A. salmonicida*, an obligate fish pathogen that is not considered further here. The group of mesophilic motile aeromonads is considered of potential human health significance; the main species are: *A. hydrophila*, *A. caviae*, *A. veronii* subsp. *sobria* (WHO 2002).

Human health effects

Clinical features

Aeromonas spp. can cause infections in humans, including septicaemia, particularly in immunocompromised patients, wound infections and respiratory tract infections. There have been some claims that the species *A. hydrophila*, *A. caviae* and *A. veronii* subsp. *sobria* can cause gastrointestinal illness but epidemiological evidence is yet not consistent (WHO 2002).

The main putative virulence factors are: exotoxins, endotoxin (LPS), presence of S-layers, fimbriae or adhesins and the capacity to form capsules (Merino 1995). The key virulence factors of *Aeromonas* have been investigated (Albert 2000) (Rahim 2004) (Erova 2006).

Dose – response data

The infectious dose is unknown.

© 2010 Global Water Research Coalition. Waterborne Pathogens: Review for the Drinking-water Industry. Prepared by Emmanuelle Guillot and Jean-François Loret. ISBN: 9781843391791. Published by IWA Publishing, London, UK.

Target populations

Individuals at the greatest risk of infection are children, the elderly and the immunocompromised (Merino 1995).

Diagnosis

Aeromonas can be isolated by culture of faeces on ampicillin containing blood agar. Confirmation is by biochemical tests (Hunter 1997).

Treatment

Treatment for infection with *Aeromonas* is generally not necessary for gastrointestinal illness. But for other pathologies, antibiotic therapy is usually implemented.

Geographical distribution

Worldwide.

Epidemiology

The health significance of detecting mesophilic aeromonads in public water supplies is not well understood. Although reports from Australia have suggested an association between gastroenteritis and aeromonads in drinking water (Burke 1984a) (Burke 1984b), other more recent works have shown little similarity between the majority of water-related *Aeromonas* strains and diarrhoea-associated isolates (Havelaar 1992) (Borchardt 2003) (Szczuka 2004). A consensus would be that only certain strains isolated from drinking water supplies are likely human enteric pathogens.

Transmission routes

The common routes of *Aeromonas* infections may be the ingestion of contaminated water or food or contact of the organism with a break in the skin through water related activities, such as swimming, diving, boating and fishing (Schubert 1991).

Waterborne outbreaks

No waterborne outbreak linked to either drinking water or recreational water contact has ever been reported (Hunter 1997). To date there is no firm evidence that direct transmission occurs via drinking water (WHO 2002).

Ecology

Regrowth in distribution systems/biofilms

Aeromonas has shown potential to grow in water distribution systems, especially in biofilms (Chauret 2001), where it may be resistant to chlorination. Biofilm production could explain the persistence of this Gram-negative pathogen organism in chlorinated tap water (Fernandez 2000).

Growth of *Aeromonas* generally occurred in the peripheral parts of distribution systems and was associated particularly with drinking water derived from anaerobic groundwater containing methane

(WHO 2002). *Aeromonas* densities usually showed a seasonal pattern, with peak values occurring in summer (WHO 2002). It has been recently reported that the isolation of *Aeromonas* in natural mineral water was the consequence of a localized development of a biofilm, with no exogenous contamination of the aquifer (Villari 2003).

The factors that affect the occurrence of *Aeromonas* in water distribution systems are not fully understood, but organic content, temperature, the residence time of water in the distribution network and the presence of residual chlorine have been shown to influence population sizes (WHO 2002).

Occurrence

Aeromonas species are commonly found in soil and water. They have been isolated in lakes, rivers, marine waters, sewage effluents as well as drinking waters (LeChevallier 1982) (Kühn 1997) (Legnani 1998) (Chauret 2001) (Sen 2004) and mineral and thermal waters (Hunter 1993) (Biscardi 2002). Reported densities in clean rivers, lakes and storage reservoir water range from 1 to 10^2 CFU/mL, in groundwater fewer than 1 CFU/mL, in post-treated drinking water from 0 to 10^2 CFU/mL and in drinking water distribution systems up to 10^2 to 10^3 CFU/mL attributed to growth in biofilms (Nichols 1996) (WHO 2002).

Aeromonas have also been found in high numbers in raw sewage (10^6 to 10^8 CFU/mL) and in sewage effluents (10^3 to 10^5 CFU/mL) (Theron 2002).

Since relatively little is known about the pathogenic mechanisms of *Aeromonas* species, several studies have demonstrated that many mesophilic aeromonads, mainly *A. hydrophila*, isolated from surface and drinking water can exhibit toxigenic factors and can thus pose a public health risk (Handfield 1996a) (Kühn 1997) (Ormen 2001). Recently, Razzolini *et al.* (Razzolini 2008) showed the presence of potentially pathogenic *Aeromonas* species in Brazilian domestic and public reservoirs and drinking fountains.

The rare species *Aeromonas culicicola*, so far only known in mosquitoes in India, has been surprisingly isolated from a drinking water supply in Spain (Figueras 2005), but since the taxonomy of *A. culicicola* is still controversial, these data will have to be confirmed.

Reservoirs

Environmental water sources.

Survival

Not extensively documented. Temperature has an influence on the survival ability of the bacteria in nutrient-poor water (Sautour 2003). Aeromonads have been found to grow between 5°C and 45°C with an optimal growth range of 22–28°C. Brandi *et al.* (Brandi 1999) reported that the survival of *Aeromonas* spp. varies considerably, depending on species and water type. Kühn *et al.* (Kühn 1997) reported that potentially pathogenic *Aeromonas* strains could persist for several months in drinking water.

Additionally Aeromonads readily multiply in domestic or industrial wastewater and are found in siphons, sinks and drainage systems (Schubert 1991).

Inactivation/removal

Published data on removal and inactivation are summarized in annex 1. It appears that if free cells of *Aeromonas* are relatively susceptible to common chlorine-based disinfectants (Knochel 1991) (Medema 1991), populations associated with biofilms may survive high chlorine levels (WHO 2002).

Recently, Silvestry-Rodriguez *et al.* (Silvestry-Rodriguez 2007) reported the efficacy of silver as a possible secondary disinfectant in water distribution systems.

Treatment can significantly reduce levels of *Aeromonas* in finished drinking water, but it appears difficult to control its growth in biofilms (Chauret 2001) (WHO 2002).

Surrogates

E. coli (or alternatively, thermotolerant coliforms) cannot be used as an index for the presence/absence of *Aeromonas* spp.

Environmental detection

Reference culture methods include membrane filtration followed by incubation on ampicillin-dextrin agar (Havelaar 1987) (HMSO 1994) (Handfield 1996b) typically at 28–30°C for 24–48 hours. Pre-enrichment with alkaline peptone water has proved successful for recovery of small numbers of *Aeromonas* from water (Moyer 1992), but identification to phenospecies or genospecies levels using classical biochemical testing can be problematic.

Pianetti *et al.* (Pianetti 2005) (Pianetti 2008a) showed that flow cytometry is a suitable method to detect strains of *A. hydrophila* from water samples in a viable but non culturable state.

A large number of phylogenetic studies on the genus *Aeromonas* has been published based on 16S rDNA RFLP analysis (Figueras 2000) or sequence analysis of gyrB and rpoD genes (Yanez 2003) (Soler 2004). Ormen *et al.* (Ormen 2005) reported a lack of agreement between biochemical and genetic identification of *Aeromonas* spp.

Recently, Donohue *et al.* (Donohue 2007) proposed a new analysis approach based on matrix-assisted laser desorption/ionization-mass spectrometry (MALDI-MS) to classify the *Aeromonas* species from environmental water samples.

As yet, few alternative immunological or molecular methods have been reported to detect and identify *Aeromonas* species in environmental water samples. Some PCR assays have been described: the detection of *A. caviae* and *A. trota* by polymerase chain reaction in seafood and water samples (Khan 1997); the detection of *Aeromonas* spp. in environmental water using 16S rDNA targeted oligonucleotide primers (Dorsch 1994), and recently the PCR identification of virulence factors of *Aeromonas* strains isolated from US drinking water utilities (Sen 2004) and RT-PCR (Yu 2008). 0ne study (Sen 2005) reported the development of a rapid identification method for *Aeromonas* species by multiplex PCR. The simultaneous detection of multiple pathogen species including *A. hydrophila* was described in marine waters by multiplex PCR (Kong 2002) and in wastewater by using oligonucleotide microarray technology (Lee DY 2008).

To date, there is no rapid and specific molecular method for the detection and quantification in environmental water samples of the most important human pathogen species, *A. hydrophila*, *A. caviae*, *A. veronii* subsp. *Sobria.*

In epidemiological studies, molecular typing methods are widely used such as RAPD, repetitive extragenic palindromic PCR (REP-PCR) (Szczuka 2004), pulsed-field gel electrophoresis (Borchardt 2003) (Villari 2003), ribotyping (Moyer 1992) and PCR-Density Gradient Gel Electrophoresis (DGGE) analysis (Tacão 2005).

Research needs

The pathogenicity, ecology and epidemiology of *Aeromonas* spp. are not yet fully understood and require further studies.

- At first, there is a need to develop rapid and specific PCR methods to detect and quantify the main pathogen species in water: *A. hydrophila*, *A. caviae*, *A. veronii* subsp. *sobria.*
- Waterborne transmission has not yet been well established. The health significance of detecting mesophilic aeromonads in public water supplies is not well understood: the occurrence and characterization of the virulent strains in drinking water should be investigated. The study currently conducted under the AWWARF project n°2957 ("Characterization of waterborne *Aeromonas* species

for their virulence potential") should address this need. Genetic diversity and stability of *Aeromonas* strains in water need to be studied, and further epidemiological studies should be carried out.

- The role of biofilm in the survival and growth of putative pathogenic *Aeromonas* strains in potable water should be investigated. The results of epidemiological studies may suggest that some water supplies are colonized with strains that are pathogenic to humans while others are not (Hunter 1997). Future research could help determine which factors in a water supply influence the strains types likely to colonize it. Transfer of antibiotics resistance and virulence factors in drinking water biofilms should be an area of future research using Aeromonas as a model organism.
- Further work is needed to understand the clinical significance of enteric isolates of *Aeromonas*, to clarify the pathogenic mechanisms of *Aeromonas* spp. and substantiate the causative role of these organisms in gastroenteritis. There is also a need for reliable data on the human infective dose for enteropathogenic aeromonad strains. Host susceptibility and bacterial virulence factor relationships should be further investigated.
- Treatment methods and/or operations for controlling *Aeromonas* regrowth in distribution systems are needed.

Burkholderia pseudomallei

Health significance	**Low**
Persistence in water supplies	**May multiply**
Resistance to chlorine	**Low**
Relative infectivity	**Low**
Important animal source	**No**

Adapted from table 7.1 Waterborne pathogens and their significance in water supplies (WHO 2004).

Microbiology

General description

Burkholderia pseudomallei is an environmental organism commonly found in soil and muddy water; it is a strictly aerobic, non-spore-forming, oxidase and catalase positive, Gram-negative bacillus.

B. pseudomallei is the aetiological agent of melioidosis, a life threatening septic infection prevalent in South-East Asia and Northern Australia. *B. pseudomallei* is an organism that has been considered as a potential agent for biological warfare and biological terrorism (Dance 2002).

Classification

Until recently this species was classified as belonging to the genus *Pseudomonas*.

Human health effects

Clinical features

Illness from melioidosis can be categorized as acute or localized infection, acute pulmonary infection, acute bloodstream infection, and chronic suppurative infection (Krishnan 2008). Unapparent infections are also possible. The incubation period (time between exposure and appearance of clinical symptoms) is not clearly defined, but may range from 2 days to as long as 26 years (Hunter 1997). Whether exposure to *B. pseudomallei* will result in disease, probably depends on the balance between the virulence of the strain, the immune status of the host (e.g. diabetes mellitus) and the size of the inoculum (Dance 2000).

Dose – response data

The infectious dose is unknown.

Target populations

Although melioidosis can occur in healthy children and adults, it occurs mainly in people whose defense mechanisms against infection are impaired by underlying conditions or poor nutrition or living conditions. Diabetes mellitus, renal failure, cirrhosis, and malignancy are known to be risk factors (Krishnan 2008).

© 2010 Global Water Research Coalition. Waterborne Pathogens: Review for the Drinking-water Industry. Prepared by Emmanuelle Guillot and Jean-François Loret. ISBN: 9781843391791. Published by IWA Publishing, London, UK.

Diagnosis

Melioidosis is diagnosed by isolating *B. pseudomallei* from the blood, urine, sputum, or skin lesions. Various serological tests, such as agglutination and complement fixation tests, are available.

Treatment

Most cases of melioidosis can be treated with appropriate antibiotics.

Geographical distribution

Tropical and subtropical areas of South-East Asia and Northern Australia (Dance 1999).

Melioidosis is endemic in South-East Asia, with the greatest concentration of cases reported in Vietnam, Cambodia, Laos, Thailand, Malaysia, Myanmar (Burma), and Northern Australia.

Epidemiology

Transmission routes

Although the epidemiology of melioidosis is not yet fully understood, humans and animals are believed to acquire the infection by inhalation of dust, ingestion of contaminated water, and contact with contaminated soil and water, especially through skin abrasions, and by contamination of war wounds in the case of military troops. Person to person transmission can occur by contact with the blood and body fluids of an infected person (Godoy 2003).

Waterborne outbreaks

Not extensively documented. In two Australian outbreaks of melioidosis, indistinguishable isolates, typed by pulsed-field gel electrophoresis, were isolated from cases and the drinking water supply (Inglis 2000) (Currie 2001).

Ecology

Regrowth in distribution systems/biofilms

No yet documented. May survive and multiply within biofilms.

Occurrence

B. pseudomallei is a widely distributed environmental saprophyte (Dance 2000). It can multiply in soil and water. The organism can predominantly be found in tropical regions, typically in soil or surface-accumulated muddy water, from where it may reach raw water sources and also drinking water supplies (WHO 2004). Isolation rates are higher during the rainy season and in still water rather than running water. Some studies reported its isolation from drinking water supplies (Inglis 2000) (Zanetti 2000) (Currie 2001).

Reservoirs

The main reservoir is believed to be environmental water and soil in tropical areas (Dance 2000). Besides humans, many animal species are susceptible to melioidosis. These include sheep, goats, horses, swine, cattle, dogs, and cats.

Survival

Little is known about the climate, physical, chemical and biological factors which control the proliferation and survival of *Burkholderia* spp. in the environment. The optimal temperature and pH values for *B. pseudomallei* are 24°C to 32°C and 5 to 8, respectively (Tong 1996), but Chen *et al.* (Chen 2003) reported that this bacterium can still grow at 4°C, which was suggested to be related with the occurrence of melioidosis in some cold areas.

Inactivation/removal

Data are limited. Two studies reported the effect of chlorine disinfection in potable water (Howard 2003) (Howard 2005). The most recent study has shown that chlorine proved the most effective disinfectant with a 99.99% reduction of a 10^6 CFU/mL pure bacterial culture, followed by 99.9% reduction by monochloramine, and 99% reduction by UV and suggested that amoebae can enhance the survival of the bacteria (Howard 2005). The published inactivation data are summarized in annex 1.

Surrogates

Because of the environmental occurrence of *B. pseudomallei*, *E. co*li (or, alternatively, thermotolerant coliforms) is not a suitable index for the presence/absence of this organism.

Environmental detection

A reference culture method is available but it is time consuming: it includes filtration through a 0.22 μm filter, then incubation on a selective broth at 40°C (Ashdown 1979) for seven days, followed by biochemical identification.

Some PCR assays have been developed to rapidly detect *B. pseudomallei* in soil (Brook 1997) or clinical samples (Lew 1994). Recently, real-time PCR assays have been used in blood samples (Tomaso 2005) or in soil (Kaestli 2007). However, no rapid molecular method has yet been developed for water samples.

In molecular epidemiology studies, molecular typing methods have been used, such as pulsed-field gel electrophoresis (Inglis 2000) (Currie 2001), ribotypes (Trakulsomboon 1997) or restriction endonuclease analysis (Yap 1995).

Research needs

The epidemiology and ecology of *B. pseudomalllei* are still poorly understood. In particular, more research is needed to investigate the role of amoebae and biofilms in the survival of *B. pseudomallei* in potable water.

To document the occurrence of this organism in water, there is a need for a rapid, sensitive and specific molecular method to detect and quantify *B. pseudomallei* in environmental water samples.

More epidemiological studies, including molecular methodologies for microbial source tracking, should be carried out, investigating the role of waterborne transmission.

Cyanobacteria

Health significance	**High**
Persistence in water supplies	**Yes**
Resistance to chlorine	**Moderate**
Relative infectivity	**Unknown**
Important animal source	**No**

Microbiology

General description

Cyanobacteria, otherwise known as blue-green algae, are among the most ancient species known on this planet. They are single-celled organisms that live in fresh, brackish, and marine water. A feature of cyanobacteria is that because they are planktonic organisms and need to float, they possess gas vacuoles, which act as buoyancy aids (Hunter 1997). Depending upon the species and environmental conditions, colonies may form filaments, sheets or even hollow balls.

Cyanobacteria are photosynthetic prokaryotes of significant ecological and biotechnological interest, since they strongly contribute to primary production and are a rich source of bioactive compounds. In eutrophic fresh and brackish waters, their mass occurrences (water blooms) are often toxic and constitute a high potential risk for human health. Of the cyanobacterial genera which include toxing-forming species, those of particular concern when mass populations occur include *Microcystis*, *Anabaena*, *Planktothrix* (formerly known as *Oscillatoria*), *Aphanizomenon*, *Cylindrospermopsis*, *Phormidium*, *Nostoc*, *Anabaenopsis* and *Nodularia*.

Cyanobacteria were included in the two first versions of the Contaminant Candidate List (CCL) (U.S. Environmental Protection Agency 1998) (U.S. Environmental Protection Agency 2005).

Guideline values for drinking water were recently introduced by the World Health Organization (WHO 1998), with a recommended limit of 1 μg of microcystin-LR equivalents per liter.

Classification

Despite being called algae, cyanobacteria are true prokaryote bacteria. They belong to the class Photobacteria, photosynthetic bacteria that contain chlorophyll. The phylogeny of the cyanobacteria is poorly understood at present. Most classification schemes are organized by cell or colony shape, but recent efforts may soon provide a truly evolutionary scheme.

Human health effects

The cause

Most illnesses related to cyanobacteria are thought to be mediated by toxins. Cyanobacterial toxins are classified according to the way in which they affect the human body.

Hepatotoxins (which affect the liver) are produced by some strains of the cyanobacteria *Microcystis*, *Anabaena*, *Oscillatoria*, *Nodularia*, *Nostoc*, *Cylindrospermopsis* and *Umezakia*. Hepatotoxins are the most widely distributed types of cyanobacterial toxins in aquatic environments; of these, the most commonly encountered in freshwaters are microcystins, which have also been identified as potent liver carcinogens (Nishiwaki-Matsuhima 1992).

© 2010 Global Water Research Coalition. Waterborne Pathogens: Review for the Drinking-water Industry. Prepared by Emmanuelle Guillot and Jean-François Loret. ISBN: 9781843391791. Published by IWA Publishing, London, UK.

Neurotoxins (which affect the nervous system) are produced by some strains of *Aphanizomenon* and *Oscillatoria*. Cyanobacteria from the species *Cylindrospermopsis raciborski* may also produce toxic alkaloids, causing gastrointestinal symptoms or kidney disease in humans (WHO 1999).

Understanding of the true range of cyanobacterial toxins, of their occurrence and health significance is still incomplete.

The cyanotoxins of highest concern to the water industry are:

- Neurotoxins: Anatoxin-a, Anatoxin-a(s), Saxitoxins, Neosaxitoxins including the new neurotoxic B-methylamino-L-alanine (BMAA) which may be associated with neurodegenerative diseases such as Alzheimer's disease (Cox 2003);
- Hepatotoxins: Microcystins, Nodularins, Cylindrospermopsins;
- Dermatotoxins: Aplysiatoxins and Lyngbiatoxins (which also affect the gastro-intestinal tract);
- Lipopolysaccharide (LPS) endotoxins which can affect any exposed tissue (WHO 1999).

Clinical features

Disease due to cyanobacterial toxins varies according to the type of toxin and the type of water or water-related exposure (drinking, skin contact, etc.). Adverse health effects associated with exposure to high concentrations of cyanobacterial toxins include stomach and intestinal illness, breathing disorders, allergic responses, skin irritation, liver damage and neurotoxic reactions such as tingling fingers and toes. Scientists are exploring the health effects associated with long-term exposure to low levels of cyanobacterial toxins. Some studies have suggested that such exposure could be associated with chronic illnesses, such as liver cancer and digestive-system cancer. Microcystins and nodularins are potent toxins, which are also tumor promoters (Falconer 1989). Low doses can disrupt the enzymes that control cell division, contributing to cancer (Ito 1997). The causes of the high prevalence of liver cancer in parts of China may be associated to repeated exposure to cyanobacterial microcystins in drinking water (Yu 1989).

Animals, birds, and fish can also be poisoned by high levels of toxin-producing cyanobacteria.

Dose – response data

The toxicity of known cyanobacterial toxins by various administration routes has been determined (Kuiper-Goodman 1999): the lethal dose$_{50}$ (LD_{50}) for microcystins by oral administration has been estimated to be between 5000–10900μg/kg, and a provisional guideline value of 1 μg/L for microcystin-LR in drinking water has been proposed by the WHO (WHO 2004).

Target populations

Children are at higher risk than adults for illness from cyanobacterial harmful algal blooms because of their low weight, with respect to the dose of toxin they can receive.

Diagnosis

Diagnosis is based primarily on history (recent contact with an algal bloom), signs of poisoning, and necropsy findings. It may be possible to identify cyanobacterial cells in the stomach content or faeces. In people who develop allergic-type symptoms, a skin test may be useful.

Treatment

No specific treatment is available. However, intense supportive therapy may be needed in some cases.

Geographical distribution

Cyanobacteria are found worldwide, including North and South America, Africa, Australia, Europe, Scandinavia and China. In warmer climates, these organisms can grow year-round.

Epidemiology

Transmission routes

Humans can be exposed to cyanobacterial toxins by drinking water that contains toxins (from a lake, reservoir or untreated water), by direct skin contact, in particular by swimming in water that contains high concentrations of cyanobacterial cells, or by inhaling water droplets from irrigation or water-related recreational activities. Transmission through consumption of filter-feeding shellfish such as mussels or fish may occur (Hunter 1991).

Waterborne outbreaks

Human populations have been adversely affected by the ingestion of water containing cyanobacterial cells and their toxins and by recreational and occupational skin contacts (Falconer 1983) (Hunter 1997) (Carmichael 2001). The most recent and serious known human poisoning episode attributed to cyanobacterial toxins occurred at Caruaru, Brazil in 1996, causing the death of over 50 patients. An outbreak of acute liver failure occurred at a dialysis center due to human intoxication by microcystins during renal dialysis treatment. It was estimated that 19.5 µg/L microcystin was in the water used for dialysis treatments, i.e.19.5 times the level set as a guideline for safe drinking water supplies by the World Health Organization (Carmichael 2001).

Several studies suggested that the high prevalence of liver cancer in parts of China may be associated to repeated exposure to cyanobacterial microcystins in drinking water (Yu 1989).

Ecology

Regrowth in distribution systems/biofilms

Toxic algae may grow in open-air finished water reservoirs (Lippy 1976).

Occurrence

Cyanobacteria are ubiquitous in aquatic environments. They are common in freshwater lakes and reservoirs and the intertidal zone of seashores, and can also be found on damp rocks, salt marshes, river beds and tree trunks (Hunter 1997).

Toxic cyanobacterial populations have been reported in freshwaters in over 45 countries, and in numerous brackish, coastal, and marine environments (Codd 2005) as well as in European freshwater bodies (Via-Ordorika 2004). Hoeger *et al.* (Hoeger 2005) confirmed that toxin-producing cyanobacteria (blue-green algae) are abundant in surface waters used as drinking water resources. Vieira *et al.* (Vieira 2005) reported the detection of toxic cyanobacteria at a concentration level up to of 20,000 cells/mL as well as Microcystins at concentrations of up to 1.25µg/L in public water supply reservoirs in Brazil.

Reservoirs

Aquatic environments. Specific reservoirs have not been identified.

Survival/growth

Most toxic cyanobacteria grow best in warm, clear, eutrophic or hypertrophic waters that are not turbulent. Explosive overgrowths of cyanobacteria are encouraged by excess nutrients in the water – primarily phosphorus and nitrogen – from sources such as agriculture, urban and rural runoff water, and wastewater. Maximum growth rates are attained by most cyanobacteria at temperatures above 25°C. Conditions limiting growth include low temperature (<10°C), low phosphate and nitrogen concentrations and high turbulence.

Inactivation/removal

Comprehensive information on management strategies for toxic blue-green algae is presented in a recent report by House *et al.* (House 2004). Rapala *et al.* (Rapala 2002) (Rapala 2006) evaluated the removal of cyanobacteria and cyanotoxins at a surface water treatment plant.

Chlorine has been shown to successfully remove a range of cyanotoxins including microcystins. Each cyanotoxin requires specific treatment parameters, particularly solution pH and free chlorine residual (Senogles-Derham 2003) (Daly 2007). Chlorination was ineffective in removing anatoxin-a (Nicholson 1994). The most effective process for destroying microcystins and nodularins seems to be ozonation (WHO1999) (Brooke 2006). Microcystins and anatoxin-a would be destroyed under conditions usually applied for ozonation prior to granular activated carbon (GAC) filtration. The saxitoxin class of compounds is very resistant to oxidation by ozone and would require further treatment such as GAC filtration (Rositano 2001). Newcombe *et al.* (Newcombe 2002) confirmed that ozonation of saxitoxin at moderate doses is ineffective. One study investigated the potential limitation of a water treatment system involving ozonation coupled with several filtration steps, when faced with various bloom conditions (Hoeger 2002). Recently, Sakai *et al.* (Sakai 2007) studied the effects of low or medium pressure ultraviolet lamp irradiation on pure cultures of *Microcystis aeruginosa* and *Anabaena variabilis*.

Surrogates

As cyanobacteria are common environmental organisms, *E. coli* (or alternatively, thermotolerant coliforms) cannot be used as an index for the presence/absence of cyanobacteria.

Environmental detection

- *Methods of detection of cyanobacterial toxins:*

Concentration of cyanobacterial toxins from water samples may be required and can be performed using concentration procedures such as Solid Phase Extraction (SPE) (Environment Agency 1998) or using more recent immunoaffinity methods (McElhiney 2002) (Arando-Rodriguez 2003). Most methods for the detection of cyanobacterial toxins in water samples have been directed at microcystin. The traditional method for detection and quantification of cyanobacterial toxins is the mouse bioassay. Törökné *et al.* (Törökné 2007) evaluated the sensitivity of two rapid microbiotests (Rapidtoxkit and Thamnotoxkit) based on the ingestion of cyanobacterial toxins by crustacean larvae.

Sensitive biological assays, such as the protein phosphatase inhibition assay (Heresztyn 2001), have been proposed as alternatives to the mouse assay. A rapid bioluminescence assay has also been developed (Lawton 1990). A wide range of chromatographic methods is being used. Currently, routine analysis of microcystins is most commonly carried out using high-performance liquid chromatography with photodiode array detection (HPLC-PDA) (Environment Agency 1998) (McElhiney 2005). The gold standard method for cyanotoxin analysis uses liquid chromatography (LC) coupled with mass spectrometry (MS), but this methods is laborious and time-consuming. A promising alternative for microcystin detection is the use of immunological assays (Metcalf 2003) (Sheng 2006). Recently, Pyo *et al.* (Pyo 2009) developed a microchip based enzyme-linked immunosorbent assay to detect low concentrations of cyanobacterial toxins in water, but one drawback is that these methods do not provide a direct indication of the toxicity, which can be possible with in vitro enzyme inhibition assays.

- *Cyanobacteria detection methods:*

Detection and identification of cyanobacteria are traditionally done by microscopic examination of water samples, but the morphological characteristics do not allow differentiation between toxic and non toxic cyanobacteria. In vivo fluorescence methods using phycocyanin can be used for the monitoring of potentially toxic cyanobacteria in freshwater (Gregor 2007) (Brient 2008).

Genetic methods such as PCR, and more recently DNA microarrays, are increasingly being developed to detect the presence of genes responsible for cyanobacterial toxin production (Pan 2002) (Fergusson 2003) (Hisbergues 2003) (Burns 2004) (Pearson 2008) (Rantala 2008). Other articles reported the development of species-specific PCR assays using specific sequences in the RNA polymerase gene (rpoC1) to detect the toxic cyanobacterium *Anabaena circinalis* (Fergusson 2000) and *Cylindrospermopsis raciborskii* (Wilson 2000). Baker *et al.* reported the development of PCR assay followed by RFLP or sequencing of the PCR products for identification of cyanobacteria in environmental samples (Baker 2001). Other studies developed molecular probes for the detection of toxigenicity in species capable of producing microcystins (Tillett 2001). Baker *et al.* (Baker J 2002) described the analysis of a cyanobacterial bloom by use of a PCR-based method for direct detection and identification of strains present and determination of their toxigenicity. A multiplex PCR assay for the rapid identification of potentially toxic environmental Microcystis from environmental samples has recently been described (Ouahid 2005). Several studies used a real-time PCR approach to quantify hepatotoxin-producing cyanobacteria (Foulds 2002) (Kurmayer 2003) (Furukawa 2006) (Koskenniemi 2007) (Rasmussen 2008).

In addition, improved molecular approaches to study cyanobacterial diversity at the strain level have been widely described, including PCR- RFLP (Neilan 1995), multiplex- Randomly Amplified Polymorphic DNA (Neilan 1995) and PCR-denaturing gradient gel electrophoresis (DGGE) (Lyautey 2005) (Zwart 2005).

Although a large number of molecular methods have been developed to detect and identify potential toxigenic cyanobacteria in water, all these methods are at the experimental stage and no validated protocol yet exists. It is impossible to compare results obtained by using different methods. Further evaluation and comparative studies need to be carried out in order to validate such alternative methods.

Research needs

Because toxigenic cyanobacteria can grow in drinking water reservoirs and recreational water, the USEPA included cyanobacteria in the first two versions of the CCL (1998 and 2005) and considered cyanobacterial toxins as a priority for investigation. Substantial research programs need to be conducted on the following themes:

- At first, there is a need to develop and validate molecular techniques for the detection and quantification of toxigenic cyanobacteria in water samples in order to acquire data on the occurrence and epidemiology of these organisms.
- It is also necessary to further develop accurate, sensitive and specific detection methods – including the sample treatment step – for individual toxins.
- Few studies have explored the links between harmful cyanobacterial algal blooms and human health. Particularly, the potential human health impact associated with the consumption of treated water from source waters with toxic blue-green algae remains unknown. Information on health effects is needed to document episodic and low dose exposure to cyanotoxins via recreational and drinking waters.
- Further research programs should be carried out in order to investigate the conditions favoring the occurrence of potentially toxic cyanobacteria and cyanotoxins, understand the dynamics

of cyanobacterial bloom events and associated problems, and thus facilitate the improved management of fresh water resources.
- In order to assess the health risk related to cyanotoxins more data are needed on their occurrence and distribution as well as on their persistence and degradation in aquatic environments.
- Additional research is required to determine whether standard and advanced water-treatment practices successfully remove cyanobacteria and their toxins from drinking water.
- Finally, risk management strategies for cyanobacterial cells and toxins in waterbodies need to be developed.

Legionella

Health significance	**High**
Persistence in water supplies	**Multiply**
Resistance to chlorine	**Low**
Relative infectivity	**Moderate**
Important animal source	**No**

Adapted from table 7.1 Waterborne pathogens and their significance in water supplies (WHO 2004).

Microbiology

General description

Legionellae were first recognized as human pathogens in 1976 after an outbreak of pneumonia among veterans attending a convention in Philadelphia (Fraser 1977).

Legionella spp. are heterotrophic bacteria found in a wide range of water environments; they can proliferate at temperatures above 25°C. They are Gram-negative, rod-shaped, non-spore forming bacteria that require L-cysteine for growth and primary isolation.

Classification

The genus *Legionella*, a member of the family *Legionellaceae*, has at least 48 species of which 19 have been associated with disease in humans (Muder 2002). The type species is *Legionella pneumophila*. Some species can be further differentiated into serotypes: there are at least 15 for *L. pneumophila* but so far no more than two for any other species.

Human health effects

Clinical features

Legionella infections can lead to two forms of disease, namely legionellosis or Legionnaires' disease, an acute purulent pneumonia with an incubation of 3–6 days, and Pontiac fever, a self-limiting non-pneumonic disease with high attack rate and onset (5 h to 3 days), consisting of fever, headache, nausea, vomiting, aching muscles and coughing.

L. pneumophila serogroup 1 is most frequently related to human disease. A study revealed that *L. pneumophila* serogroup 1 is specifically present in the Paris area (Lawrence 1999). *L. micdadei* is the second most frequent species causing legionellosis in the United-States and *L. longbeachae* infections are the most frequent in Australasia including New-Zealand (WHO 2002). To date legionellosis has been detected almost exclusively in humans, but some animals (guinea-pigs, rats, mice, marmosets, and monkeys) are susceptible to experimental infection (WHO 2002).

Dose – response data

The infectious dose is unknown, but it is supposed to depend on the susceptibility of the exposed population. Infectivity may be substantially enhanced if amoebae are inhaled. Vacuoles in infected

© 2010 Global Water Research Coalition. Waterborne Pathogens: Review for the Drinking-water Industry. Prepared by Emmanuelle Guillot and Jean-François Loret. ISBN: 9781843391791. Published by IWA Publishing, London, UK.

amoebae may contain many hundreds of *Legionella* cells providing a large inoculum in a restricted area of the respiratory tract (WHO 2002).

A model to qualitatively assess the risk of legionellosis has been proposed in which legionellosis risk can be described in terms of three factors: pathogen proliferation potential, exposure potential and population susceptibility (Cooper 2004) (McCoy 2004).

A simple quantitative microbial risk assessment (QMRA) based on the maximum risk curve has been applied for *Legionella* in drinking water distribution systems (Storey 2004a). More recently, exponential dose-response models for humans, derived from experiments on guinea-pigs, have been proposed to model the rates of infection and of Legionnaire's disease in humans (Armstrong 2008).

Target populations

Host factors influence the likelihood of Legionnaires' disease: males are more frequently affected than females, and most cases occur in the 40–70 year age group. Risk factors include smoking, alcohol abuse, cancer, diabetes, chronic respiratory or kidney disease and immunosuppression, as in transplant recipients. The disease is extremely rare among children under 15 years.

Diagnosis

Different laboratory tests can be used in the diagnosis of *Legionella* infection: culture on selective media, urinary antigen test, direct fluorescent antibody (DFA) stain of sputum or other samples from lung and antibody testing (serology). PCR detection and other molecular methods have been increasingly used for clinical diagnosis of legionellosis.

Treatment

Several antibiotics are effective *in vivo* against *Legionella*. The newer macrolides and quinolones are considered drugs of choice. Pontiac fever patients recover without specific therapy.

Geographical distribution

Worldwide.

Epidemiology

Transmission routes

Inhalation of *Legionella*-containing aerosols is considered as the usual mode of infection. Aspiration following ingestion of contaminated water is another route of infection. There is no evidence of person-to person transmission (Leclerc 2002).

Waterborne outbreaks

As legionellosis is a notifiable disease in most developed countries, a lot of epidemiological data are available. Legionellosis outbreaks are regularly reported and the per annum incidence rate of the disease is currently of 10 cases per million persons in Europe (http://www.ewgli.org).

Although legionellosis has been mostly sporadic and, until recently, frequent in hospitals, recent large outbreaks of Legionnaires' disease in Australia and Europe have spurred renewed interest in Legionnaires' disease.

Legionella outbreaks and single cases have been traced to cooling towers, evaporative condensers, hot water from showers, spas and whirlpools, nebulizers and other aerosol-producing medical devices

(Leclerc 2004). The information available suggests that the number of legionellae in the water cooling systems associated with outbreaks is always above 10^5 and usually greater than 10^6 CFU/L when determined by culture.

The most spectacular outbreak of legionellosis occurred in July 2001 in Murcia, Spain, with 449 confirmed cases and six deaths. The source of the outbreak was proven to be improperly controlled contaminated hospital cooling towers in the north of the city (Garcia-Fulgueiras 2003).

Previously, an outbreak occurred in April 2000 in Victoria, Australia, causing 93 confirmed cases and three deaths. The cause was an improperly treated cooling tower at the newly opened Melbourne Aquarium (McCoy 2004). This was the largest outbreak of legionnaire's disease ever recorded in Australia.

In 1999, an outbreak of Legionnaires' disease affected many visitors to a flower show in the Netherlands causing 21 deaths. The source was a contaminated whirlpool spa on exhibition at the flower show (Den Boer 2002). Another outbreak associated with spa pools was reported in New-Zealand (Ruscoe 2006). Recently, Simmons *et al.* (Simmons 2008) reported a Legionnaires' disease outbreak in Auckland caused by contaminated roof-collected rainwater.

In 2005 a community outbreak affecting 55 persons and associated with a cooling tower occurred in a large area in Catalonia, Spain (Sala Ferré 2008).

In France, 19 cases of legionellosis were identified among visitors to Paris during the World Soccer cup in June 1998. Three patients died. More recently in May 2005, an outbreak of 34 cases of Legionnaires' disease was detected in a suburb of Lyon. No deaths were reported. The source of infection could not be identified (Schmitt and Bitar 2005).

Outside hospitals, there have been no reports of outbreaks or recurrent cases of disease following consumption or use of drinking-water kept cool and not subjected to prolonged periods of stagnation (WHO 2002).

Several cases of Pontiac fever have been reported among employees of wastewater treatment plants in Denmark and Finland (Gregersen 1999) (Kusnetsov 2008). In France, wastewater treatment plants have also been involved in two legionellosis outbreaks: one indirectly generating 86 cases with 18 deaths, due to the contamination of a cooling tower located near a lagoon equipped with surface aerators (InVS 2004) and the other directly generating 9 cases (Van Cauteren 2008).

Ecology

Regrowth in distribution systems/biofilms

Aquatic biofilms provide comfortable ecological niches in which *Legionella* can settle and proliferate at temperatures between 20°C and 50°C. Legionellae frequently colonize water distribution systems where biofilms may be the reservoir for recontamination of water (Walker 1993). *Legionella* spp. are able to colonize all parts of internal distribution systems of hospitals, hotels and buildings (Yu 2000) (Perola 2005). According to some authors, the biofilms may support the survival and growth of legionellae outside a host cell (Rogers 1992). Other works found that amoebae were required for multiplication of the bacteria within biofilms (Murga 2001) (Kuiper 2004) (Wadowsky 2004). A recent study showed that *L. pneumophila* could persist for long periods (>4months) in biofilms in the viable but nonculturable state (Alleron 2008).

Concentrations have been found to be as much as 10 times higher in biofilms from faucets than from water collected from the faucets (Ta 1995). There is some evidence that pipe material can also affect colonization of legionellae (WHO 2002).

Occurrence

Legionella can commonly be found in natural aquatic environments and water supplies that meet drinking water standards (Leclerc 2004). The ubiquitous nature of legionellae in water means that water supplies, regardless of their source, may contain *Legionella* spp.

It has often been recovered in surface water and also in groundwater (Brooks 2004). One recent study pointed out the presence of *Legionella* species in Antarctic lakes, which suggests that these bacteria can adapt to extreme conditions (Carvalho 2008). Costa *et al.* (Costa 2005) reported that some strains were persistent in groundwater for at least 12 years. By using PCR-based methods, *Legionella* spp. can be detected in large numbers in all drinking water supplies, up to 10^5 cells/L, whatever the origin of the water (ground or surface). The *L. pneumophila* / *Legionella* spp. ratio in distributed waters is however generally low, below 5% (Diederen 2007) (Wullings 2006). *Legionella* concentrations are particularly high in sediments from distribution systems, up to 10^4 cells/mL of sediment (Loret 2008d).

These organisms have also been found in high numbers in sewage and treated wastewaters (Palmer 1993) (Gregersen 1999) (Kusnetsov 2006) (Medema 2002) as well as in sewage-contaminated coastal waters (WHO 2002).

Reservoirs

Aquatic biofilms and protozoa are natural reservoirs of legionellae. But other natural niches may exist and have not yet been identified.

Survival

The ubiquitous nature of *Legionella* reflects its ability to survive under varied water conditions, including temperatures from 0 to 63°C, and pH from 5 to 8.5. Temperature is a factor that significantly influences *Legionella* survival and growth. *L. pneumophila* multiplies at temperatures between 20°C and 50°C with maximal growth between 30°C and 40°C (Leclerc 2004).

It is now clear that legionellae are natural intracellular parasites of aquatic protist hosts, mostly amoebae such as *Acanthamoeba*, *Hartmannella*, *Vahlkampfia* and *Naegleria* which are "amplifiers" for legionellae in the environment. As natural reservoirs, protozoan phagocytes amplify *L. pneumophila* in fresh and potable water supplies. Batch experiments with tap water showed that *L. pneumophila* did not multiply in the absence of protozoa (Wadowsky 2004).

Free-living amoebae appear to be very important in legionellosis. Intracellular growth even enhances the infectivity of *L. pneumophila* to human-derived cells (Fritsche 1998) (Leclerc 2004).

Protozoa (*Acanthamoeba*) can enable legionellae to survive drinking water treatment processes including disinfection and be carried by water distribution mains into buildings (Kilvington 1990). This added resistance may play a large role in the survival of this bacterium and in its ecology and persistence in water distribution systems, cooling towers and hot tubs or other environments. A recent study suggested that a *L. pneumophila* serogroup1 isolate could persist in a domestic shower system despite repeated cycles of chlorination (Cooper 2008).

In low-nutrient environments, *L. pneumophila* is able to enter a viable but non culturable state, a process that is now well documented as starvation survival. This state can also result from other stresses such as temperature, antibiotics, salt and chlorine (Roszak 1987). The addition of amoebae to the dormant bacteria appears to result in the resuscitation of *Legionella* to a culturable state (Steinert 1997).

Legionella's ability to develop symbiotic relationships with other members of heterotrophic flora bacteria is thought to be important for their survival and proliferation in water. Only few studies have attempted to characterize the interactions between water bacteria and *Legionella*.

Legionellae have been shown to survive all phases of wastewater treatment and they apparently survive after discharge into the ocean (Palmer 1993).

Inactivation/removal

As for other bacteria, physical removal treatments such as coagulation, flocculation, sedimentation and filtration will reduce the number of *Legionella* in finished water. The effectiveness of conventional treatment is however limited by the ability of the bacterium to colonize filter media, especially GAC, together with amoebae (Loret 2008d).

Published data on inactivation are summarized in annex 1. UV light systems at doses commonly used in drinking water treatment may be useful for localized disinfection, keeping in mind that there are no distal residual effects (WHO 2002). *Legionella* strains are killed within 3 min on exposure to hot water at 70°C and exposure to ultraviolet light at 90 $mW.s/cm^2$ (Miyamoto 2000).

Legionella species are susceptible to ozone. The major problem with ozone is the poor penetration to all parts of the water distribution system.

Long-term disinfection has been obtained with continuous copper-silver ionization techniques, but there are no publications indicating the effectiveness of ionization in cooling towers. Recently, Pianetti *et al.* (Pianetti 2008b) evaluated the inactivation efficacy of combinations of copper and silver ions and free chlorine. A recent study reported the relative efficacy of electrochemical disinfection on *L. pneumophila* in tap water (Delaedt 2008).

In comparison to indicator organisms such as *E. coli*, a higher Ct value is necessary to achieve a comparable level of reduction in *Legionella* using chlorine and chlorine dioxide.

It has been shown that monochloramine as a disinfectant in municipal and hospital systems (rather than the more traditional free chlorine) is effective in controlling *Legionella.*

Recently chlorine dioxide, applied as a continuous treatment, was identified as the most effective for controlling *L. pneumophila* in a domestic water system (Thomas V 2004), but the same study showed that amoebae resisted all the treatments applied and probably acted as reservoirs for *L. pneumophila*, allowing quick re-colonization of the system once the treatments were interrupted.

Several studies showed the potential protection of *Legionella* by biofilms and amoebic hosts from the effects of disinfectants, low pH and heat (Miyamoto 2000) (Storey 2004b).

Surrogates

Escherichia coli (or thermotolerant coliforms) is not a reliable index for the presence/absence of *Legionella* (WHO 2004).

No suitable indicators have been identified to signal increasing concentrations of *Legionella* spp. in building plumbing systems. There is some evidence that increasing *Legionella* concentrations are accompanied or preceded by an increase in other bacteria, resulting in an elevated HPC (Heterotrophic Plate Count) measurement (i.e. > 100 CFU/mL) (WHO 2002). However, the correlation between HPC and *Legionella* is not consistent.

Environmental detection

- Culture methods are available: isolation usually requires concentration of the bacterial flora by centrifugation or membrane filtration, followed by heat (50°C for 30 min.) or acid (pH 2.2 for 5 min.) treatment before inoculation on to media containing antibiotics. Kusnetsov *et al.* (Kusnetsov 1994) provided a study of the comparison of various culture methods for the detection of *Legionella* spp. in water samples.

  The ISO 11731 method is consistent with the original method developed by the CDC and with standard methods used in Australia and Singapore (AU/NZ, 3896). A similar method is used in France (AFNOR T90-431) (McCoy 2004).

  All these culture methods require 10 days to yield confirmed results and cannot detect VBNC bacteria. Recently, Gião *et al.* (Gião 2008b) validated the use of the SYTO 9/propidium iodide fluorochrome uptake assay to rapidly detect viable but noncultivable *L. pneumophila* in the environment. Amoebal co-culture constitutes an efficient means to isolate *Legionella* as well as other amoeba-resistant bacteria (Greub 2004).

Saint (Saint 1998) described a convenient alternative to standard confirmation procedures for *Legionella* and *L. pneumophila* employing the EnviroAmp *Legionella* system and seroagglutination.

- *Legionella* spp. can also be detected by immunological methods: Yamamoto *et al.* (Yamamoto 1993) used a direct immunofluorescence assay for the detection of *L. pneumophila* (serogroups 1 to 6) in cooling tower waters. A rapid and simple immunochromatographic assay kit (Duopath Legionella) has been evaluated (Inoue 2007) (Koide 2007). But these approaches lack sensitivity for the detection of *Legionella* in water samples and do not distinguish between living and dead cells.
- Molecular methods have been developed to detect *Legionella* spp. or/and *L. pneumophila* in water samples, including fluorescence in situ hybridization (Manz 1995) (Buchbinder 2002) (Declerck 2003a) or the most widely used approach based on PCR-based detection (Catalan 1994) (Maiwald 1994) (Ballard 2000) (Levi 2003). A PCR assay was reported for the identification and discrimination of *L. longbeachae* serogroups 1 and 2 (Saint and Ho 1999). PCR also appears to be a reliable method for cooling tower waters (Koide 1993) (Yamamoto 1993) (Chang 1995); Majid *et al.* (Majid 2007) described a PCR-Elisa method and Horng *et al.* (Horng 2006) a multiplex nested PCR followed by detection using immunochromatography (PCR-ICT hybrid assay). However cooling water often contained PCR-inhibiting substances that could result in false negative PCR results for *Legionella* (Declerck 2003b). Catalan *et al.* (Catalan 1997) developed a nested-PCR method to detect *L. pneumophila* in wastewater.
  Recently, two real-time PCR assays were developed: one for the direct confirmation of *Legionella* primary cultures (Giglio 2005), the other for rapid detection of *L. pneumophila* in water (Fiume 2005). Some PCR-diagnosis kits are available including Aqua Screen® Water Test System (Minerva Biolabs GmbH) and AGFOOD kit (Biotools), but all these PCR-based detection methods cannot differentiate between dead or living cells and are essentially qualitative.
- An interesting study showed the ability of a *L. pneumophila* specific PCR-based method to detect not only culturable bacteria but also viable but non culturable (VBNC) forms and dead bacterial DNA (Dusserre 2008). Recently a new approach combining the use of photoactivated ethidium monoazide and PCR or real-time PCR was proposed to specifically detect viable *Legionella* cells (Chang 2009).
- An increasing number of quantitative PCR approaches have been developed including a competitive PCR approach (Declerck 2006) Several recent studies reported the development of real-time PCR methods to quantify *Legionella*: a real-time LightCycler PCR for the detection of legionellae in hospital water (Wellinghausen 2001), a quantitative real-time PCR assay combined with an immuno-magnetic separation system (Yanez 2005) and quantitative real-time PCR assays tested on a large number of water samples (Joly 2006) (Behets 2007) (Yaradou 2007). These promising approaches should allow quick monitoring of water samples for the risk assessment of environmental contaminations. It must be noted that a real-time PCR assay for quantification of *Acanthamoeba* (a well known natural reservoir for *Legionella*) has also been described (Riviere 2005).
- A new approach based on a flow-through chemiluminescence microarray readout system was developed to simultaneously detect *Escherichia coli* O157:H7, *Salmonella typhimurium* and *Legionella pneumophila* (Wolter 2008).
- *Typing methods:*Both environmental strains and clinical isolates can be successfully subtyped by molecular techniques such as ribotyping, macrorestriction analysis by pulsed-field gel electrophoresis or RAPD (Van Belkum 1996) (WHO 2002) (Lepeuple 2004).

Research needs

For more than two decades since it was first recognized, legionellosis has represented a continuing public health challenge. There are still many gaps in our knowledge of the ecology and epidemiology of this organism.

Analytical methods

Several quantitative-real-time PCR methods have been developed these last years, but they need to be validated and standardized in order to be recognized as alternatives to the reference culture method.

There is also a need for molecular approaches which could determine viability and thus distinguish between dead and living bacteria, including culturable and VBNC forms.

Ecology

Further research should be carried out on the normal behavior of *Legionella* species and their natural environmental habitat and ecological niches.

For example, the need for protozoa for the proliferation of *L. pneumophila* in aquatic habitats is still not fully understood and remains controversial. Little is known about the range of this parasitic association and there may be many undiscovered protist hosts.

The interactions between *Legionella* and other bacteria in such diverse habitats as free water and biofilms remain also unknown.

Occurrence:
There is a need to acquire more data on the occurrence and level of contamination of the different species of *Legionella,* by using molecular methods.

Epidemiology

The role of biofilms and protozoa in the epidemiology of legionellosis should be further investigated. What is the impact of the transmission mode (within biofilms, within protozoa or as free-living bacteria) on the outcome of the disease following exposure to *Legionella pneumophila*?

The risk associated with waste water treatment or re-use should be better evaluated.

The host susceptibility factors for legionellosis should be investigated

Treatment

Work is needed to improve the control of *Acanthamoeba* spp. and other amoebae, in order to better control the presence of *Legionella* in water supplies. More broadly, as for mycobacteria, water treatment approaches that minimize selection of treatment resistant pathogens, the survival of protozoa, biofilm formation and production of disinfection by-products need to be investigated.

Non-tuberculosis mycobacteria (NTM)

Health significance	**Low**
Persistence in water supplies	**Multiply**
Resistance to chlorine	**High**
Relative infectivity	**Low**
Important animal source	**No**

Adapted from table 7.1 Waterborne pathogens and their significance in water supplies (WHO 2004).

Microbiology

General description

The tuberculous or "typical" species of *Mycobacterium*, such as *M. tuberculosis*, *M. bovis*, *M. africanum* and *M. leprae*, have only human or animal reservoirs and are not transmitted by water. By contrast *Mycobacteria* spp. that have been implicated in waterborne transmission are generally known as atypical mycobacteria or mycobacteria other than tuberculosis (NTM) and are natural inhabitants of a variety of water environments. Environmental opportunistic mycobacteria include both slowly and rapidly growing mycobacteria. By far, the most important slowly growing species are *Mycobacterium avium* and *M. intracellulare*, called the *M. avium* complex (MAC) (also called *M. avium-intracellulare*) responsible for the majority of non- *M. tuberculosis* mycobacteria infections in developed countries (Horsburgh 1996) (WHO 2004).

MAC was included in the first two versions of the USEPA Drinking water contaminant candidate list (U.S. Environmental Protection Agency 1998) (U.S. Environmental Protection Agency 2005).

They are aerobic, Gram-positive, rod-shaped, and acid-fast bacteria. They differ from other bacteria by having a complex cell wall with high lipid content, which is used in identification of the organisms using acid-fast staining.

Classification

Mycobacterium is the single genus in the family Mycobacteriaceae, order Actinomycetales. Over 70 *Mycobacterium* species have been defined, at least 30 of which cause disease in human and animals. Among environmental opportunistic mycobacteria, the main potential human pathogens species are: *M. avium*, *M. intracellulare*, *M. chelonae*, *M. kansasii*, *M. marinum*, *M. fortuitum* and *M. ulcerans*.

Human health effects

Clinical features

Atypical *Mycobacterium* spp. are opportunistic bacterial pathogens which can cause a range of diseases involving the skeleton, lymph nodes, skin and soft tissues, as well as the respiratory, gastrointestinal and genitourinary tracts. Manifestations include pulmonary disease, Buruli ulcer, osteomyelitis and septic arthritis in people with no known predisposing factors (WHO 2004).

© 2010 Global Water Research Coalition. Waterborne Pathogens: Review for the Drinking-water Industry. Prepared by Emmanuelle Guillot and Jean-François Loret. ISBN: 9781843391791. Published by IWA Publishing, London, UK.

Pulmonary disease resembling tuberculosis may be associated with *M. kansasii, M. avium-intracellulare*; lymphadenitis may be associated with *M. scrofulaceum, M. avium* complex; skin ulcers and soft tissue wound infections may be associated with *M. fortuitum, M. cheloni, M. ulcerans* and *M. marinum.* Current research suggests a possible role for MAC organisms in the development of Crohn's disease (Hermon-Taylor 2000), an inflammatory bowel disease similar to Johne's disease in sheep, cattle and goats.

Dose – response data

The infectious dose is unknown.

Target populations

For most immunocompetent people, exposure to MAC has no effect on health, or possibly causes localized infection, but it constitutes a major cause of disseminated infections in immunocompromised patients, commonly causing death in HIV-positive persons (WHO 2004).

Diagnosis

Diagnosis is difficult and time-consuming. It may be suggested by seeing acid-fast bacilli in sputum or during the histological examination of biopsy specimens, then it is confirmed by culture on appropriate media, which can take up to two months (Hunter 1997).

Treatment

Treatment of the infection depends upon the sensitivity of the infecting organism to specific antibiotics. As many as 4–6 drugs may be used to treat some infections and treatment may require 6 months to 2 years. Certain lymph node infections and skin lesions can be surgically removed.

Geographical distribution

Worldwide.

Epidemiology

Transmission routes

Transmission of nontuberculous mycobacteria to man occurs by various mechanisms depending on the species (Dailloux 1999). The process by which these bacteria are transmitted is not completely understood. Principal routes of infection appear to be inhalation, contact and ingestion of contaminated water as well as consumption of contaminated food, but there is no evidence for person-to-person transmission.

Waterborne outbreaks

Epidemiological studies suggest that natural or drinking waters are the principal source of human contamination (von Reyn 1994). Mycobacterial infections linked to contaminated hospital water, particularly hot water systems, have been recognized for many years (Aronson 1999). Infections are generally sporadic, but in all these cases, there is only circumstantial evidence of a causal relationship between the occurrence of bacteria in drinking-water and human disease (WHO 2002). Recently however, drinking water has been identified as the source of exposure to *M. avium* after comparing clinical isolates and strains isolated from point of use water taps using molecular methods (Hilborn 2008).

Infections have also been linked to contaminated water in spas (Lumb 2004).

Ecology

Mycobacteria can be recovered from a wide variety of environmental niches (natural water, drinking-water, biofilms, aerosols, soils, foods, plants and animals).

Regrowth in distribution systems/biofilms

MAC and other mycobacteria have been shown to be present in biofilms (Carter 2003) (September 2004) (WHO 2004) where they can proliferate with densities usually between 10^3 and 10^4 CFU/cm^2 (Schulze-Robbecke 1992). Some species, such as *M. Kansasii*, can colonize cold water distribution systems whilst *M. xenopi* and *M. avium* are commonly associated with hot water systems (WHO 2004). High numbers of atypical mycobacteria may occur in distribution systems after events that dislodge biofilms, such as flushing or flow reversals. Torvinen *et al.* (Torvinen 2007) reported that temperature is a key factor in the survival of *M. avium* in drinking water biofilms. Resistance to ozone and chlorine-based disinfectants is undoubtedly one reason why MAC and other mycobacteria grow and persist in drinking water distribution systems. In fact disinfection of water can lead to selection of *M. avium* and *M. intracellulare* and other mycobacteria by killing competitor organisms (Falkinham 2001).

One study showed that the survival of *M. avium* in a model drinking water distribution system was dependent upon a complex interaction between pipe surface, nutrient levels and disinfectants (Norton 2004). This study suggested that reducing the biodegradable organic material in drinking water, control of corrosion, maintenance of an effective disinfectant residual, and management of hot water temperatures can help limit the occurrence of *M. avium* complex in drinking water biofilms.

Occurrence

Data are limited because isolation of atypical mycobacteria from water is laborious and time-consuming.

MAC and some other mycobacteria have been isolated from natural waters throughout the world, including lakes, rivers, ponds and streams, sea water and wastewater. They also have been frequently recovered from drinking water systems before and after treatment, from distribution systems and from raw source waters (Falkinham 2001) (LeChevallier 2001) (Le Dantec 2002) (Thomas 2008). Although the data confirm that members of the genus *Mycobacterium* are isolated in drinking water, the numbers and frequencies of recovery of *M. avium* and *M. intracellulare* are low.

Environmental mycobacteria have been isolated in bottled table waters in Greece (Papapetropoulou 1997). MAC has been recovered from pools and spas in the Nertherlands (Havelaar 1985).

Reservoirs

Nontuberculous mycobacteria are acquired from the environment (soil, water, dust, tissues of domestic animals...), but specific reservoirs of these organisms leading to human disease have not been defined (Horsburgh 1996).

Although environmental opportunistic mycobacteria have been isolated from wild and domestic birds and animals, the zoonotic potential of MAC infections remains unclear (Falkinham 2002).

Survival

Not extensively documented.

The physiological characteristics of non-tuberculous mycobacteria provide an understanding of their ecological distribution and survival: 1) they can grow over a wide pH range and they grow best at low pH values (around 5); 2) they grow microaerobically; 3) *M. avium* strains can grow over a wide range of temperatures, up to 45°C or more (up to 57°C), unlike *M. intracellulare* which can grow only up to 42°C; 4) they grow equally well in water with high levels of salt and are found in large numbers in brackish swamps and estuaries; 5) they are hydrophobic and collect largely at air-water interfaces;

6) they are relatively resistant to heavy metals and may be isolated from water with high levels of metal pollution; 7) their high resistance to disinfection by chlorine undoubtedly contributes to their persistence in drinking water systems (Leclerc 2004).

One study reported that the remarkable stress survival exhibited by *M. avium* further suggests persistence in a large range of environments (Archuleta 2002). More recently, Archuleta *et al.* (Archuleta 2005) found that *M. avium* enters a state of metabolic dormancy in response to starvation in the environment.

M. avium is capable of survival and growth in phagocytic protozoa and amoebae (*Acanthamoeba polyphaga* and *A. castellanii*) (Adékambi 2006), and such growth can increase its virulence and provide protection from antimicrobial agents (Steinert 1998). One study reported that protozoa intracellular growth of *M. avium*, *M. intracellulare*, and *M. scrofulaceum* was 40-fold faster than that of water-grown isolates (Strahl 2001).

Inactivation/removal

Published data on inactivation are summarized in annex 1.

MAC are relatively resistant to chlorine, chloramine, chlorine dioxide, and ozone (Taylor 2000). One study reported the persistence of nontuberculous mycobacteria in a drinking water system despite the addition of ozonation and filtration treatment (Hilborn 2006). Falkinham *et al.* (Falkinham 2003) suggested that simultaneous exposure of *M. avium* and other mycobacterial cells to chlorine at high temperatures might result in increased eradication of *M. avium* from public water supplies. Other opportunistic environmental mycobacteria, including *M. kansasii*, *M. marinum*, *M. fortuitum*, *M. phlei*, and *M. chelonae*, have been shown to be relatively resistant to chlorine at concentrations used in municipal water systems for disinfection. However, Vicuña-Reyes *et al.* (Vicuña-Reyes 2008) reported that chlorine dioxide effectively inactivates *M. avium* at doses currently applied in drinking water treatment.

M. avium were killed within 3 min when exposed to hot water at 70°C and exposed to ultraviolet light at 90 mW.s/cm^2 (Miyamoto 2000). Shin *et al.* (Shin 2008a) recently suggested that *M. avium* is one of the rare waterborne micro-organisms that are highly resistant to both chemical and UV disinfection in water. Nevertheless, UV doses applied in drinking water treatment are sufficient to effectively inactivate *M. avium.*

Surrogates

Escherichia coli (or thermotolerant coliforms) is not a reliable index for the presence/absence of atypical mycobacteria (WHO 2004).

No suitable indicators have been identified to signal increasing concentrations of MAC organisms in water systems. There is some evidence that the presence of *M. avium* is associated with turbidity in raw water (Falkinham 2001) but further investigation on this issue is needed.

Currently the presence of mycobacteria in water is not regulated by any country or international organization.

Environmental detection

Culture-based methods for isolation of MAC and other slow-growing mycobacteria are difficult as these organisms grow very slowly and tend to be dominated by other, more rapidly growing organisms.

Chemical decontamination methods were developed to reduce overgrowth problems but they may destroy some mycobacterial cells. This step is not necessary for drinking water samples. Several studies have been conducted to try and determine the optimum decontamination method but there is no clear consensus as to the best method.

Alternative purification methods based on IMS have been proposed for the specific enrichment of target cells (WHO 2004). Amoebal co-culture has also proved an efficient means to isolate mycobacteria as well as other amoeba-resistant bacteria (Greub 2004).

Reference culture methods to isolate NTM from water include four main steps: concentration by filtration or centrifugation, decontamination of the sample, growth on a selective media and enumeration of samples. Thompson *et al.* (Thompson 2008) suggested that for the analysis of drinking water samples, filtration is a more sensitive method than centrifugation for the concentration of mycobacteria.

These culture-based methods are laborious and time-consuming (weeks or months), they cannot detect VBNC bacteria and they do not allow species identification.

To overcome these limitations, molecular methods have been increasingly used for several years. Detection and identification of NTM to the species level have been reported using a culture enrichment step and PCR-restriction fragment length polymorphism (PCR-RFLP) in tap water (Chang 2002) and in bottled table waters (Papapetropoulou 1997).

FISH has been developed to visualize mycobacteria *in situ* in culture. This technique constitutes a promising approach particularly for biofilm samples but still needs extensive development (Stender 1999) (Hongmanee 2001) (Lehtola 2006b) before being operational on environmental water samples.

Specific PCR tests have been developed: a PCR detection of *M. ulcerans* in the environment (Stinear 2000), a real-time PCR method for the quantification of *M. ulcerans* DNA (Rondini 2003) and a PCR detection of *M. avium* complex (MAC) (MacGregor 1999). Two detection tests specific for *M. avium paratuberculosis* from water samples have been developed: a real-time PCR assay for quantitative detection (Rodriguez-Lazaro 2005) and an IMS-PCR assay (Whan 2005). Recently, a nested PCR-restriction fragment length polymorphism protocol was described to detect and differentiate nontuberculous mycobacteria species in hospital tap water (Shin 2008b).

Currently all these available alternative molecular methods are at the experimental stage and validated protocols do not yet exist.

Several commercial tests for rapid species-level identification of mycobacteria have been developed (INNO-Lipa Mycobacteria kit from Innogenetics and Accuprobe from Genprobe) but these tests lack specificity and can give false-positive results.

Fingerprinting methods such as pulsed-field gel electrophoresis (PFGE), RAPD, or Inter-insertion sequence PCR, have proven very useful in epidemiological studies (WHO 2004). A new approach for genotyping mycobacteria in clinical samples using DNA chip technology has been reported (Troesch 1999). Extensive research work must be carried out before this promising approach can be operational in environmental water samples.

Research needs

The following issues can constitute priorities for future research on atypical mycobacteria:

Analytical methods

Develop specific tests and testing strategies for atypical mycobacteria.

- Culture-based methods: Improvements in the methodology for isolation and enumeration of mycobacteria in water and biofilms samples are necessary. For example by developing a more selective medium or isolation conditions which will enable only MAC or other mycobacteria to grow and that will not require a decontamination step.
- Molecular methods: Develop and validate specific PCR tests for rapid detection and quantification of pathogenic environmental mycobacteria species and strains.

Virulence factors

Virulence factors are poorly understood for mycobacteria. The lack of knowledge of virulence markers hinders the identification of the sources of MAC infection.There is a need to identify genetic or phenotypic markers of pathogenic strains. Genetic markers are the basis for modern molecular detection methods and their knowledge will allow the development of pathogen-specific molecular tests.

Epidemiology

The host susceptibility factors for MAC infection should be investigated.

Biofilm

Biofilm might play a role in the survival of virulent mycobacteria strains. Further studies should be carried out on this topic in order to identify approaches for inhibiting *M. avium-intracellulare* (and other mycobacteria) biofilm formation and colonization, for example, by identifying *M. avium* genes involved in biofilm formation and investigating the effect of disinfectants on the establishment of biofilm.

Additional studies are needed on the role of amoebae and protozoa in the survival, proliferation and the virulence of atypical mycobacteria in biofilms.

Studies on the transfer of resistance to antibiotics and virulence factors in drinking water biofilms should be an area of future research.

Waterborne transmission

Limited data exist to evaluate the risk of *M. avium* complex infection from water.

- Further epidemiological studies using molecular biology techniques such as pulsed field gel electrophoresis should be carried out in order to investigate the role of water as the principal source of contamination.
- More systematic studies of drinking water systems over long period of times and of the efficacy of water treatment steps need to be carried out.
- Additional research is needed on the occurrence, survival and reservoirs of MAC in water.

Inactivation

- There is a need to develop disinfection strategies, which are effective on MAC.
- Additional research on the susceptibility of atypical mycobacteria to treatment processes is needed. More broadly, as for *Legionella*, water treatment approaches that minimize selection of treatment resistant pathogens, the survival of protozoa, biofilm formation and production of disinfection by-products need to be investigated.
- Furthermore, data are limited on the effectiveness of control measures that could be applied to reduce the health risk associated with these organisms. As the eradication of nontuberculous mycobacteria in water is not possible, there is a need for preventive measures, which could minimize their presence and proliferation, particularly in biofilm of drinking-water systems.

Risk assessment

Dose-response data, in particular for MAC, are needed to establish public health goals for potable water supplies (WHO 2004).

Viruses

Adenovirus

Health significance	**High**
Persistence in water supplies	**Long**
Resistance to chlorine	**Moderate**
Relative infectivity	**High**
Important animal source	**No**

Adapted from table 7.1 Waterborne pathogens and their significance in water supplies (WHO 2004).

Microbiology

General description

Adenoviruses are medium-sized (90–100 nm), nonenveloped icosohedral viruses containing double-stranded DNA. First isolated from the adenoids of children in 1953, they are the only human enteric viruses with double stranded DNA.

They were included in the first two versions of the U.S. Environmental Protection Agency drinking water contaminant candidate list (CCL) (U.S. Environmental Protection Agency 1998) (U.S. Environmental Protection Agency 2005).

Classification

Human adenoviruses belong to the family of Adenovirus and the genus of Mastadenovirus. They have been classified into six subgenera (A to F) and 47 immunologically distinct types (serotypes).

Human health effects

Clinical features

Adenoviruses can cause a variety of clinical syndromes depending on the infecting serotype:

- Subgenera A-E are generally associated with respiratory infections, conjunctivitis, pharyngitis, pneumonia, acute and chronic appendicitis, exanthematous disease, bronchiolitis, acute respiratory disease, febrile illness (sore throat, glands) (Baum 1995).
- Subgenus F, which includes types 40 and 41, is consistently associated with gastroenteritis. The two serotypes, 40 and 41 are referred to as enteric adenoviruses. They are considered second only to rotaviruses as the primary causes of gastroenteritis in children (Uhnoo 1984).

Adenoviral infections typically last from a couple of days to a week. Severe respiratory infections can last longer and cause persistent symptoms, like a cough. Pneumonia can last from 2 to 4 weeks.

For some adenovirus serotypes, the clinical spectrum of disease associated with infection varies depending on the site of infection; for example, infection with adenovirus 7 acquired by inhalation is associated with severe lower respiratory tract disease, whereas oral transmission of the virus typically causes no or mild disease.

© 2010 Global Water Research Coalition. Waterborne Pathogens: Review for the Drinking-water Industry. Prepared by Emmanuelle Guillot and Jean-François Loret. ISBN: 9781843391791. Published by IWA Publishing, London, UK.

Dose – response data

The infectious dose in humans is unknown but presumably is 10–100 virus particles. A dose-response model developed from a study involving human inhalation of aerosols has been proposed for adenovirus type 4 (Crabtree 1997).

Target populations

Adenoviruses are opportunistic pathogens, as they tend to affect children, the elderly and the immunocompromised. The virus is also associated with crowding and stress.

Diagnosis

Antigen detection, polymerase chain reaction assay, virus isolation, and serology can be used to identify adenovirus infections (from throat, conjonctival swabs, or stools). Adenovirus typing is usually accomplished by hemagglutination-inhibition and/or neutralization with type-specific antisera.

Since adenoviruses can be excreted for prolonged periods, the presence of the virus does not necessarily mean that it is associated with disease.

Treatment

Most infections are mild and require no therapy or only symptomatic treatment. Because there is no virus-specific therapy, serious adenovirus illness can be managed only by treating symptoms and complications of the infection.

Vaccines have been developed for adenovirus serotypes 4 and 7, but only for purposes of preventing Acute Respiratory Disease (ARD) among military recruits.

Geographical distribution

Worldwide.

The prevalence of adenoviruses is high in both developed and developing countries.

Epidemiology

Transmission routes

All types of adenoviruses are transmitted by faecal-oral route, by direct person-to-person contact, and occasionally by indirect waterborne or foodborne transmission.

For non enteric adenoviral infections, airborne spread through coughing and sneezing is probably common.

Waterborne outbreaks

Although the potential health risks constituted by adenoviruses in water sources and supplies are widely recognized (American Water Works Association 1997) (U.S. Environmental Protection Agency 1998) (WHO 2002), no food or waterborne outbreak associated with drinking water has ever been reported.

The only waterborne outbreaks reported in the literature have been associated with inadequately chlorinated swimming pools. These outbreaks are generally associated with types 3 and 4 and result in pharyngoconjuctivitis (Papapetropoulou 1998) (Harley 2001).

Ecology

Regrowth in distribution systems/biofilms

They are environmentally inert and only replicate in their hosts. Their persistence in biofilms has never been documented.

Occurrence

Human adenoviruses (Ads) are ubiquitous and can be found in the environment where contamination by human faeces and sewage has occurred.

Using molecular techniques, adenoviruses have been frequently found in treated and untreated waters throughout the years (Castignolles 1998) (Chapron 2000) (Lee 2002) (Van Heerden 2003) (Jiang 2004) (Lee C 2004) (Lee 2004) (Choi 2005), in recreational water, sewage (Puig 1994) (Schvoerer 2000) and coastal water and shellfish (Pina 1998b).

Data are limited about the prevalence of enteric adenoviruses 40 and 41 in water sources. Levels of contamination in water are not documented.

The public health significance of adenoviruses presence in water is not well established (Jiang 2006).

Survival

Not extensively documented.

Adenoviruses are unusually stable to chemical or physical agents and adverse pH conditions, allowing for prolonged survival outside the body.

Adenoviruses have been found to be quite persistent in both tap and sea water. They can survive for days to weeks. Enriquez *et al.* (Enriquez 1995) reported that adenoviruses (Ad 40 and Ad 41) were much more stable than either poliovirus type 1 or hepatitis A. Infectivity is undiminished by refrigeration (4°C) for 70 days (American Water Works Association 1997).

Reservoirs

Recently, Lorenzo-Morales (Lorenzo-Morales 2007) suggested that *Acanthamoeba* should be considered as a potential reservoir for some adenovirus serotypes.

Inactivation/removal

Published data on removal and inactivation are summarized in annex 1.

The most effective germicides (with a minimum 15 min. contact time) are phenol (5%) and sodium hypochlorite (household bleach diluted to 200 ppm or 10%).

The virus is also effectively killed by heat, at 56°C (American Water Works Association 1997).

Inactivation can be achieved with chlorination or ozonation levels comparable to those used in water treatment. Adenovirus type 40 is readily inactivated by 0.1 mg/L of free chlorine (American Water Works Association 1997) (Thurston-Enriquez 2003) (Thurston-Enriquez 2005a), or 0.3 mg/L of ozone (Thurston-Enriquez 2005b), but a recent study (Sirikanchana 2008) revealed that monochloramine disinfection does not provide adequate control of adenoviruses in drinking water at high pH and low temperature.

In a recent study conducted at different drinking water treatment plants Albinana-Gimenez (Albinana-Gimenez 2009) demonstrated that conventional treatment can provide more than 5 log reduction of adenoviruses.

Adenoviruses are much more resistant to ultraviolet light (UV) than any other pathogen of concern because their double-stranded DNA genome allows them to use the host cell enzymes during replication to repair UV damage in the DNA (Qin 1996) (Nwachuku 2005) (Eischeid 2009). The UV doses for a 90% inactivation of adenoviruses type 2, 15, 40 and 41 are in the range of 25 to 56 mW-s/cm (Hijnen 2006). Linden *et al.* (Linden 2007) reported enhanced UV inactivation using polychromatic UV lamps. A recent study revealed that the genomes of adenoviruses are equally damaged by medium pressure and low pressure UV, but that medium pressure UV was more effective to inactivate adenoviruses in cell culture (Eischeid 2009).

Surrogates

Adenoviruses have been detected in drinking water supplies that met accepted specifications for treatment, disinfection and conventional indicator organisms.

Because of the higher resistance of the viruses to disinfection, *E. coli* (or, alternatively coliforms) is not a reliable index of the presence/absence of adenoviruses in drinking-water supplies (WHO 2004).

Environmental detection

Most serotypes grow well in standard cell culture line except fastidious adenovirus serotypes 40 and 41 for which no standard culture method is available.

Recently Cromeans *et al.* (Cromeans 2008) described plaque assays for adenoviruses 40 and 41 which will require optimization before being applied to environmental water samples.

The virus may be identified by electron microscopy (Thomas P 1999) and ELISA (Uhnoo 1984) but this method is not sensitive enough for environmental water samples.

The basic method for concentration of adenoviruses from water is the same as for enteroviruses. A recent study (Lambertini 2008) proposed a new method based on glass wool filters to concentrate waterborne viruses including adenoviruses from large-volume samples (1,000 liters) of drinking water.

A large number of molecular sensitive and specific tests have been developed to detect all types of adenoviruses in water: PCR or nested-PCR based detection (Puig 1994) (Castignolles 1998) (Pina 1998b) (Cho 2000) (Schvoerer 2000) (Lee 2002).

Several studies reported the detection of infectious adenoviruses in water by a combination of cell culture and PCR (Greening 2002) (Ko 2003) (Lee 2004).

Several quantitative real-time PCR assays have been described to quantify human adenovirus genomes in environmental waters (Jiang 2005) (Choi 2005), in addition to enterococci (He 2005), but Choi *et al.* (Choi 2005) reported that samples with positive results obtained by real-time PCR gave negative results by tissue culture, suggesting that the adenoviruses from the analyzed samples were non infectious. A TaqMan real-time RT-PCR method in conjunction with cell culture infectivity has been developed to detect low levels of infectious adenoviruses in water (Ko 2005). Recently a new integrated cell culture quantitative PCR assay was developed to quantify adenoviruses in disinfection studies (Gerrity 2008).

All the molecular methods available are at the experimental stage and validated protocols do not yet exist. To date analysis of adenoviruses in water is likely to be restricted largely to research purposes.

Calicivirus

Health significance	**High**
Persistence in water supplies	**Long**
Resistance to chlorine	**Moderate**
Relative infectivity	**High**
Important animal source	**Potentially**

Adapted from table 7.1 Waterborne pathogens and their significance in water supplies (WHO 2004).

Microbiology

General description

Caliciviruses are small, non-enveloped, single-stranded RNA organisms about 7.5Kb long. They are spherical and about 27–40 nm in diameter and generally display a typical surface morphology resembling cup-like structures.

Caliciviruses are included in the Contaminant Candidate List (CCL) (U.S. Environmental Protection Agency 1998) (U.S. Environmental Protection Agency 2005).

Classification

Human caliciviruses (HuCVs) have been grouped into two genera: *Norovirus* and *Sapovirus* (Berke 1997).

Sapovirus spp. demonstrate typical calicivirus morphology and were called classical caliciviruses in the past. Sapporo virus is the prototype strain for *Sapovirus*. (Parshionikar 2003).

Norovirus spp. (previously termed Norwalk-like viruses or small round structured viruses) generally fail to reveal the typical morphology. They have been divided into genogroups I and II, each with a number of distinct genetic clusters. Norwalk virus is a prototype strain for genogroup I, while Snow Mountain virus is a prototype strain for genogroup II (Ando 2000).

The Norwalk virus was the first human calicivirus identified by electron microscopy during a gastroenteritis outbreak in Norwalk, Ohio, in 1972 (Kapikian 1972).

Human health effects

Clinical features

Human caliciviruses, especially the *Norovirus* genus, are a major cause of acute nonbacterial gastroenteritis worldwide and are believed to be one of the major causes of waterborne disease.

The illness leads to vomiting and/or diarrhea, and abdominal cramps. Fever and respiratory symptoms have been reported in a small number of cases. The disease is marked by an abrupt onset, 24–28 h after exposure, but is typically mild and self-limiting, only lasting 12–60h (Kaplan 1982). However, children and persons with compromised immunity may need hospitalization for dehydration.

Since some cases involve vomiting only and no diarrhea, the condition is also known as "winter vomiting disease". Infections by HuCVs induce a short-lived immunity (WHO 2004).

© 2010 Global Water Research Coalition. Waterborne Pathogens: Review for the Drinking-water Industry. Prepared by Emmanuelle Guillot and Jean-François Loret. ISBN: 9781843391791. Published by IWA Publishing, London, UK.

Dose – response data

The infectious dose is low, about 10–100 viral particles (Green 1997). A dose-response model for norovirus, taking into account the capacity of the virus to form aggregates, has recently been described. According to this model, the probability of infection for a single Norwalk virus particle exceeds that reported for any other virus studied to date (Teunis 2008).

Target populations

Norovirus causes illness in people of all age groups, whereas *Sapovirus* predominantly causes illness in children.

Diagnosis

Detection in stools has been done by electron microscopy (EM), immune electron microscopy (IEM) but these methods are not very sensitive and require highly skilled microscopists as well as expensive equipment. In 1995, reverse transcriptase polymerase chain reaction (RT-PCR) was introduced, and has since become the laboratory standard.

Treatment

There is no specific antiviral treatment. Oral rehydration therapy is the recommended treatment prescribed to avoid severe dehydration.

Geographical distribution

Worldwide.

Epidemiology

Noroviruses are the viruses most commonly associated with food and waterborne outbreaks of gastroenteritis (Green 1997). 50% to 80% of gastroenteritis outbreaks in the United States, Japan and some European countries have been attributed to noroviruses (Hedberg 1993) (Dedman 1998) (Hedlund 2000) (Lopman 2003) (Blackburn 2004).

Transmission routes

The virus is transmitted by faecal-oral route through contaminated food, water and by person-to-person contact. Contaminated surfaces, such as door handles, telephone handsets, and water taps, have been shown to carry caliciviruses in detectable amounts, thereby forming an important mechanism for virus transmission (Brown 1999). Transmission by aerosol produced by vomiting is probably common (Hedberg 1993).

Waterborne outbreaks

The first well-documented outbreak of *Norovirus* gastroenteritis linked to drinking water occurred in a Washington State elementary school in the USA in May 1978 (Taylor 1981). Since that time, many waterborne outbreaks of *Norovirus* associated to drinking water have been reported worldwide (Kaplan 1982b) (Hedberg 1993) (Kukkula 1997) (Kukkula 1999) (Schvoerer 1999) (Hafliger 2000) (Huffman 2003) (Nygard 2003a) (Parshionikar 2003) (Maunula 2005) (Godoy 2006). Waterborne outbreaks associated with swimming in contaminated recreational water have also been reported (Baron 1982) (Kappus 1982) (Sartorius 2007).

Moreover, numerous outbreaks due to HuCV-contaminated foods, such as shellfish, salads, and deli sandwiches have been reported (Johansson 2002) (Makary 2008).

Ecology

Regrowth in distribution systems/biofilms

As with all viruses, HuCVs are environmentally inert and only replicate in their hosts (human).

Recently, Lehtola *et al.* (Lehtola 2007) reported the persistence of caliciviruses, as measured by reverse transcription-PCR in drinking water-associated biofilms.

Occurrence

The distribution of HuCVs in the environment reflects the distribution of human faecal contamination.

The occurrence of caliciviruses in water has been studied only since molecular detection methods have been available.

Noroviruses are frequently detected in various types of environmental water samples, such as sewage (Lodder 1999), river water (Gilgen 1997), well water (Beller 1997) (Nygard 2003b), seawater (Katayama 2002), and even mineral waters (Beuret 2000) (Beuret 2002) and tap water (Kukkula 1999) (Haramoto 2004).

RT-PCR based -detection cannot determine whether or not the virus is infectious and thus as suggested by Gassilloud *et al.* (Gassilloud 2003b), positive results must be interpreted with caution in terms of public health.

Levels of contamination are not documented.

Reservoirs

The main reservoir of infection appears to be humans but some data suggest that calves and pigs may be reservoir hosts of Norovirus as well (van Der Poel 2000) (van Der Poel 2003).

It is believed that caliciviruses emerged from ocean reservoirs, with subsequent zoonotic and interspecies movement (Smith 1998).

Survival

A high stability of noroviruses in the environment has been reported, particularly at low temperatures (Ferrero Poschetto 2007) (Belliot 2008) (Lee 2008). One study (Kadoi 2001) reported the survival of a cultivable model virus (feline calicivirus strains) in marine water: infective viruses can survive at 10°C or lower temperature during 30 days. Ngazoa *et al.* (Ngazoa 2008) reported that norovirus genome may persist for long periods of time in different types of water.

Inactivation/removal

Published data on removal and inactivation are summarized in annex 1. Knowledge on efficient inactivation methods for HuCVs is limited and based on studies with cultivable model viruses, mainly feline calicivirus (FCV) (Doultree 1999) (Thurston-Enriquez 2003) (Duizer 2004).

Caliciviruses remained infective to volunteers following: pH 2.7, 20% ether at 4°C for 18 hours, temperatures of 60°C for 30 minutes, and chlorine levels consistent with those found in drinking water systems (Keswick 1985). Other work showed that feline calicivirus was surprisingly resistant to chlorine, requiring 1000 mg/L of freshly reconstituted granular hypochlorite for complete inactivation (Doultree 1999). Duizer *et al.* (Duizer 2004) reported incomplete inactivation of FeCV by 70% ethanol and 300 ppm free chlorine, while Belliot *et al.* (Belliot 2008) reported a reduction of over 4 logs with 60%

ethanol or isopropanol. Studies conducted with purified viruses, however, demonstrated that chlorine concentrations and contact times in drinking water production are sufficient to effectively inactivate FCV as well as norovirus (Thurston-Enriquez 2003) (Urakami 2007) (Shin 2008). Several studies have also demonstrated the efficacy of chlorine dioxide against caliciviruses at doses used for drinking water treatment (Thurston-Enriquez 2005a) (Zoni 2007).

Both FCV and norovirus have shown to be inactivated by ozone at doses usually applied for drinking water treatment (Thurston-Enriquez 2005b) (Shin 2003).

In several studies, FCV appears to be UV resistant (Nuanualsuwan 2002) (De Roda Husman 2004), but recent studies demonstrated that feline and canine caliciviruses as well as murine noroviruses can be controlled by UV disinfection (Hijnen 2006) (Lee 2008).

There is a strong need for further studies on disinfection of these persistent organisms.

Surrogates

Due to higher resistance of the viruses to disinfection, *E. coli* (or, alternatively, thermotolerant coliforms) is not a reliable index of the presence/absence of HuCVs in drinking-water supplies (WHO 2004).

Environmental detection

HuCVs cannot be propagated in available cell culture systems.

Their detection depends mainly on reverse transcription-PCR, but RT-PCR fails to determine virus infectivity and does not give the level of contamination.

Conventional filter adsorption-elution methods are the most widely used for concentrating the virus from large volumes of water (Sobsey 1980) (Schwab 1995) (Gilgen 1997) (Gassilloud 2003a). Haramoto *et al.* (Haramoto 2004) reported a concentration method using a cation-coated filter. A procedure for concentration of enteric viruses from coastal water using a negatively charged membrane was described (Katayama 2002). Lambertini *et al.* (Lambertini 2008) described a new method based on glass wool filters to concentrate waterborne viruses including noroviruses from large-volume samples (1000 L) of drinking water.

A large number of studies reported the development of RT-PCR assays to detect *Norovirus* from sewage (Lodder 1999), environmental water samples (Schwab 1995) (Gilgen 1997) (Myrmel 1999) (Parshionikar 2003) (La Rosa 2007) and mineral water (Beuret 2003). A triplex real-time reverse transcription-PCR (RT-PCR)-based method that detects and distinguishes between noroviruses belonging to genogroups I, II, and III has been described (Wolf 2007).

Immunomagnetic separation combined with RT-PCR was developed to detect *Norovirus* (genogroup I) in water (Myrmel 2000). A novel, magnetic-bead-based histo-blood group antigen assay was evaluated for the recovery of low numbers of norovirus particles in environmental waters (Cannon 2008). Several studies described TaqMan real-time transcription PCR assays for rapid detection (Haramoto 2004) and typing of NoV of norovirus (Trujillo 2006), but these tests are not quantitative.

Recently, Victoria *et al* (Victoria 2008) evaluated HA (mixed cellulose esters) negatively charged membranes to concentrate viruses coupled with quantitative PCR detection.

A single analytical process using ultrafiltration combined with high-density oligonucleotide array has been developed to simultaneously detect waterborne pathogens including bacteria, viruses (Enterovirus, Hepatitis A, Norovirus) and parasites (Guillot 2003) (see Enterovirus section).

The method currently used for typing *Norovirus* is sequencing and subsequent phylogenetic analysis of RT-PCR products (Schvoerer 1999). One study reported a simple method for genotyping noroviruses using oligonucleotide array (Vinje 2000).

At the present time, molecular tests for Noroviruses are only available in specialist laboratories and cannot be used for routine testing of water samples. Commercial test kits are not yet available. Evaluation studies and standardization of the procedures are necessary to get the new methods approved by health authorities; this is the basic requirement for routine application.

Enterovirus

Health significance	**High**
Persistence in water supplies	**Long**
Resistance to chlorine	**Moderate**
Relative infectivity	**High**
Important animal source	**No**

Adapted from table 7.1 Waterborne pathogens and their significance in water supplies (WHO 2004).

Microbiology

General description

Enteroviruses (EVs) are a large group of small (20 to 30 nm in diameter) non-enveloped single-stranded RNA viruses with an icosahedral symmetry.

Enteroviruses have been the most studied in the past because they are easily isolated by standard cell culture methods.

EVs (Coxsackievirus and Echovirus) were included in the first two versions of the U.S. Environmental Protection Agency drinking water contaminant candidate list (CCL) (U.S. Environmental Protection Agency 1998) (U.S. Environmental Protection Agency 2005).

Classification

The genus *Enterovirus* is a member of the Picornaviridae family. This genus consists of 69 serotypes that infect humans: poliovirus types 1–3, coxsackievirus types A1–A24, coxsackievirus types B1–B6, echovirus types 1–33 and the numbered enterovirus types EV 68–EV 71. Members of the genus are collectively referred to as enteroviruses.

Other species of the genus infect animals other than humans, for instance the bovine group of enteroviruses (WHO 2004).

Human health effects

Clinical features

Enteroviruses are one of the most common causes of human infection. However most infections are unapparent and lead to a non-specific febrile illness referred to as the "summer flu" (Dagan 1996). Some serotypes, however, may cause serious clinical syndromes such as poliomyelitis, encephalitis, meningitis, myocarditis, and pericarditis (Melnick 1990). Enterovirus 71 infections lead to blisters in the oral cavity, palms and soles, ulcers and fever (hand-foot-mouth disease).

Most infections, particularly in children, are asymptomatic, but still lead to the excretion of large numbers of the viruses, which may cause clinical disease in other individuals (WHO 2004).

© 2010 Global Water Research Coalition. Waterborne Pathogens: Review for the Drinking-water Industry. Prepared by Emmanuelle Guillot and Jean-François Loret. ISBN: 9781843391791. Published by IWA Publishing, London, UK.

Dose – response data

The infectious dose is low: 1–10 infectious particles. Different dose-response models based on experiments on humans or animals have been described for echovirus 12, poliovirus 1 and 3 (Regli 1991) and coxsackievirus type B4 (Mena 2003).

Target populations

Everyone is at risk of EV infection. Infants, children, and adolescents are more susceptible to EV infection, due in part to a lack of prior immunity and to poor hygiene habits.

Diagnosis

Diagnosis can be performed by cell culture from the throat, swabs and faeces. Serological diagnosis may also be used but is of limited value due to large number of strains.

Treatment

No specific treatment is available for enteroviral infections. Treatment focuses on management of complications.

No vaccine is available for non-polio enteroviruses.

Geographical distribution

Worldwide.

Epidemiology

Some EV serotypes have been responsible for large and serious epidemics throughout the world in the recent years: Echovirus caused several large epidemics of aseptic meningitis in Japan, Europe, the Middle East and Cuba (Kirschke 2002). EV 71 has caused major outbreaks of hand-foot-and mouth disease worldwide since 1995 (Ho 1999).

Rapid genetic evolvements of some new strains of EVs as well as development of anti-viral drug resistance have been reported (Nel 2004b).

Transmission routes

EVs are transmitted via the faecal-oral route. Person to person contact and inhalation of airborne viruses or viruses in respiratory droplets are considered to be the predominant routes of transmission of enteroviruses in communities. Persons may also become infected by contact with respiratory secretions (e.g., saliva, sputum, or nasal mucus). Transmission from drinking water could also be important, but this has not yet been confirmed (WHO 2004).

Waterborne outbreaks

Despite the wide occurrence of enteroviruses in the water environment, few water-related outbreaks have been reported. The limited knowledge on the role of waterborne transmission could be related to a number of factors, including the wide range of clinical symptoms, frequent asymptomatic infections, the diversity of serotypes and the dominance of person to person spread (WHO 2004).

The only outbreak of poliomyelitis linked by an epidemiological study to drinking water occurred in Taiwan in 1982, affecting 1031 persons (Kim-Farley 1984).

Coxsackievirus has been associated with numerous common source outbreaks, including at least two documented waterborne outbreaks (Ikeda 1993).

Echovirus-30 was responsible for a large waterborne outbreak of aseptic meningitis in Belarus in summer/autumn of 1997 (Amvrosieva 2001).

A school outbreak caused by echovirus 3, where contamination of drinking and swimming water was demonstrated was reported in South Africa (Yeats 2005).

Ecology

Regrowth in distribution systems/biofilms

As for all viruses, EVs are environmentally inert and only replicate in their hosts.

One study reported the tendency for virus accumulation within drinking water biofilms (Quignon 1997); the role of biofilm in the survival and persistence of viruses remains unclear.

Occurrence

The virus is shed in the faeces of persons with both asymptomatic and symptomatic infection.

Enteroviruses are widely present in water environment: numerous studies have documented the presence of enteroviruses in raw water (Jiang 2004) (Pusch 2005), groundwater (Abbaszadegan 1999) (Fout 2003), wastewater (Payment 1981), recreational water (Denis-Mize 2004) (Gersberg 2006) as well as in treated drinking water (most often "bacteriologically safe") (Keswick 1984) (Cho 2000) (Gofti-Laroche 2001) (Lee and Kim 2002) (Patti 2003) (Numanovic 2004).

Vivier *et al.* (Vivier 2004) reported the monitoring of drinking water for the presence of enteroviruses by cell culture-PCR, over a period of 1 year: Enteroviruses (predominantly coxsackie B viruses) were detected in 11% and 16% of the drinking water samples (that met bacteriological standards) from two treatment plants.

EVs have also been isolated from swimming pools (Keswick 1981).

In addition, EVs are found in food and have been isolated from shellfish.

Levels of contamination are not extensively documented (because the molecular methods increasingly used do not readily allow quantification). Levels of 1–100 PFU/L of enteric viruses in contaminated surface water, of 1–10 PFU /100 L in less polluted surface water, 1–10PFU/ 1000 L in treated drinking water were reported (Payment 1981) (Hurst 1991).

In the future, further research should be done to determine the public health significance of EV occurrence in water.

Reservoirs

Humans are the only reservoir of EVs.

Survival

Not extensively documented.

These viruses are quite resistant to environmental stresses and stable under acidic conditions (pH 3). Their persistence was shown in water supplies over 1 month at 20°C (WHO 1993).

Ultraviolet radiation reduces survival, and survival in fresh waters is longer than in marine water (Hunter 1997).

Skraber *et al.* (Skraber 2004) compared the survival in river water of infectious polioviruses to somatic coliphages and poliovirus genomes considered as potential indicators. At three different temperatures

(4°C, 18°C and 25°C), somatic coliphages and poliovirus genomes persisted longer than infectious polioviruses. According to these results and previously reported results (Hot 2003), the authors suggested that poliovirus genomes detected by RT-PCR may be useful for predicting the presence of waterborne viruses.

Inactivation/removal

Published data on removal and inactivation are summarized in annex 1. Hurst *et al.* (Hurst 1991) reported the removal of enteric viruses from various stages of conventional water treatment.

EVs have been shown to be particularly resistant to chlorine and UV inactivation (Payment 1985) (Batigelli 1993) (Blackmer 2000). But polioviruses, coxsackieviruses and echoviruses are inactivated by ozone (Roy 1982) (Emerson 1982) (Harakeh 1985), and a 4 log reduction of poliovirus type 1 can be achieved with a UV dose of 30 mW.S/cm̀ (Hijnen 2006).

Surrogates

Enteroviruses have been detected in drinking water supplies that met accepted specifications for treatment, disinfection and conventional indicator organisms.

Because of the higher resistance of the viruses to disinfection, *E. coli* (or, alternatively coliforms) is not a reliable index of the presence/absence of enteroviruses in drinking-water supplies (WHO 2004).

Environmental detection

A standard cell culture method is available for most enteroviruses (Fout 1996), but this approach is time consuming (6–15 days) and not very sensitive.

Although cell culture still remains the "gold standard" for enterovirus isolation, RT-PCR – which is more rapid and sensitive – has become widely used.

Concentration:

The method of choice to concentrate viruses from water is the adsorption/elution technique (APHA 1998) using positively charged filters (Sobsey 1980) (Fout 1996) with improvements for groundwater (Dahling 2002) or electronegative filters (Beuret 2003). Bigliardi *et al.* (Bigliardi 2004) described a new concentration method using tangential flow ultrafiltration. A study (Hsu 2007) evaluated the performance of six concentration methods based on different adsorption and elution procedures.

Recently, Lambertini *et al.* (Lambertini 2008) proposed a new method based on glass wool filters to concentrate waterborne viruses including adenoviruses from large-volume samples (1000 L) of drinking water.

Purification:

Schwab *et al.* (Schwab 1996) developed a broad-spectrum immunocapture method for concentration and purification of enteric viruses before RT-PCR detection. Four commercial kits for viral RNA extraction from water concentrates were evaluated: NucliSens and Centricon-QIAamp proved to be very sensitive, efficient and robust methods (Burgener 2003).

Detection:

Numerous RT-PCR methods used for water samples have been reported to detect polioviruses (Tansuphasiri 2000), enteroviruses (Abbaszadegan 1999) (Gofti-Laroche 2001) (Shieh 2008) or simultaneously several viruses: adenoviruses and enteroviruses (Cho 2000), noroviruses and enteroviruses (Beuret 2003), enteroviruses and hepatitis A viruses (Tsai 1994) (Li 2002), adenoviruses, enteroviruses and hepatitis A viruses in sewage and shellfish (Formiga-Cruz 2005) and enteroviruses, hepatitis A virus, Norwalk virus, reoviruses, and rotaviruses (Fout 2003). Rutjes *et al.* (Rutjes 2005) evaluated the detection of enterovirus RNA from large-volume water samples by using the NucliSens miniMAG system combined with a commercial real-time NASBA assay.

A single analytical process using ultrafiltration combined with high-density oligonucleotide array (GeneChip technology from Affymetrix, Santa Clara, Ca.) to simultaneously concentrate and detect waterborne pathogens including bacteria, viruses (Enterovirus, Hepatitis A, Norovirus) and parasites in treated and distributed water samples has been described (Guillot 2003). This process is at the experimental stage and requires significant improvements, particularly in terms of specificity and sensitivity.

However, conventional RT-PCR does not determine the infectivity of the viruses and does not allow easy quantification.

An integrated cell culture-RT-PCR procedure has been used to detect infectious enteroviruses in water (Egger 1995) (Schwab 1995) (Chapron 2000) (Greening 2002) (Lee and Kim 2002) (Vivier 2004) (Shieh 2008).

Few studies reported the quantitative detection of enterovirus in water (Donaldson 2002) (Fuhrman 2005). Gregory *et al.* (Gregory 2006) described a rapid one-step quantitative reverse transcriptase PCR assay with competitive internal positive control for the detection of enteroviruses in environmental samples.

Typing methods such as PCR-Restriction Fragment Length Polymorphism (Kuan 1997) (Vivier 2001) and PCR and sequencing (Oberste 1999) have been used.

Currently all the available molecular methods are at the experimental stage and validated protocols do not yet exist. In the future, evaluation studies and standardization of these new methods will be necessary to get them approved by the authorities.

Hepatitis A

Health significance	**High**
Persistence in water supplies	**Long**
Resistance to chlorine	**Moderate**
Relative infectivity	**High**
Important animal source	**No**

Adapted from table 7.1 Waterborne pathogens and their significance in water supplies (WHO 2004).

Microbiology

General description

Hepatitis A virus (HAV) is a small (27–28 nm) spherical, non-enveloped (naked) positive strand RNA virus, about 7.5 kb in length. HAV has a single antigenic serotype.

Classification

HAV is classified under the genus Hepatovirus within the *Picornaviridae* family.

Human health effects

Clinical features

Hepatitis A is an acute, usually self-limiting infection of the liver caused by hepatitis A virus (HAV). Symptoms typically include fever, malaise, anorexia, nausea and abdominal discomfort, followed by dark urine and jaundice (Koff 1992). Complications of hepatitis A include relapsing hepatitis, cholestatic hepatitis and fulminant hepatitis. Chronic infection with HAV does not occur. In children below six years of age, HAV infection is usually asymptomatic, with only 10% developing jaundice (Yang 1988).

The average incubation period is 28 days, but may vary from 15–50 days. Approximately 10–12 days after infection the virus can be detected in blood and faeces. Virus excretion may begin up to a week before the symptoms are apparent, making control difficult.

Infection with HAV induces lifelong immunity.

Dose – response data

The infectious dose in humans is unknown but presumably is 10–100 virus particles (WHO 2002).

Target populations

The severity of disease and mortality may be dependent on the general state of health (immunodeficiency and malnutrition…) and increase in older age groups.

© 2010 Global Water Research Coalition. Waterborne Pathogens: Review for the Drinking-water Industry. Prepared by Emmanuelle Guillot and Jean-François Loret. ISBN: 9781843391791. Published by IWA Publishing, London, UK.

Diagnosis

Hepatitis A is diagnosed by finding IgM-class anti-HAV in serum collected during the acute or early convalescent phase of disease. Commercial kits are available. Detection of the virus or viral antigens in the stools is of limited value for routine diagnosis.

Treatment

No specific treatment is currently available.

Vaccines are available for persons 2 years of age and older.

Geographical distribution

HAV has a worldwide distribution.

Epidemiology

An estimated 1.5 million clinical cases of hepatitis A occur every year.

The prevalence of clinical disease has typical geographically based characteristics. In countries where inadequate sanitary conditions prevail, the virus persists in the environment and almost 100% of the population acquires infection in childhood. In developed, industrialized countries HAV has ceased to circulate in the environment and the general population. Here, infections predominantly occur in adults traveling to endemic areas or exposed to infected individuals or members of high risk groups (e.g. children in day care centres, drug users) (WHO 2000) (Grabow 2002).

Transmission routes

Transmission occurs primarily through the faecal-oral route, and is closely associated with poor sanitary conditions. The most common modes of transmission include close personal contact with an infected person (usually among household contacts or sexual partners) and ingestion of faecally contaminated water and food. Infection can also be spread in association with injecting and non-injecting drug use.

Asymptomatic and non-jaundiced HAV-infected persons, especially children, are an important source of HAV transmission (Staes 2000). Outbreaks by person-to-person spread commonly occur among families, institutions and troops (day-care centres, schools and institutions for mentally ill people and food establishments) (Banker 2003).

Waterborne outbreaks

Food and waterborne outbreaks of HAV in developed countries have been recognized for over 40 years, but are infrequently reported (Fiore 2004). Epidemiological evidence to link hepatitis infection to food and water sources is sparse because of the long incubation period (Seymour 2001). In developing countries they are uncommon because of high levels of immunity in the resident population (Fiore 2004).

Waterborne outbreaks have often been associated with water supplies that complied with generally accepted guidelines for indicators and treatment procedures (Grabow 2002).

In 1972, an outbreak of HAV affected 49 children and one adult in a rural elementary school in Alabama, probably associated with contaminated school drinking water (Baer 1977). In 1989, an outbreak of hepatitis A affecting 47 people in central Italy was linked to a single brand of bottled mineral water (Stroffolini 1990). In 2006, Zhang *et al.* (Zhang 2008) reported an outbreak of hepatitis A in recently vaccinated students due to consumption of ice snacks made with untreated well water.

Coursaget *et al.* (Coursaget 1998) reported for the first time a waterborne outbreak of acute hepatitis simultaneously due to HAV and HEV observed in 1993 in Djibouti.

In the United States, the highest incidence of waterborne infectious hepatitis occurred between the 1950s and the 1970s with 47 outbreaks reported (Craun 1996). Since the early 1990s, very few outbreaks of HAV have been reported to the CDC (De Serres 1999).

The largest ever recorded outbreak of HAV infection occurred in Shanghai in 1988 involving 250,000 people and was traced to consumption of raw clams (Halliday 1991). Another large outbreak of HAV infection linked to raw seafood consumption occurred in 2004 in Italy, affecting 882 persons (Pontrelli 2008).

Outbreaks associated with recreational water contact were also reported (Bryan 1974) (Mahoney 1992).

Ecology

Regrowth in distribution systems/biofilms

As all viruses, HAV only replicates in its hosts (human or animal).

The presence and persistence of HAV within biofilms have not been investigated.

Occurrence

The ecology in the environment mirrors that of the distribution of human faecal contamination (both asymptomatic and symptomatic infection).

Using molecular biology techniques, HAV has been found in environmental water (Grabow 2002), HAV has been detected in sewage and wastewater (Jothikumar 1998) (Espigares 1999), in polluted rivers and in drinking water (Morace 1993) (Jothikumar N 2000) (Schvoerer 2000) (Pina 2001) (Taylor 2001).

Further study is required to determine the public health significance of HAV occurrence in water.

Levels of contamination in environmental waters are insufficiently documented.

Survival

Not extensively documented. The survival of infectious HAV greatly depends on the temperature. Under favorable conditions (ambient temperatures) HAV may survive in the environment for months (Mbithi and Springthorpe 1991). It could survive in wastewater and groundwater for 90 days or more at 10°C. Crance *et al.* (Crance 1998) reported that infectious HAV remain stable at 4°C throughout 92 days in synthetic seawater. Hepatitis A virus and poliovirus were shown to survive in excess of one year in mineral water stored at 4°C (Biziagos 1988).

Reservoirs

Humans are the only known reservoir of the organism, but a variety of non-human primates are susceptible to HAV under experimental laboratory conditions. Transmission of the virus from chimpanzees to humans during close contact is well documented (Zuckerman 1993), but to date there is no evidence that animals may serve as a significant reservoir for HAV. This question needs to be answered.

Inactivation/removal

To date HAV removal or inactivation by water treatment and disinfection processes has not been extensively documented. The published data are summarized in annex 1.

The virus is relatively stable at low pH levels (as low as 3) and moderate temperatures, but is inactivated by high temperatures (at 85°C/185°F for 1min).

Only 2% glutaraldehyde, a quaternary ammonium formulation containing 23% HCl (toilet bowl cleaner), and sodium hypochlorite (greater than 5,000 μg/mL of free chlorine) reduced the virus titer by more than 99.9% (Mbithi 1990).

HAV is resistant to disinfection by some chemical disinfectants (phenolics, iodine-based products, alcohols, and solutions of acetic, peracetic, citric, and phosphoric acids).

HAV appears to be relatively resistant to chlorine (Grabow 1983) (Peterson 1983) (Li 2004b). In the United States a level of 5 mg chlorine.L^{-1} with a contact time of 1 min is recommended for inactivation of hepatitis A virus. HAV is readily inactivated by ozone (Hall 1993).

Surrogates

Due to the higher resistance of the viruses to disinfection, *E. coli* (or, alternatively, thermotolerant coliforms) is not a reliable index of the presence/absence of HAV in drinking-water supplies (WHO 2004).

Nevertheless, one study (Gersberg 2006) reported a significant correlation between the densities of faecal bacterial indicators and the levels of HAV in recreational marine waters.

Environmental detection

HAV is not readily detectable by routine cell procedures. It needs a prolonged incubation in cell culture requiring at least 4–5 weeks.

The method of choice to concentrate viruses from water is the adsorption/elution technique (APHA 1998) using positively or negatively charged membrane filters (Sobsey 1980) (Fout 1996), followed by organic flocculation for secondary concentration. Villar *et al.* (Villar 2006) evaluated the performance of negatively charged membrane adsorption-elution concentration methods to monitor hepatitis A virus in tap, mineral and coastal water samples.

An amplified enzyme-linked immunosorbent assay (A-ELISA) for detecting and quantifying hepatitis A virus in estuarine water samples was reported but this approach was time consuming and not very sensitive (Nasser 1987).

Thus, the most feasible current approach to detect HAV from water samples is based on molecular techniques (mainly RT-PCR). Numerous RT-PCR methods have been reported to detect HAV alone or in combination with other viruses in environmental waters: detection of HAV in wastewater (Divizia 1998) (Espigares 1999), detection of HAV in bottled water (Perelle 2008), simultaneous detection of rotavirus, HAV, and small round structured viruses in river water (Gilgen 1997), detection of enterovirus and HAV (Li 2002); detection of HAV and HEV in drinking water (Jothikumar 2000), detection of HAV and rotavirus in spring water samples (Brassard 2005), detection of poliovirus, HAV and rotavirus in sewage and ocean water (Tsai, 1994), detection of HAV, rotavirus and enterovirus in shellfish and sediment (Le Guyader 1994), detection of enteroviruses, HAV, Norwalk virus, reoviruses, and rotaviruses (Fout 2003) and detection of adenoviruses, enteroviruses and HAV in sewage and shellfish (Formiga-Cruz 2005).

A combination of immunocapture followed by RT-PCR from environmental water samples has also been described (Monceyron 1994) (Jothikumar 1998) (Abd El Galil 2004) (Abd El Galil 2005).

A single analytical process using ultrafiltration combined with high-density oligonucleotide array has been developed to simultaneously detect waterborne pathogens including bacteria, viruses (Enterovirus, Hepatitis A, Norovirus) and parasites (Guillot 2003) (see Enterovirus section).

RT-PCR is highly sensitive and specific but it does not determine the infectivity of the viruses or the level of contamination. However, Bhattacharya *et al.* (Bhattacharya 2004) showed that RT-PCR could discriminate between infectious and non-infectious hepatitis A viruses following UV inactivation treatment. The procedure may be adapted to other viruses and inactivating agents.

An integrated cell culture/polymerase chain reaction (ICC/RT-PCR) technique was developed for the detection of infectious HAV alone or in combination with enteroviruses in environmental waters (Reynolds 2004). Though promising, this approach is still not ideal as it is time consuming (at least 48h), expensive and not readily applicable for routine diagnostic work.

Several methods based on quantitative real-time RT-PCR were reported for environmental water samples (Brooks 2005) (Gersberg 2006) (Sanchez 2006) (Dubois 2007).

Nested PCR amplification and partial sequencing have been used in molecular epidemiology studies (Pina 2001) (Sanchez 2004).

All available molecular methods are currently at the experimental stage mainly for research purposes and validated protocols do not yet exist.

Hepatitis E

Health significance	**High**
Persistence in water supplies	**Long**
Resistance to chlorine	**Moderate**
Relative infectivity	**High**
Important animal source	**Yes**

Adapted from table 7.1 Waterborne pathogens and their significance in water supplies (WHO 2004).

Microbiology

General description

Hepatitis E (HEV) is a new waterborne infectious viral agent, first characterized in the 1990s.

HEV is a small (27–34 nm) spherical, non-enveloped positive strand RNA virus, with a single serotype, about 7.2 kb in length; it was described for the first time in 1983 (Balayan 1983).

Classification

HEV was previously classified into the family *Caliciviridae*. However, it is now reclassified into the Hepatitis E-like viruses as an unassigned genus (Berke 2000). There are indications of antigenic variation, and possibly even differences in serotypes of the virus (WHO 2004), but to date only a single serotype has been recognized.

Human health effects

Clinical features

The disease caused by HEV is called hepatitis E, or enterically transmitted non-A non-B hepatitis. Other names include faecal-oral non-A non-B hepatitis, and A-like non-A non-B hepatitis.

Hepatitis caused by HEV is clinically indistinguishable from hepatitis A disease. The incubation period, however, is longer (6–8 weeks). Symptoms include discomfort, anorexia, abdominal pain, arthralgia, fever and vomiting. Patients generally excrete HEV for 1–2 weeks. The disease is usually self-limited and resolves in 2 weeks, leaving no sequelae.

HEV infection in children is mostly asymptomatic or causes a very mild illness without jaundice (anicteric) that goes undiagnosed (Grabow 2002).

Dose – response data

The infectious dose is unknown.

© 2010 Global Water Research Coalition. Waterborne Pathogens: Review for the Drinking-water Industry. Prepared by Emmanuelle Guillot and Jean-François Loret. ISBN: 9781843391791. Published by IWA Publishing, London, UK.

Target populations

The disease is most often seen in young to middle aged adults (15–40 years old) with a relatively low mortality rate of 0.1–1%.

Pregnant women are especially susceptible to severe disease, and excessive mortality has been reported in this group (up to 30%) (Balayan 1997) (WHO 2002).

Diagnosis

Diagnosis is performed by detecting specific IgM antibodies in the patient's serum.

Treatment

Most cases of hepatitis E only require non-specific supportive treatment consisting of rest and analgesics. No available therapy can alter the course of acute infection. There is currently no vaccine available.

Geographical distribution

Hepatitis E is mostly confined to tropical and subtropical areas (China, Asia, Africa, Mexico and Central America). Hepatitis E in developed countries is generally restricted to persons who travel to an endemic area (Balayan 1993) (Langer 1996), but evidence indicates that HEV might also be prevalent at a very low level in Europe (Clemente-Casares 2003).

Increased travel for business and tourism to HEV endemic areas and increasing globalization of food markets by industrialized countries have the potential of introducing HEV into new geographical areas.

Epidemiology

The risk factors for HEV infection are related to poor sanitation in large areas of the world and HEV shedding in faeces.

Transmission routes

The virus is primarily transmitted through the faecal-oral route.

HEV is transmitted mainly by faecally contaminated water. Foodborne transmission has been suggested but not demonstrated conclusively.

Person-to-person transmission is uncommon (Aggarval 1994).

There is no evidence for sexual transmission or for transmission by transfusion.

Waterborne outbreaks

Large outbreaks associated with sewage-contaminated drinking water have been reported in Asia (mainly India and China), Africa, the Middle East, and North America (Mexico) (Mast 1996).

One of the first known waterborne outbreaks was in Delhi, India in 1955–1956; approximately 29,300 people became ill (Vishwanathan 1957). In 1991 the largest waterborne outbreak of HEV occurred in Kanpur, India, with an estimated total of over 79,000 cases of disease (Nail 1992). More recently, several waterborne hepatitis E outbreaks were reported in India (Sarguna 2007) (Sailaja 2009) and in Sudan (Guthmann 2006). To date no U.S. outbreak has been reported but several sporadic cases have occurred (Kwo 1997).

Ecology

Regrowth in distribution systems/biofilms

As with all viruses, HEV is environmentally inert and only replicates in hosts (human or animal).

The presence and persistence of HEV within biofilms have not been investigated.

Occurrence

The ecology of this virus in the environment mirrors that of the distribution of animal or human faecal contamination.

For lack of practical detection techniques, data on the occurrence of HEV virus in water are scarce. A recent study in Spain demonstrated the presence of HEV in waste water, but the virus was not detected in raw nor treated surface water (Albinana-Gimenez 2007).

Reservoirs

The reservoir of HEV has not yet been established, although several findings have suggested that at least some strains of HEV may be zoonotic (Clayson 1996) (Meng 2000).

Some animals including domestic animals (mainly pigs), cattle, goats and even rodents may serve as reservoirs of HEV infections in humans and many human infections may originate from water sources polluted by animal waste. The zoonotic aspect of HEV needs to be more investigated and confirmed in the future.

Survival

Not documented.

The only clue to the behavior of HEV in the water environment is provided by the detection of the virus in raw and treated wastewater (Jothikumar 1993) (Pina 1998a), which suggests that the virus can survive in hostile environments.

Inactivation/removal

HEV removal and inactivation by water treatment have not yet been studied, but data on waterborne outbreaks suggest that HEV may be as resistant as other enteric viruses to disinfection processes.

Surrogates

Due to the likelihood that the virus has a higher resistance to disinfection, *E. coli* (or, alternatively, thermotolerant coliforms) is not a reliable index of the presence/absence of HEV in drinking-water supplies (WHO 2004).

Environmental detection

No method is currently available for routine analysis of water and food.

HEVs are undetectable by conventional cell culture methods.

Some studies reported the development of molecular detection methods to isolate HEV from water or food (shellfish).

As for most viruses, the environmental detection of HEV requires two steps: concentration and detection.

The use of a membrane filter-based adsorption and elution method (Grimm 2002) or a filtration column filled with granular activated carbon (GAC) (Jothikumar 1995) has been proposed for concentration of HEV viruses from large volume water samples (at least 100 L).

Some RT-PCR-based detection methods have been reported for environmental samples (Jothikumar 1993) (Jothikumar 1995) (Grimm 2002). Jothikumar *et al.* (Jothikumar 2000) developed a duplex RT-PCR for simultaneous detection of Hepatitis A and E viruses isolated from drinking water. Few studies reported a quantitative real-time PCR assay for HEV (Orru 2004)(Jothikumar 2006)(Qiao 2008).

Currently the few available molecular methods are in the experimental stage and restricted to research purposes. These protocols should be optimized and validated before being recognized as official methods.

Rotavirus

Health significance	**High**
Persistence in water supplies	**Long**
Resistance to chlorine	**Moderate**
Relative infectivity	**High**
Important animal source	**No**

Adapted from table 7.1 Waterborne pathogens and their significance in water supplies (WHO 2004).

Microbiology

General description

First detected in 1973 in Australia from children with acute diarrhoea (Bishop 1973) rotaviruses are non-enveloped double-stranded RNA viruses approximately 70 nm in diameter. Using electron microscopy, intact virus particles have the appearance of a wheel – hence the name "rota".

Classification

Rotaviruses are classified with the *Reoviridae* family. Seven serological groups have been identified, three of which (groups A, B, and C) infect humans with group A being the most important human pathogen, whereas others infect a wide spectrum of animals (Cukor 1984) (Jiang 2002).

Human health effects

Clinical features

The illness starts abruptly with fever, vomiting and diarrhoea, which normally last 24–48 hours, although diarrhoea can last up to five days. The incubation period for rotavirus disease is approximately 1–2 days. Immunity after infection is incomplete, but repeated infections tend to be less severe than the original infection. For persons with healthy immune systems, rotavirus gastroenteritis is a self-limited illness, lasting for only a few days.

Rotaviruses are excreted in very large quantities in the faeces of infected subjects at a rate of 10^{11} virus particles per gram.

Dose – response data

As little as one infectious particle of rotavirus can trigger disease in animal (Graham 1987) or human models (Ward 1986). A β-Poisson dose-response model based on experimentation on humans has been described (Teunis 2000).

Target populations

Humans of all ages are susceptible to rotavirus infection but rotavirus most commonly affects infants and young children.

© 2010 Global Water Research Coalition. Waterborne Pathogens: Review for the Drinking-water Industry. Prepared by Emmanuelle Guillot and Jean-François Loret. ISBN: 9781843391791. Published by IWA Publishing, London, UK.

Diagnosis

Diagnosis may be made by rapid antigen detection of rotavirus in stool specimens. Strains may be further characterized by enzyme immunoassay (several commercial kits are available for group A rotavirus) or reverse transcriptase polymerase chain reaction (RT-PCR), but such testing is not commonly done.

Treatment

Treatment is non-specific and consists of oral rehydration therapy to prevent dehydration. About one in 40 children with rotavirus gastroenteritis requires hospitalization for intravenous fluids.

In 1998, a live, attenuated rotavirus vaccine (Rotashield, Wyeth Laboratories, Marietta, PA) was licensed in the United States and recommended for routine immunization of U.S. infants. However, 9 months later, the use of Rotashield was suspended because some children who received the vaccine developed bowel obstruction. Efforts are ongoing to develop other rotavirus vaccines, and within the next few years, a rotavirus vaccine may be available (Parashar 2003).

Geographical distribution

Worldwide.

Epidemiology

Human rotaviruses (HRVs) are the most common cause of severe diarrhoea among children, worldwide (Cukor 1984). It is estimated that rotavirus is responsible for the deaths of approximately 500,000 children each year, over 80 percent of them in developing countries (Parashar 2003). In developed countries deaths are relatively scarce, but rotavirus gastroenteritis is the most frequent reason for admission of young children to hospital (Glass 1997).

Group A rotavirus is endemic worldwide. It is the leading cause of severe diarrhoea among infants and children. In temperate areas, it occurs primarily in the winter, but in tropical areas it occurs throughout the year. Group B rotavirus is responsible predominantly for adult diarrhoeal disease and therefore designated as adult diarrhoea rotavirus or ADVR. Group C rotavirus has been associated with rare and sporadic cases of diarrhoea in children in many countries.

Transmission routes

The primary mode of transmission is the faecal-oral route. Person-to-person transmission and the inhalation of airborne HRVs or aerosols containing the viruses would appear to play a much more important role than the ingestion of contaminated food or water (Craun 1991) (WHO 2004).

Person-to-person spread through contaminated hands is probably the most important means by which rotaviruses are transmitted in very small communities such as pediatric and geriatric wards, day care centers and family homes.

Waterborne outbreaks

Foodborne and particularly waterborne spread are probably a significant route of transmission in developing countries, but in developed countries reports are rare. Several waterborne outbreaks of rotavirus have been reported: in a school in Rio de Janeiro in 1980 (Sutmoller 1982), in 1981 in Colorado (Hopkins 1984), and two very large outbreaks in China affecting more than 12,000 adults (Hung 1984). More recently, a large infantile gastroenteritis outbreak occurred in Albania; it was primarily caused by rotavirus (Villena 2003) (Divizia 2004).

All of the outbreaks of rotavirus have been associated with a direct faecal contamination of the water supply or improper treatment.

Ecology

Regrowth in distribution systems/biofilms

As all viruses, HRVs are environmentally inert and only replicate in their hosts (human).

The presence and persistence of HRVs within biofilms have not been investigated.

Occurrence

The distribution in the environment reflects the distribution of human (both symptomatic and asymptomatic) faecal contamination.

Rotaviruses have been found in treated and tap water (Keswick 1984) (Gratacap-Cavallier 2000) (van Zyl 2004), in river water in Switzerland (Gilgen 1997) and in Spain (Abad 1998). No studies reported their presence in groundwater.

They are present in large amounts in wastewater and sewage (Smith 1982) (Hejkal 1984) (Dubois 1997) (Pusch 2005).

Levels of contamination are not documented.

Further research is needed to determine the public health significance of their occurrence in water.

Survival

Not extensively documented.

Rotaviruses can survive for weeks in potable and recreational waters and for at least 4 hours on human hands (Ansari 1991).

Temperature is an important factor, with low temperatures favoring viral survival in natural waters. Raphael *et al.* (Raphael 1985) found no significant drop in rotavirus titer after 64 days at 4°C in raw water, treated tap water or filtered water. However, a 99% drop in titer was observed after 10 days at 20°C.

The virus is stable to storage in ambient tropical temperatures for more than 2 months (Fischer 2002).

Reservoirs

Humans are the only reservoir for HRVs.

However, animal rotaviruses have been detected in drinking water. The role of water in the spread of animal strains to human populations and the emergence of reassortant strains have thus been suggested and require further investigation (Gratacap-Cavallier 2000).

Inactivation/removal

Published data on removal and inactivation are summarized in annex 1. Rotaviruses are inactivated below pH 3 and above pH 10. They are stable to freeze-thawing, sonication, incubation at 25°C overnight or at 37°C for 1 h and to treatment with acid, ether and chloroform (Estes 1979).

Rotaviruses are relatively resistant to commonly used hard-surface disinfectants and hygienic hand-wash agents (Ansari 1991).

Rotaviruses have been reported to be more resistant to chloramines and ultraviolet light disinfection than enteroviruses and more resistant to inactivation by preformed chloramines and ozone than poliovirus type 1 (Abad 1994) (Crabtree 1996). Rotavirus type SA 11 is efficiently controlled by UV disinfection (Hijnen 2006).

Surrogates

Due to a higher resistance of the viruses to disinfection, *E. coli* (or, alternatively, thermotolerant coliforms) is not a reliable index of the presence/absence of HRVs in drinking-water supplies (WHO 2004).

Environmental detection

Rotaviruses known as fastidious viruses do not grow in routine cell culture.

The standard methods for the diagnosis of specific infectious rotaviruses include electron microscopy (Brandt 1983), immunological methods such as latex agglutination or enzyme immunosorbent assays (Doern 1986) but these approaches are laborious and not sensitive enough to be applied to water samples.

The basic method for concentration of rotavirus form large volume water samples includes adsorption and elution from positively charged microporous filters (Sobsey 1980) with some modifications (Guttman-Bass 1983) (Mehnert 1993b) (Gilgen 1997). Recently, Gutiérrez-Aguirre *et al.* (Guiérrez-Aguirre 2008a) optimized the concentration of rotavirus from water samples using methacrylate monolithic chromatographic supports.

Immunofluorescence assays have been successfully developed to detect rotavirus in sewage (Smith 1982) (Agbalika 1985) (Mehnert 1993a) (Guttman-Bass 1987).

Water samples are mostly analyzed by RT-PCR (Tsai 1994) (Jothikumar 1995) (Gratacap-Cavallier 2000) (Gofti-Laroche 2001) (van Zyl 2004) (Kittigul 2005), often combined with the detection of other viruses: detection of hepatitis A virus and rotavirus in spring water samples (Brassard 2005), detection of enterovirus, rotavirus, hepatitis A virus, and small round structured viruses in water samples (Gilgen 1997).

RT-PCR-based detection is highly sensitive and rapid (a few hours), but it cannot differentiate between infectious and non-infectious viruses.

To determine the infectivity, some approaches have been proposed combining cell culture, immunofluorescence and flow cytometry (Abad 1998) (Bosch 2004), or cell culture and ELISA (Sellwood 1995). Though they seem promising, these approaches are still not ideal as they are time consuming (8–10 days), expensive and not readily applicable to routine diagnostic work.

Kurokawa *et al.* (Kurokawa 2004) used real-time PCR for rapid detection of three viruses (Rotavirus, Adenovirus and Norovirus) in faecal samples. Recently, two studies developed real-time RT-PCR detection of multiple rotavirus genotypes (Gutiérrez-Aguirre 2008b) (Jothikumar 2009), but as yet, no rapid quantitative method has been validated for environmental water samples.

Currently all the available molecular methods are at the experimental stage for research purposes and validated protocols do not yet exist.

Virus research needs

Much remains unknown about waterborne viruses, namely in terms of ecology and epidemiology. For lack of appropriate diagnosis technology, enteric viruses have not been frequently identified as the aetiologic agents of waterborne disease outbreaks. There is a lack of information on the environmental survival and transport, the occurrence and impact on health as well as the efficiency of water treatment processes.

Research themes of interest to water utilities include:

Analytical methods

A standardized approach to virus detection in water is essential for understanding the "true" environmental prevalence of enteric viruses.

First there is an urgent need to develop rapid and reliable analytical methods to detect waterborne viruses in water.

Quantitative cell culture techniques must be developed for assaying the infectivity of caliciviruses rotaviruses and adenoviruses types 40 and 41. These methods will serve as reference for infectivity for these viruses.

Rapid, simple, quantitative methods (TaqMan RT-PCR) to quantify viral contamination levels in water (treated and untreated) need to be developed for some viruses.

These molecular methods must eventually be evaluated and validated in order to be used as alternatives to the reference culture method, if any.

RT-PCR based detection methods are useful for rapid screening of water to provide evidence of virus occurrence, but they give no information on the infectivity of the virus. Positive results would then have to be followed up with other tests, such as the integrated cell culture-PCR assay to determine the health significance of the findings.

Further research using new molecular methods is necessary in order to acquire water-related data, to compare performance with existing cell culture methods and to correlate any positive results with human health risks by carrying out epidemiological studies.

Occurrence

Data on the presence and distribution of waterborne viruses in water are limited.

Little information is available on the types and levels of viruses found in groundwater.

There is a lack of quantitative data on virus occurrence in surface water, groundwater and treated water.

Further research must be carried out to determine the public health significance of the occurrence of waterborne viruses in water.

Epidemiology

The role of waterborne transmission needs to be more clearly defined and documented, particularly for adenoviruses, enteroviruses and rotaviruses.

Infective dose data and susceptibility of the human population remain unknown, particularly for adenoviruses, HAV and HEV.

There are few data available on the occurrence of viral waterborne outbreaks, particularly in Europe. Further epidemiological studies using molecular techniques for viruses detection should be carried out.

Ecology

Reservoirs:

Investigation must be done to determine whether animal to human transmission is possible for caliciviruses, rotaviruses and HEV.

Environmental survival and transport:

There is little information on the environmental persistence of viruses. The potential incidence of environmental factors such as humidity, temperature, seasonal variation and other geographic factors should be further evaluated.

The possible role of biofilm in the survival of viruses in drinking water systems by providing protection from disinfection should also be investigated.

The transport and fate of viruses in groundwater systems should be further investigated.

Treatment

The effectiveness of water treatment processes requires validation. Studies on the performance of water treatment processes to remove viruses from drinking water should be carried out in full-scale conventional water treatment plants and not only in pilot tests.

How viruses go through physical and chemical treatment barriers and enter finished drinking water supplies needs to be understood.

The effectiveness of water treatment and disinfection processes must be evaluated for virus aggregates.

Indicators

There is a need for new, relatively simple and easy to-use-indicators, to assess the presence of infectious viruses in water. Suitable indicator organisms based on model systems using similar non-pathogenic cultivable human or mammalian viruses may be the most promising alternative.

The presence of waterborne viruses may reliably be monitored by practical indicator systems consisting of appropriate combinations of suitable indicators.

Protozoa

Cryptosporidium

Health significance	**High**
Persistence in water supplies	**Long**
Resistance to chlorine	**High**
Relative infectivity	**High**
Important animal source	**Yes**

Adapted from table 7.1 Waterborne pathogens and their significance in water supplies (WHO 2004).

Microbiology

General description

Cryptosporidium is now one of the most commonly identified intestinal pathogens throughout the world (WHO 2004). It was discovered to infect humans only in 1976, and waterborne transmission was confirmed for the first time in 1984.

Cryptosporidium is an obligate, intracellular, coccidian parasite with a complex life cycle including sexual and asexual replication, which infects the gastrointestinal tract of animals and humans. Thick-walled spherical oocysts with a diameter of 4–6 µm are shed in faeces. The oocyst containing four sporozoites constitutes the environmentally resistant transmission stage of the parasite. The whole life cycle of *Cryptosporidium* can be completed within a single host.

Classification

Cryptosporidium is an apicomplexan protozoan parasite. All species of *Cryptosporidium* are classified taxonomically within the family Cryptosporidiidae, suborder *Eimeriorina*, order *Eucoccidiorida*, subclass *Coccidiasina* and class *Coccidia*.

Eleven species are currently recognized within the genus *Cryptosporidium. C. parvum* has been segregated into genotype 1 (recently named *C.hominis)* (Morgan-Ryan 2002) which is restricted to humans and genotype 2 (*C. parvum*) which infects humans and various mammalian hosts. While *C. parvum* and *C. hominis* are the primary species known to infect humans, other studies suggest that *C. muris, C. meleagridis, C. felis, C. canis* and a cervine genotype may also cause diarrhea in humans (Carey 2004) (Fayer 2004).

The taxonomy of *Cryptosporidium* species is unclear and is still being developed using molecular techniques. New species and genotypes of *Cryptosporidium* are being identified for which the infectivity for humans is not clear.

A user-friendly database called CryptoDB representing a collaborative effort to locate all genome data for *C. parvum* and *hominis* is accessible through a free searchable website (http://CryptoDB.org).

Human health effects

Clinical features

Cryptosporidiosis is the illness caused by exposure to *Cryptosporidium*. The average incubation period varies widely but is usually about 7 days. In normally healthy individuals, cryptosporidiosis is usually

© 2010 Global Water Research Coalition. Waterborne Pathogens: Review for the Drinking-water Industry. Prepared by Emmanuelle Guillot and Jean-François Loret. ISBN: 9781843391791. Published by IWA Publishing, London, UK.

characterized by an acute self-limiting diarrhoeal illness, commonly lasting one to two weeks, from which the patient recovers fully. Other symptoms are nausea, abdominal cramps and vomiting and mild fever. In patients who are immuno-suppressed, including those with AIDS, the disease is much more severe and persistent. Illness can last for several months or until deaths. Non-gastrointestinal illness, such as cholecystitis, hepatitis and respiratory disease may also occur in such individuals.

The peak intensity of oocysts shedding, with an average concentration of 10^6/g, coincides with the peak intensity of clinical symptoms.

Dose – response data

The infectious dose is not clearly known. Experiments in human volunteers indicated a median infective dose of 132 oocysts and a minimum infective dose of less than 30 (DuPont 1995). Studies using animal models showed that as few as 1 to 10 oocysts may initiate an infection (Leclerc 2004). A mathematical model based on data from the Milwaukee outbreak suggested that some individuals developed cryptosporidiosis following ingestion of only one oocyst (Haas 1994). Exponential dose-response models based on experimentation on humans have been described (Haas 1996) (Teunis 1999). The variability in infectivity between different *Cryptosporidium* isolates and between hosts has been investigated, and dose-response models incorporating the variability of infectivity of the isolates or the variability of the immune status of the hosts have been proposed (Teunis 2002a) (Teunis 2002b).

Target populations

Anyone can have cryptosporidiosis but immuno-compromised persons (e.g. AIDS patients) as well as young children and the elderly are particularly at risk.

Diagnosis

Diagnosis is usually based on microscopic demonstration of oocysts in faeces. Oocysts are usually stained by an acid-fast staining procedure. Other methods include direct or indirect immuno-fluorescence microscopy and ELISA using commercially available kits (Quintero-Betancourt 2002). None of these techniques are species-specific nor can they determine whether oocysts are infectious or not. Only molecular techniques, including PCR, can help determine the possible source and risk to human health.

Treatment

As yet there is no effective specific medical treatment.

Geographical distribution

Worldwide.

Epidemiology

Transmission routes

Cryptosporidium is transmitted by the faecal-oral route. Consumption of contaminated drinking water or food is source of infection. However, there are many other possible ways of becoming infected with *Cryptosporidium* such as person-person contact, animal-person contact, contaminated swimming pools and other recreational waters (rivers and lakes), or foreign travel.

Waterborne outbreaks

Numerous reports of outbreaks of cryptosporidiosis related to drinking water have been described mainly in North America, the UK, and Japan (Fayer 2004a) (Hlavsa 2005) (Karanis 2007a) (Eisenstein 2008). The first occurred in 1984 in Texas, with sewage-contaminated groundwater as the source of the infection (D'Antonio 1985). The largest documented waterborne outbreak occurred in Milwaukee, USA, in 1993 when more than 400,000 people were infected by the drinking-water supply (MacKenzie 1995). The total cost of illness associated with this outbreak has been estimated at 96.2 million US dollars (WHO 2004). In European countries other than the United Kingdom, few *Cryptosporidium* outbreaks have been reported (Semenza 2007). The first outbreak occurred in 1995 in the Netherlands, but the source of transmission was not identified (Van Asperen 1996). A second outbreak due to the contamination of a water tank was reported in Italy in 1997 (Pozio 1997). In France, *Cryptosporidium* was involved in four (out of ten) waterborne outbreaks investigated between 1998 and 2006 (Beaudeau 2008). Most outbreaks were caused by *C. parvum* genotype 1 (recently called *C. hominis*), even in areas where the majority of sporadic cases were caused by *C. parvum* genotype 2.

Waterborne outbreaks of cryptosporidiosis have been attributed to contaminated drinking water, from both surface-water and ground-water sources, and to recreational water, including swimming pools. Outbreaks caused by drinking-water have been attributed to contamination of the source water by heavy rainfall or melting snow, to sewage contamination of wells, to inadequate treatment or treatment deficiencies and to combinations of these factors. But many outbreaks have been associated with potable water, even when processing systems had been operated in accordance with conventional standards of water treatment, and while current microbiological standards were met.

Swimming-pool associated outbreaks have also been widely reported (WHO 2002) (Takagi 2008).

In addition, foodborne transmission and traveller's diarrhoea related to cryptosporidiosis have been documented.

Ecology

Regrowth in distribution systems

Oocysts are environmentally inert and cannot multiply outside their hosts. The role of biofilm in the persistence of oocysts remains insufficiently documented, although oocysts of *Cryptosporidium* have been shown to be able to attach to and persist in biofilms (Piriou 2003) (Helmi 2008).

Occurrence

Cryptosporidium oocysts are ubiquitous in surface waters worldwide. Reported concentrations are generally in the range 0.01–100 /L (WHO 2002). The sources of environmental contamination can be both human sewage and animals. Groundwaters that mix with surface water or other sources of contamination (e.g. surface run-off) may contain low levels of *Cryptosporidium* (< 1/L). They also have been detected in treated water samples in the range 0.001–0.3/L. Surveys conducted in various regions of the United States demonstrated the presence of *Cryptosporidium* oocysts in 67–100% of wastewaters, 24–100% of surface waters, and 17–26.8% of drinking waters (Rose 1991a).

A genotyping study of *C. parvum* and *C. hominis* human and bovine isolates from seven countries in Europe, Middle East, Africa, North America and Australasia demonstrated a quasi-complete phylogenetic segregation among countries for both species (Tanriverdi 2008).

Although more and more studies have collected data on *Cryptosporidium* species occurrence and infectivity over the past few years (Xiao 2001) (Le Chevallier 2003), information is still lacking about the health significance of the presence of *Cryptosporidium* oocysts in water, in particular whether human pathogenic species are present and whether the oocysts are viable and infectious.

Reservoirs

Many animal reservoirs, including farm livestock, pets and wildlife have been identified (Graczyk 1997a). They spread oocysts to humans and the water environment, particularly through agricultural and human effluents.

Survival

The robust nature of *C. parvum* oocysts enables them to survive for months in surface water.

Carey *et al.* (Carey 2004) and Peng *et al.* (Peng 2008) summarized the effects of environmental selected factors on the viability and infectivity of *C. parvum* oocysts; the main data are reported below.

Oocysts are able to survive for several months in water kept at 4°C. Temperature, however, appears to be the most lethal factor affecting oocysts in the environment. An exponential relationship between the die-off rate of *C. parvum* and temperature (positive or negative) has been demonstrated (Peng 2008). At extreme temperatures, oocyst viability and infectivity are adversely affected. One study (King 2005) reported that oocysts incubated at 4°C and 15°C remained infective over the 12-week holding period, whereas a 4 log(10) reduction in infectivity was observed for both 20 and 25°C incubation treatments at 12 and 8 weeks, respectively, for all water types examined. The authors also showed that temperature inactivation at higher temperatures was a function of increased oocyst metabolic activity, as measured by oocyst ATP content.

Solar radiation has been reported as a critical process affecting oocyst survival in the environment (King 2008).

Rapid freezing inactivates oocysts whereas slow freezing, which is typically observed in the natural environment, does not.

C. parvum oocysts are susceptible to inactivation by dessication.

Oocysts were also able to survive for extended periods in seawater during 8 to 12 weeks at varying temperatures and salinities, similar to those in which Eastern oysters survive.

Peng *et al.* (Peng 2008) suggested that oocyst degradation in faeces is more rapid than in water, probably because of other faecal microorganisms associated with the degradation process.

Inactivation/removal

Published data on removal and inactivation are summarized in annex 1.

Chemical treatment:
C. parvum oocysts are resistant to inactivation by common disinfectants such as free chlorine and monochloramine at the concentrations used in water treatment. One study reported that the Ct values for the inactivation of *C. parvum* (for a 3-log reduction in oocyst viability) under chlorinated recreational water conditions were higher than the current recommendations of the Center for Disease Control and Prevention (Shields 2008).

Several studies showed that ozone more effectively inactivates *Cryptosporidium* oocysts (Wohlsen 2007) (Pereira 2008). Effective oocyst inactivation was achieved ($>2\ \log_{10}$ units) with 2 mg/L of ozone (for 10 min) at 20°C (Keegan 2003). But the effectiveness of ozone decreases at lower temperatures. Sequential disinfection of oocysts with different chemical agents can result in greater inactivation levels (Carey 2004). However, the infectivity of oocysts is unaffected by exposure to coagulants and subsequent exposure to disinfectants (Keegan 2008).

Physical removal or inactivation:
Despite their small size, *Cryptosporidium* oocysts are readily removed by conventional processes used in drinking water treatment plants such as coagulation, settling, rapid filtration and slow sand filtration. A well-operated treatment plant using chemical coagulation, settling and filtration should achieve about

99.9% oocyst removal (MicroRisk 2007). But the level of removal by conventional treatment has been shown to be dependent on initial pathogen concentration (efficacy was lower with lower initial pathogen spike doses) (Assavasilavasukul 2008).

One study demonstrated that membrane filtration of the public water supply was effective in reducing the risk for sporadic human *Cryptosporidium* infection in a population in England between 1996–2002 (Goh 2004).

Micro and ultra filtration can remove over 99.99% as long as the integrity of the system is maintained.

Investigations based on infectivity assays have shown that UV light irradiation inactivates oocysts. Effective oocyst inactivation – as measured by cell culture -TaqMan PCR - was achieved ($>2\ \log_{10}$ units) with low-pressure UV lamps (20 mJ/cm^2) (Keegan 2003) or with pulsed-UV light (15 mJ/cm) (Lee S.U. 2008). *Giardia muris* cysts and *C. parvum* oocysts exposed to medium-pressure UV doses of 60 mJ/cm^2 or higher did not exhibit resistance to and/or reactivation following treatment (Belosevic 2001). It was demonstrated that *C. hominis* oocysts (the predominant genotype in human cryptosporidiosis infections) displayed similar levels of infectivity and had the same sensitivity to UV light as *C. parvum* (Johnson 2005). The efficacy of low and medium pressure lamps for oocyst inactivation is in the same order of magnitude (Hijnen 2006).

Freezing and thawing at more than 72°C for 1 min or 45°C for 10–20 min effectively inactivate oocyst (Carey 2004).

Recently, Olvera *et al.* (Olvera 2008) investigated the potential of ultrasonic treatments (1MHz ultrasound) in inactivation of *C. parvum*.

Surrogates

Owing to the exceptional resistance of the oocysts to disinfectants, *E. coli* (or, alternatively, thermotolerant coliforms) cannot be relied upon as an index for the presence/absence of *Cryptosporidium* oocysts in drinking-water supplies.

Continuous monitoring of turbidity or particle count can give an early warning of particle breakthrough and alert operators to an increased risk of the presence of oocysts in the treated water.

Legislation:

The UK legislation requires that oocysts must not be present in drinking water at a concentration higher than 1 per 10 L.

Environmental detection

Concentration, purification, and detection are the three key steps in all methods that have been approved for routine monitoring of waterborne oocysts.

Methods for total enumeration (dead and live oocysts).

- Approved methods USEPA 1622 (*Cryptosporidium*) and USEPA 1623 (*Cryptosporidium* and *Giardia*): Both methods include filtration (Pall Gelman Envirochek^{TM} HV filter or Idexx/Genera Filta-Max^{TM} filter), Immuno-magnetic Separation (IMS) and Immuno-fluorescence Assay (IFA). Confirmation using differential interference contrast microscopy (DIC) and nucleic acid stain (DAPI) can be added (USEPA 1999a) (USEPA 1999b).
- Drinking water inspectorate DWI approved method (UK): This method is similar to the USEPA methods, using Idexx/Genera Filta-Max ^{TM} filter for concentration (DWI 1998).
- French standardization association (AFNOR) approved method NFT90-455: this method is similar to the USEPA methods.

  In a study designed to evaluate the recovery efficiencies of *C. parvum* oocysts from various filtration cartridges (Lee 2004), the Filta-Max filter proved to perform the best.

- Alternative methods to concentrate water samples have been described, including cross-flow filtration, continuous-flow centrifugation (Higgins 2003) and vortex-flow filtration.

  Graczyk *et al.* (Graczyk 1997b) investigated the use of the cellulose-acetate membrane filter dissolution method to concentrate and recover viable and infective oocysts. Several studies described the simultaneous concentration of bacteria, viruses and parasites from drinking water samples by using ultrafiltration-based techniques (Guillot 2003) (Hill 2005) (Polaczyk 2008).
- Flow cytometry has been used for selective separation of *Cryptosporidium* oocysts and *Giardia* cysts from samples as an alternative to IMS purification (Medema 1998) and to manual epifluorescence microscopy examination (Montemayor 2007). Several protocols using flow cytometry to quantify *Giardia* cysts and *Cryptosporidium* oocysts in water samples have been described (Gratacap-Cavallier 2000) (Hsu 2005) (Barbosa 2008a).
- Recently, an immunological approach, a surface plasmon resonance-based inhibition assay using a sensor chip, was developed for real-time detection of *C. parvum* oocysts (Kang 2008).

Conventional approaches however cannot discriminate among the different species of *Cryptosporidium.* Only molecular biology techniques are able to specifically detect human pathogenic species and genotypes.

- Since 1991, many PCR-based detection methods, including IMS-PCR have been described (Laxer 1991) (Johnson 1995) (Rochelle 1997a) (Hallier-Soulier 2000). Jiang *et al.* (Jiang 2005) developed and evaluated different direct extractions of *Cryptosporidium* DNA from water concentrates. An immunocapture-based polymerase chain reaction (PCR) assay for simultaneous detection of *C. parvum* oocysts and *Giardia intestinalis* cysts in surface water has been reported (Rimhanen-Finne 2002). Several studies described protocols based on PCR-RFLP to detect and discriminate the human pathogenic species *C. parvum* and *C. hominis* in water samples (Xiao 2004) (Ochiai 2005).
- As an alternative to PCR-based approaches, a loop-mediated isothermal amplification (LAMP) procedure was evaluated for the first time for sensitive detection of *Cryptosporidium* oocysts in faecal and water samples (Karanis 2007b).
- A single analytical process using ultrafiltration combined with high-density oligonucleotide array has been developed to simultaneously detect waterborne pathogens including bacteria, viruses and parasites (*Cryptosporidium* and *Giardia*) (Guillot 2003) (see Enterovirus section).
- Several methods using real-time PCR to detect and/or quantify *C. parvum* oocysts in water samples have been published (Higgins 2001) (Fontaine 2002) (Fontaine 2003) (Ramirez 2006). A multiplex real-time PCR (qPCR) assay for simultaneous quantification of *G. lamblia* and *C. parvum* in water and sewage has been reported (Guy 2003).

Methods for determining oocyst viability:

- In vitro excystation: the method is not very sensitive and is difficult to apply on a routine basis.
- Incorporation of vital dyes: DAPI/PI procedure or SYTO9 staining. To date interpretation remains difficult and results are not reproducible (WHO 2002).

Several molecular approaches have been described:

- FISH (Fluorescent in situ hybridization) technique (Vesey 1998) (Smith 2004) (Lemos 2005) (Taguchi 2006). This technique however does not appear sensitive enough for application to environmental water concentrates that contain autofluorescent algae and mineral particles.
- The combination of in vitro excystation with PCR amplification (Wagner-Wiening 1995). In vitro excystation /PCR tests show the same limitations as excystation itself.
- RT-PCR: The detection of mRNA transcripts from an induced heat shock protein 70 (*hsp70*) gene has been shown to be sensitive enough to detect small numbers of viable oocysts in environmental water concentrates (Stinear 1996) (Kaucner 1998) (Hallier-Soulier 2003). One study described the use of nucleic acid sequence-based amplification (NASBA) with subsequent electrochemiluminescent

detection for the specific and sensitive detection of viable oocysts of *C. parvum* in environmental samples (Baeumner 2001). As the RT-PCR -based methods above, this approach targets *C. parvum* heat shock protein hsp70 mRNA. Jenkins *et al.* (Jenkins 2000) described an RT-PCR assay directed at mRNA encoding the enzyme amyloglucosidase (CPAG) to differentiate viable from non-viable *C. parvum* oocysts. As for PCR assays available, all these molecular methods which determine viability are at the experimental stage for research purposes and validated protocols do not yet exist.

Methods for determining oocyst infectivity:

- Animal infectivity: this assay has long been considered the method of choice for determining the infectivity of oocysts, but is impractical for use on a routine basis. It requires extensive space and equipment.
- Cell culture based methods: Foci Detection Method (FDM) combines cell culture and immunofluorescence assay (Slifko 1997). This method is time-consuming and tedious.
- An in vitro infectivity assay using cell culture combined with RT-PCR to detect the infectious *C. parvum* oocysts has been described (Rochelle 1997b).
- An approach called cell culture quantitative sequence detection (CC-QSD) which combines quantitative TaqMan PCR method with cell culture method has been developed (Keegan 2003) (Di Giovanni 2005) (Keegan 2008). To date, these promising molecular approaches need optimization and standardization before routine use on environmental water samples.

Several typing methods are available to discriminate human and animal *Cryptosporidium* species and genotypes such as PCR-RFLP and RAPD. Straub *et al.* (Straub 2002) developed an oligonucleotide microarray to differentiate between closely related *C. parvum* isolates and *Cryptosporidium* species.

Research needs

There are obvious needs to be addressed from the clinical standpoint – e.g. therapeutic drugs, vaccine development, infective dose – as well as in epidemiology concerning the actual incidence of cryptosporidiosis, but we will focus on needs with respect to waterborne oocysts. There are a number of areas relating to *Cryptosporidium* that require further research including aspects of monitoring, analysis, water treatment and disinfection efficiency, environmental survival and distribution, transport and fate of oocysts in soils and groundwater systems.

Analytical method

Clearly the most critical requirement is for standardized, improved oocyst recovery and detection methodology. The numerous published PCR-based methods identify the lack of an optimized standard method for clinical, environmental, and public health investigations of cryptosporidiosis.

The ability to distinguish viable and/or infectious oocysts would greatly facilitate interpretation of oocyst presence in waters.

There is a need to develop reliable, routine tests which specifically detect human pathogenic species (*C. parvum* and *C. hominis*) and genotypes and determine oocyst viability. RT-PCR constitutes a promising approach but requires optimization, simplification and validation.

Standardized methods for rapid and reliable quantification of *Cryptosporidium* species are also required.

Taxonomy

Further investigations including international collaboration efforts are needed to resolve the current difficulties concerning taxonomy, genotyping, phylogenetic analysis and characterization of *Cryptosporidium*.

In fact research on *Cryptosporidium* has recently resulted in a considerable progress due to available molecular methods for studying the diversity within the genus. Numerous genotyping studies yielded an enormous amount of data concerning this parasite. However there is an urgent need to review and update the sequence data for *Cryptosporidium* available in public databases. Considerable work must be done for the resolution of the taxonomy and the number of valid species in the genus *Cryptosporidium*.

Ecology

The potential incidence of environmental factors (humidity, temperature, seasonal variation, other geographic factors) should be further determined.

The role of biofilm in the survival of pathogenic *Cryptosporidium* remains insufficiently investigated. The impact of predation is another issue that has not been sufficiently addressed in survivability experiments.

Little information is available on the significance to the environmental ecology of the state in which oocysts occur in water, i.e. suspended or attached to particles.

Concerning groundwater, transport and fate of *Cryptosporidium* in groundwater systems should be further investigated.

Occurrence and distribution

Few studies reported the species distribution of *Cryptosporidium* parasites in environmental samples.

Studies using molecular typing tools to identify *Cryptosporidium* isolates from human, animal or environmental samples including water and sludge need to be carried out. The objective will be to determine which *Cryptosporidium* species or genotypes are present in Europe, the USA and other parts of the world, how they are spatially distributed, how they circulate in ecosystems, and whether geographic origin is predictive of genotypes.

In addition the health significance of the presence of oocysts in water is not documented for lack of adequate methods for viability determination. It would be critical to determine if *Cryptosporidium* oocysts detected in particular in groundwater and in treated water samples are viable and potentially infectious and whether human pathogenic species are present. National databases should be established to provide comprehensive information on the occurrence of oocysts in both sources and treated water.

Treatment

Additional research into improving water treatment and sewage treatment practices is needed, particularly in testing the efficiency of ozone in oocyst inactivation using *Cryptosporidium* human pathogenic species. Studies on the performance of treatment processes to remove oocysts from drinking water should be carried out in full-scale conventional water treatment plants and not only in pilot tests.

In addition, further evaluation and development of the use of bacterial spores or other surrogates to assess treatment performance are necessary.

Giardia

Health significance	**High**
Persistence in water supplies	**Moderate**
Resistance to chlorine	**High**
Relative infectivity	**High**
Important animal source	**Yes**

Adapted from table 7.1 Waterborne pathogens and their significance in water supplies (WHO 2004).

Microbiology

General description

Giardia is an ubiquitous parasite affecting humans and a wide range of domestic and wild mammals.

G. duodenalis (syn. *G. intestinalis*; *G. lamblia*) is the most common intestinal parasite of humans worldwide (Marshall 1997).

Giardia has a relatively simple life cycle consisting of a flagellate trophozoite that multiplies in the gastrointestinal tract and an infective thick-walled cyst that is shed intermittently but in large numbers in faeces. The cyst is the infective form, which is responsible for the transmission of the parasite and the infection of humans and animals.

G. duodenalis cysts are elliptical, 8–12 µm long and 7–10 µm wide. The cyst wall is 0.3–0.5 µm thick and has a fibrillous structure. Two to four nuclei are found in each cyst, together with axonemes of the flagella of the trophozoite (WHO 2004).

Classification

The taxonomy of *Giardia* at the species level remains unclear because only limited morphological differences are present between different species. In addition the high degree of genetic heterogeneity found in human and animal isolates makes speciation difficult. It is still unclear how and whether this heterogeneity is related to host specificity and pathogenicity of *Giardia* (WHO 2002).

Six species in the genus are currently considered valid: *G. duodenalis* (synonym *G. lamblia* or *G. intestinalis*) in a wide range of mammals including humans, *G. agilis* in amphibians, *G. muris* in rodents and reptiles, *G. ardeae* in the grey blue heron, *G. psittaci* in birds, and *G. microti* in muskrats and voles (Sulaiman 2004).

Two major groups of *G. duodenalis* have been recognized in humans, assemblages A and B (Thompson 2000).

Human health effects

Giardia has been known as a human parasite for 200 years. Giardiasis is considered to be a re-emerging infection because of its association with outbreaks of diarrhoea in child-care centres (Thompson 2000).

Clinical features

The symptoms of giardiasis are acute diarrhoea, abdominal cramps, bloating and excessive flatulence. In most cases, the infection is acute and self-limiting, with a duration of 2–4 weeks. The incubation period can be anything from 1 to 75 days, but on average it is 7–10 days.

© 2010 Global Water Research Coalition. Waterborne Pathogens: Review for the Drinking-water Industry. Prepared by Emmanuelle Guillot and Jean-François Loret. ISBN: 9781843391791. Published by IWA Publishing, London, UK.

The severity of the illness can vary considerably and only about a quarter of infected people show symptoms of the illness. Malabsorption of food can lead to considerable loss of weight and in children it can be a cause of failure to thrive.

Excretion of cysts varies between 10^6 and 10^8 per gram of stools but a significant proportion of the stool samples do not contain detectable levels of *Giardia*.

Dose – response data

The infectious dose has been estimated to be as low as 10 cysts for *Giardia* (Adam 2001). Exponential dose-response models based on experimentation on humans have been described (Rose 1991b) (Regli 1991).

Target populations

Giardiasis affects all age groups, although the highest incidence is in children at the age of 1–4 years, particularly in developing countries. It is also a common cause of traveller's diarrhoea.

Diagnosis

Diagnosis is usually made by light microscopic examination of a faeces sample. Procedures using the enzyme immunoassay (EIA) or a direct fluorescent-antibody method have been developed (Marshall 1997). None of these techniques are species-specific nor can they determine whether cysts are infectious or not.

Treatment

Giardiasis is readily treated by a number of different drugs (Minenoa 2003), but if left untreated it can persist for a number of years (Hunter 1997). The two drugs that seem to be most effective are metronidazole and tinidazole.

Geographical distribution

Worldwide.

Epidemiology

Transmission routes

Transmission is through the faecal-oral route.

Consumption of contaminated drinking water or food is source of infection. However, there are many other possible ways of becoming infected with *Giardia* such as person-person contact, animal-person contact, contaminated swimming pools and other recreational waters (rivers and lakes), or foreign travel – giardiasis is also known as "travellers diarrhoea".

Several authors suggest a zoonotic potential for giardiasis (Thompson 2000) (van Keulen 2002) but the infectivity of animal strains for humans has not yet been demonstrated.

Waterborne outbreaks

Waterborne outbreaks of giardiasis have been reported for some 30 years. They are a major public health problem in many industrialized nations, including the United Kingdom, Sweden, Australia, New Zealand, Canada and the United-States. About 40% of water-associated outbreaks of parasitic protozoan diseases worldwide are linked to *G. duodenalis* (Karanis 2007a). The first outbreak was documented in 1965 at Aspen, CO (Craun 1986). The first UK waterborne outbreak occurred in the Bristol area in 1985 affecting 108 persons (Jephcott 1986). Similar to cryptosporidiosis, waterborne outbreaks of giardiasis have been attributed to contaminated drinking water, from both surface-water and ground-water sources, and to recreational water, including swimming pools. Recently, an outbreak of giardiasis associated with a neighborhood interactive water fountain was reported for the first time (Eisenstein 2008).

Outbreaks caused by drinking-water have been attributed to surface water treatment failure, contamination of the source water, and leakage into the distribution system or to combinations of these factors. However many outbreaks have been associated with potable water even when processing systems had been operated in accordance with conventional standards of water treatment, and while current microbiological standards were met (WHO 2002).

Swimming pool associated outbreaks have also been widely reported (Hunter 1997).

In addition, foodborne transmission and traveller's diarrhoea related to giardiasis have been documented.

Ecology

Regrowth in distribution systems/biofilms

Cysts are environmentally inert and cannot multiply outside their hosts. Cysts of *Giardia* are thought to attach to and persist in biofilms. As for *Cryptosporidium*, the role of biofilm in the persistence of *Giardia* remains poorly investigated.

Occurrence

As *Cryptosporidium* oocysts, *Giardia* cysts are ubiquitous in surface waters worldwide. Reported concentrations are generally in the range of 0.01–100 /L (WHO 2002). The sources of environmental contamination can be both human sewage and animals. They are widely found in lakes and rivers especially where there is wildlife which uses these water sources. Concentrations as high as 88,000 cysts/L have been found in raw sewage and as high as 240 /L in surface waters (Wallis 1996). *Giardia* cysts have been detected in 81% of raw water samples and 17% of filtered water samples in the United Sates (Marshall 1997). Groundwaters that mix with surface water or other sources of contamination (e.g. surface runoff) may contain low levels of *Giardia* (< 1/L).

Giardia cysts are commonly found in untreated and in treated sewage. An investigation study of four wastewater treatment plants in Italy revealed that *Giardia* cysts were ubiquitous, whereas *Cryptosporidium* oocysts were quite rare (Cacciò 2003).

It is not clear whether or not seasonality is a general feature of *Giardia* contamination: available results are contradictory (Cacciò 2003). However, most studies on *Giardia* contamination of water have been limited to estimating the prevalence and little information has been published on the contaminating species. The distribution of *G. duodenalis* genotypes and subgenotypes in raw urban wastewater in Milwaukee was investigated (Sulaiman 2004).

The public health importance and contamination sources of *Giardia* cysts found in water are largely unclear.

Reservoirs

Giardia spp. affect a wide range of animals, including domestic pets, farm animals and wild mammals, but there is no unequivocal evidence that organisms from these sources have caused infections in humans.

Survival

Giardia cysts have been shown to survive in water for up to 2 months at temperatures as low as 8°C (Meyer 1980). *G. muris* cysts remained viable for 1 to 3 months depending on the temperature and type of water sample (deRegnier 1989). Under natural conditions, the die-off rate for *Giardia* is higher and more temperature-dependent than for *Cryptosporidium*, varying between 0.015 $\log_{10}$–units per day at 1°C and 0.28 $\log_{10}$ per day at 23°C (WHO 2002).

Inactivation/removal

Published data on removal and inactivation are summarized in annex 1.

Chemical treatment:
The effectiveness of standard chemical disinfectants such as chlorine and chloramine against *Giardia* cysts is limited. 99.99% reduction can be achieved with a Ct of 180–530 mg.min/L with chlorine, depending on the temperature and pH of the water. At Ct values of 4.7–28 mg.min/L chlorine dioxide reduces *Giardia* by 99% (WHO 2002).

Ozone is the most effective agent against *Giardia* cysts: at 20°C the Ct for 99% inactivation of *G. intestinalis* cysts is 0.6 mg.min/L (Finch 1993), but the effectiveness of ozone decreases at lower temperatures.

Physical removal or inactivation:
Cysts are particulate and are fairly readily removed by conventional processes used in drinking water treatment plants such as coagulation, settling, rapid filtration and slow sand filtration. A well-operated treatment plant based on chemical coagulation, settling and filtration should achieve at least 99.9% cyst removal (MicroRisk 2007). Level of removal by conventional treatment, however, has been shown to be dependent on initial pathogen concentration (Assavasilavasukul 2008).

Micro and ultra filtration can remove over 99.99% as long as the integrity of the system is maintained.

Irradiation with UV light is currently the most promising form of disinfection or inactivation for *Giardia*. Work is in progress to identify the conditions in which UV radiation may be used with confidence. Belosevic *et al.* (Belosevic 2001) found that *G. muris* cysts and *C. parvum* oocysts exposed to medium-pressure UV doses of 60 mJ/cm^2 or higher did not exhibit resistance to and/or reactivation following treatment. Recently, Li *et al.* (Li 2008) suggested that some *G. lamblia* trophozoites may survive or are reactivated after exposure to UV fluences up to 10 mJ/cṁ, a UV dose of 100 mJ/cṁ being required for a total inactivation.

Surrogates

Due to the exceptional resistance of the cysts to disinfectants, *E. coli* (or, alternatively, thermotolerant coliforms) cannot be relied upon as an index for the presence/absence of *Giardia* cysts in drinking-water supplies.

Continuous monitoring of turbidity or particle count can give an early warning of particle breakthrough and alert operators to an increased risk of the presence of cysts in the treated water.

Environmental detection

Methods for the analysis of a water sample for *Giardia* cysts are broadly the same as for *Cryptosporidium* oocysts. Concentration, purification, and detection are the three key steps in all methods that have been approved for routine monitoring of waterborne parasites.

Methods for total enumeration (dead and live cysts):
Approved method USEPA 1623 (*Cryptosporidium* and *Giardia*). This method includes filtration (Envirocheck cartridge), Immunomagnetic Separation (IMS) and Immunofluorescence Assay (IFA). Confirmation using differential interference contrast microscopy (DIC) and nucleic acid stain (DAPI) can be added (USEPA 1999). Recently, Shaw *et al.* (Shaw 2008) suggested modifications of EPA Method 1623 to address fully clumped samples.

- The French standardization association (AFNOR) approved method NF T 90–455: this method is similar to the USEPA 1623 method.
- Flow cytometry has been used for selective separation of *Cryptosporidium* oocysts and *Giardia* cysts from samples as an alternative to IMS purification (Medema 1998). Hsu *et al.* (Hsu 2005) described a protocol using flow cytometry to quantify *Giardia* cysts and *Cryptosporidium* oocysts in water samples.
- Alternative methods to concentrate water samples have been described, including cross-flow filtration, continuous-flow centrifugation (Higgins 2003) and vortex-flow filtration. The use of a portable continuous flow centrifuge (PCFC) has been proposed as an alternative concentration step to US-EPA Method 1623 (Zuckerman 2006).

However, the above approaches cannot specifically detect human pathogenic species and genotypes.

- Since 1991, several PCR-based detection methods, including IMS-PCR, have been described (Mahbubani 1991) (Rochelle 1997a) (Mahbubani 1998).

  A PCR- and gene probe-based detection (Mahbubani 1992) and a PCR-RFLP assay (Sulaiman 2004) could differentiate *G. duodenalis* from other *Giardia* spp.

  An IMS-PCR assay for simultaneous detection of *C. parvum* oocysts and *G. intestinalis* cysts in surface water has been reported (Rimhanen-Finne 2002).
- A single analytical process using ultrafiltration combined with high-density oligonucleotide array has been developed to simultaneously detect waterborne pathogens including bacteria, viruses and parasites (*Cryptosporidium* and *Giardia*) (Guillot 2003) (see Enterovirus section). A multiplex real-time PCR (qPCR) assay for simultaneous quantification of *G. lamblia* and *C. parvum* in water and sewage has been reported (Guy 2003).

Methods for determining oocyst viability:

- In vitro excystation (Hoff 1985): the method is not very sensitive and difficult to apply on a routine basis.
- Incorporation of vital dyes: according to Taghi-Kilani *et al.* (Taghi-Kilani 1996) SYTO-9 staining was correlated to animal infectivity. The determination of *Giardia* cyst viability in water samples has been reported by combination of immunofluorescence, fluorogenic dye staining (DAPI/PI) and differential interference contrast microscopy (Thiriat 1998). Another study described a viability assay using fluorogenic dyes, fluorescein diacetate (FDA) and labelled monoclonal antibodies (Jarmey-Swan 2000), but to date methods available cannot be used on a routine basis because microscopic interpretation remains difficult (WHO 2002).
- Electrorotation as a rapid method for determining the viability of *Giardia* has been reported (Dalton 2001).
- Recently, Barbosa *et al.* (Barbosa 2008b) described a flow cytometry protocol for the detection and viability assessment of *G. lamblia*.

These different approaches need optimization and standardization before routine use on environmental water samples.

Molecular approaches based on RT-PCR and FISH have also been described:

- RT-PCR: Methods based on detection of giardin mRNA (Mahbubani 1991) and of heat shock-induced mRNA (Abbaszadegan 1997) have been reported.
  Simultaneous detection of viable *C. parvum* oocysts and *Giardia* cysts from large-volume water samples with wound fiberglass cartridge filters and reverse transcription-PCR has also been described (Kaucner 1998).
- The detection and viability determination of *G. lamblia*, *C. parvum*, and *C. hominis* were carried out using fluorescent in situ hybridization (FISH) and monoclonal antibodies (Lemos 2005).

Methods for determining oocyst infectivity:

- Animal infectivity: this assay has long been considered the method of choice for determining the infectivity of cysts (Belosevic 1983), but it is impractical for use on a routine basis. It requires extensive space and equipment.

At the present time, there is an absence of adequate reference methods to detect and quantify the presence of viable /infectious cysts of *Giardia* human pathogenic species.

Several typing methods are available to discriminate human and animal *Giardia* species and genotypes such as PCR-RFLP (Homan 1998) and RAPD (Paintlia 1999) (Pelayo 2003) and a sequence characterization of the triosephosphate isomerase (TPI) gene (Sulaiman 2004).

Research needs

Giardia-related issues are similar to those of *Cryptosporidium* and research has often focused on both parasites. This explains why, as for *Cryptosporidium*, there are a number of areas relating to *Giardia* that require further research including aspects of monitoring, analysis, water treatment and disinfection efficiency, environmental survival and distribution, the zoonotic potential transmission, transport and fate of cysts in soils and groundwater systems.

Clearly the most critical requirement is for standardized, improved cyst recovery and detection methodology.

The ability to distinguish viable and/or infectious cysts would greatly facilitate interpretation of *Giardia* cyst presence in waters. There is a need to develop reliable, routine tests which specifically detect the human pathogenic species (*G. intestinalis*) and which determine the viability of the cysts. RT-PCR constitutes a promising approach but requires optimization, simplification and validation.

Molecular typing methods which can discriminate between the different strains of *G intestinalis* from human or animal sources would be useful for epidemiological studies.

Methods for rapid and reliable quantification of *Giardia* cysts species are also required.

Taxonomy:

Further investigation is needed to resolve the current difficulties concerning taxonomy, genotyping, and characterization of *Giardia*. Considerable work must be done for the resolution of the taxonomy and the number of valid species in the genus *Giardia*.

Zoonotic transmission:

Little is known about the relative infectivity and the intensity of the infection caused by the different strains of *G. intestinalis* of human or animal origin. The zoonotic transmission of giardiasis, although it is often addressed in the literature, remains to be demonstrated.

Ecology:

The potential incidence of environmental factors (humidity, temperature, seasonal variation, other geographic factors) should be further determined.

The role of biofilm in the survival of pathogenic *Giardia* remains poorly investigated. The impact of predation is another issue that has not been addressed in survivability experiments.

Little information is available on the significance to the environmental ecology of the state in which cysts occur in water, i.e. suspended or attached to particles.

Concerning groundwater, transport and fate of *Giardia* in groundwater systems should be further investigated.

Occurrence and distribution:

Little is known about the species distribution of *Giardia* parasites in environmental samples.

Studies using molecular typing tools to identify *Giardia* isolates from human, animal or environmental samples including water and sludge need to be carried out.

In addition, the health significance of the presence of cysts in water is not documented for lack of adequate methods for viability determination.

Treatment:

Additional research to improve water treatment and sewage treatment practices is needed, particularly in testing the efficiency of ozone and UV in cyst inactivation. Studies on the performance of treatment processes to remove cysts from drinking water should be carried out in full-scale conventional water treatment plants and not only in pilot tests.

Cyclospora

Health significance	**High**
Persistence in water supplies	**presumably long**
Resistance to chlorine	**High**
Relative infectivity	**presumably high**
Important animal source	**No**

Adapted from table 7.1 Waterborne pathogens and their significance in water supplies (WHO 2004).

Microbiology

General description

Cyclospora is a newly described waterborne and foodborne pathogen. The first known human cases of illness caused by *Cyclospora* infection were reported in 1979 (Ashford 1979), but it is only since about 1990 that the role of the organism in human disease has been identified.

Cyclospora is a single-cell, obligate, intracellular, coccidian protozoan parasite. It produces thick-walled spheroid oocysts of 8 to 10 μm in diameter, and each oocyst has two ovoid sporocysts with two sporozoites each. The oocysts stain red with modified acid-fast stains and autofluoresce under UV light.

The life cycle has not yet been completely elucidated. One of the fundamental features of the biology of this organism is the fact that *Cyclospora* oocysts are non infectious in freshly excreted stool. The oocysts are thought to require from days to weeks outside the host, under favorable environmental conditions, to sporulate, and thus become infectious (Herwaldt 2000).

Classification

Cyclospora is a member of the subphylum Apicomplexa, class Sporozoasida, subclass Coccidiasina, family Eimeriidae. Molecular phylogenetic analysis suggests that the genus is closely related to the genus *Eimeria* (WHO 2002).

Of the 17 known *Cyclospora* species, *C. cayetanensis* is the only specie of the genus thus far associated with human illness (Shields 2003).

In the past, it used to be called by such names as cyanobacterium-like, coccidia-like, and *Cyclospora*-like bodies (CLBs).

Human health effects

Clinical features

Cyclospora is a microscopic parasite that infects the small intestine of humans with a disease that is called cyclosporiasis. Sporozoites are released from the oocysts when ingested and penetrate epithelial cells in the small intestine of susceptible individuals. Clinical symptoms include water diarrhoea, abdominal cramping, weight loss, anorexia, myalgia and occasional vomiting and/or fever and last an average of 7 weeks. In AIDS patients, infections tend to last up to 4 months. Relapsing illness often occurs (Marshall 1997).

The incubation period is variable but averages 7 days with low or moderate numbers of unsporulated oocysts being excreted for up to 60 days.

Asymptomatic infection occurs and probably is most common in settings where *Cyclospora* is endemic.

© 2010 Global Water Research Coalition. Waterborne Pathogens: Review for the Drinking-water Industry. Prepared by Emmanuelle Guillot and Jean-François Loret. ISBN: 9781843391791. Published by IWA Publishing, London, UK.

Dose – response data

While unknown, the infectious dose is presumed to be low, as for *Cryptosporidium.*

Target populations

Anyone can suffer *Cyclospora* infection.

Diagnosis

Light microscopy, acid-fast staining, sporulation assay can be used. Molecular methods are increasingly used: a nested-PCR-RFLP protocol has been developed to specifically detect *C. cayetanensis* in stool (Orlandi 2000).

Treatment

Cyclospora infection is treatable with a combination of two sulfa-based antibiotics: trimethoprim and sulfamethoxazole.

Geographical distribution

In the past, *Cyclospora* infection was usually found in people who lived or travelled in developing countries. However, nowadays *Cyclospora* infection occurs worldwide but appears to be most common in tropical and subtropical areas (Herwaldt 2000). It is endemic to North, Central and South America, Southeast Asia, the Caribbean islands, India and parts of Eastern Europe (Sterling 1999).

Epidemiology

A seasonal distribution of infection, coinciding with wet or warm months of the year has been suggested (Herwaldt 2000).

Transmission routes

It is transmitted by the faecal-oral route.

The parasite can be transmitted by ingestion of contaminated water or food (fresh fruit and vegetables) with either sporulated oocysts or unsporulated oocysts that have time to sporulate before consumption.

Direct person-to-person transmission by faecal exposure is unlikely, because excreted oocysts must sporulate to become infective.

Waterborne outbreaks

Several outbreaks of *Cyclospora* infection (or isolated cases) have been linked to waterborne transmission (Slifko 2000) (Karanis 2007a). The first reported cases were among the staff of a hospital in Chicago, USA, in 1990. The infections were associated with drinking tap water that had possibly been contaminated with stagnant water from a rooftop storage reservoir (Centers for Disease Control 1991). To date, the only case of *Cyclospora* transmission through water was reported in 1994 in Nepal, where drinking-water consisting of a mixture of river and municipal water was associated with infections of 12 to 14 soldiers.

The United States and Canada experienced a number of large foodborne outbreaks in the 1990s (Herwaldt 2000). From 1990 to 2000, there were 11 foodborne outbreaks of cyclosporosis in North America that affected at least 3600 people caused by produce imported from countries where *Cyclospora* is endemic (Mansfield 2004). In 1996, the largest ever reported outbreak of cyclosporiasis, affecting more than 1,400 persons in North America was associated with eating fresh raspberries from Guatemala. Most likely, faecally-polluted water used for spraying biocides on fruit was the indirect source (Leclerc 2004).

Ecology

Regrowth in distribution systems/biofilms

Does not multiply outside host.

Its possible persistence in biofilm has not yet been documented.

Occurrence

Data are limited because of the lack of sensitive and specific environmental screening methods.

Dowd *et al.* (Dowd 2003) reported the presence of *C. cayetanesis* in source water used for consumption around the city of Guatemala using a molecular approach. In a waterborne outbreak in Nepal, *Cyclospora* oocysts were found in the drinking water supply although chlorine residuals remained at acceptable levels (0.3-0.8 ppm) (Rabold 1994). *C. cayetanesis* oocysts were also detected in wastewater using microscopic and molecular techniques (Sturbaum 1998).

Reservoirs

The primary source of contamination is human faeces containing oocysts.

Other sources may exist but have not yet been identified. Host range remains unknown including if infected animals are a source of infection for humans.

Cyclospora-like oocysts have been recovered in several animals such as ducks, chickens, dogs, including non-human primates (Sterling 1999), but their phylogenetic relationship to *C. cayetanensis* remained uncertain due to the limited availability of molecular techniques to differentiate and speciate these isolates.

Survival

Little is known.

To maintain transmission, *Cyclospora* must survive in the environment long enough both to sporulate and to be ingested thereafter by a susceptible host. A moist environment is probably more conducive to survival than a dry one. *Cyclospora* spp. are likely to survive longer at lower temperatures when suspended in water.

Preliminary studies have shown that oocysts subjected to $-20°C$ for 24h and exposure to $60°C$ for one hour cannot be induced to sporulate. Oocyst storage at $4°C$ or $37°C$ for 14 days retards sporulation (Smith 1997).

Inactivation/removal

Little information is available regarding the ability of water treatment processes to remove or inactivate *Cyclospora* spp. oocysts. Due to size (8-10µm in diameter), it is likely that physical removal will be similar to that achieved with *Giardia* and *Cryptosporidium* (WHO 2002).

According to WHO (WHO 2002), the oocysts are resistant to disinfection and are not inactivated by chlorination practices generally applied in the production of drinking water (WHO 2002).

One recent study suggested that solar disinfection may be a simple, cheap and effective means to inactivate *Cyclospora* in water, particularly in developing countries (Gaafar 2007).

Surrogates

Owing to the resistance of the oocysts to disinfectants, *E. coli* (or, alternatively, thermotolerant coliforms) cannot be relied upon as an index of the presence/absence of *Cyclospora* in drinking-water supplies.

Environmental detection

No reference methods have been specifically developed to detect *Cyclospora* in environmental samples.

Cyclospora spp. oocysts being larger than *C. parvum* and smaller than *G. intestinalis* cysts, it is therefore assumed that concentration techniques developed for *Cryptosporidium* and *Giardia* will prove effective for sampling and recovering *Cyclospora* spp. oocysts from water concentrates (WHO 2002).

Traditional microscopic identification currently used in stool samples has been applied to environmental samples. Microscopic techniques use size and morphology to identify *Cyclospora* oocysts and take advantage of the oocyst's autofluorescence under UV light, but they are difficult, labor-intensive and cannot distinguish *Cyclospora* species.

Currently there is no commercially available polyclonal or monoclonal antibody against *Cyclospora* oocysts.

Electrorotation as a rapid method for determining the viability of *C. cayetanensis* has been reported (Dalton 2001).

There is a lack of laboratory methods for assessing the viability and infectivity of *Cyclospora* oocysts. There are neither in vitro culture methods nor in vivo amplification models.

Molecular techniques have first been developed for clinical samples and have recently be applied to environmental samples, but the current PCR-RFLP protocol (Orlandi 2000) used in stool samples cannot distinguish between *C. cayetanensis* and *Cyclospora* species from other primates. Orlandi *et al.* (Orlandi 2003) developed primers for PCR detection to differentiate between *C. cayetanensis*, nonhuman primate species of *Cyclospora*, and *Eimeria* species.

An extraction-free, filter-based protocol to prepare DNA templates for use in PCR to identify *C. cayetanensis* has been described (Orlandi 2000).

One study reported the development of a PCR-RFLP assay for detection of *C. cayetanensis* in environmental waters without microscopic confirmation (Shields 2003). Recently, a highly sensitive and specific PCR assay was developed for reliable detection of *C. cayetanensis* oocysts (Lalonde 2008). These two new protocols should be of great use for environmental microbiologists and public health laboratories.

To date, only one study reported the development of a quantitative real-time PCR assay for *C. cayetanensis* (Varma 2003). This promising protocol should be validated on environmental water samples.

Research needs

There is a need for a better understanding of the environmental biology and survival as well as the epidemiology of this newly described emerging pathogen.

Environmental survival

Little is known about the effects of various environmental conditions on the rate of sporulation and on the viability of unsporulated and sporulated oocysts. What is the impact of fluctuating environmental conditions on the rate of sporulation?

Little is known about conditions that may favor the survival of *Cyclospora*. It is unknown if *Cyclospora* is more resistant to environmental stresses than *Cryptosporidium*, which is excreted fully sporulated.

Cyclospora oocyst distribution and occurrence in the environment remain poorly documented.

Epidemiology

Transmission to humans:

What is the infectious dose ? No animal or human exposure studies have been undertaken.

Do *Cyclospora* other than *C. cayetanensis* infect humans ? The number of species of *Cyclospora* that are infective to human beings is not known, as well as whether human-derived oocysts are infectious to non-human hosts.

Other potential sources of infection such as possibly insects as transport hosts should be investigated. Epidemiological studies will help clarify these transmission issues.

The genetic and antigenic diversity within *C. cayetanensis* species should be further studied. Do the different variants behave differently (e.g. sporulation rate, infectivity and clinical manifestations)?

Another fundamental issue is how the environment, including water, becomes contaminated with *Cyclospora* and whether humans are the only sources of oocysts.

Treatment

Little is known about treatment that will effectively inactivate *Cyclospora* oocysts. The effectiveness of drinking water treatment and disinfection regarding the elimination of *Cyclospora* oocysts needs to be documented.

Analytical method

Robust and efficient detection, viability and typing methods are required to better assess risks and to further epidemiological understanding. Molecular approaches seem to be the most promising.

Microsporidia

Health significance	**High**
Persistence in water supplies	**Moderate**
Resistance to chlorine	**Moderate**
Relative infectivity	**High**
Important animal source	**Yes**

Microbiology

General description

Microsporidia are intracellular eukaryotic parasites, which have the potential for zoonotic and environmental, including waterborne, transmission. They were included in the first two versions of the Contaminant Candidate List (CCL) (U.S. Environmental Protection Agency 1998) (U.S. Environmental Protection Agency 2005). These organisms are small, single-celled, obligate intracellular parasites. They produce environmentally resistant spores with a diameter of 1.0–4.5μm and a characteristic coiled polar filament for injecting the sporoplasm into a host cell to initiate infection. Their host range is extensive and includes most invertebrates and all classes of vertebrates.

Classification

The term "Microsporidia" is a non-taxonomic designation commonly used to describe a group of obligate intracellular protozoa belonging to the phylum Microspora and which were later reclassified with fungi (Weiss 2001).

Approximately 143 genera and more than 1,200 species are known. Since the first documented case in 1985, at least six genera (*Enterocytozoon*, *Encephalitozoon* (including *Septata*), *Nosema*, *Pleistophora*, *Vittaforma* and *Trachipleistophora*) including 14 species have been reported to infect humans (Franzen 1999). *Enterocytozoon bieneusi* and *Encephalitozoon intestinalis* are the most common causes of human infections.

Human health effects

Clinical features

Since the recognition of immunosuppression due to HIV infection, microsporidia have become recognized as important human pathogens. The most common symptoms are diarrhoea, weight loss, abdominal pain, vomiting and fever. Prolonged illness for up to 48 months has been reported in AIDS patients. Infections are self-limiting in immunocompetent individuals. The incubation period is still unknown. The species *Enterocytozoon bieneusi* and *Encephalitozoon intestinalis* are the most prevalent microsporidian parasites that cause gastrointestinal infection.

The pathology of microsporidia infection varies depending on the organs that are infected. Other documented infections include keratitis, conjunctivitis, hepatitis, peritonitis, myositis and central nervous system infection (Marshall 1997).

© 2010 Global Water Research Coalition. Waterborne Pathogens: Review for the Drinking-water Industry. Prepared by Emmanuelle Guillot and Jean-François Loret. ISBN: 9781843391791. Published by IWA Publishing, London, UK.

Dose – response data

The infectious dose is unknown, but it may be low because as few as 100 spores cause infection in mice (Didier 1994).

Target populations

While rare cases of acute self-limiting infections in immunocompetent individuals exist, most infections are primarily associated with immunocompromised hosts (Marshall 1997).

Diagnosis

The diagnosis of microsporidiosis relies on direct visualization of spores in clinical specimens by light or fluorescent microscopy. Trichrome stain and optical brighteners are used to detect spores in faeces, urines, respiratory secretions and other aspirates. Electron microscopy remains an important method but its sensitivity is relatively poor. Molecular diagnosis techniques have been used as alternatives (da Silva 1996).

Treatment

The most effective drugs for treating microsporidiosis in humans currently include albendazole, which is effective against the *Encephalitozoon* species but not against *Enterocytozoon bieneusi*, and fumagillin, which has broader anti-microsporidia activity but is toxic in mammals, suggesting a need to identify better drugs (Didier 2004).

Geographical distribution

Worldwide. Microsporidiosis infections have been reported in persons from all continents.

Epidemiology

Transmission routes

The route of infection has not been fully characterized, but microsporidia are likely transmitted by the faecal-oral route.

Ingestion or possibly inhalation of spores by a susceptible host from a contaminated environmental source as well as direct contact with infected individuals or animals have been suggested as possible transmission routes (Didier 2004).

Waterborne outbreaks

Few waterborne outbreaks have been documented worldwide (Karanis 2007a). In France, a drinking water outbreak of microsporidiosis occurred in the summer of 1995 where approximately 200 cases of microsporidiosis were identified, mostly in AIDS patients (Cotte 1999). However, the source of the organism and faecal contamination of the drinking water supply were not demonstrated. Foodborne associations have been suggested but not yet demonstrated.

Ecology

Occurrence

Not extensively documented for lack of environmental detection methods.

Microsporidia spores can be introduced into the environment from stools, urine or respiratory secretions of infected hosts.

Avery *et al.* (Avery 1987) detected microsporidia from concentrated ditch water.

By using molecular methods, several studies have detected these organisms in water. *E. intestinalis* spores have been found in tertiary sewage effluent, and in surface and ground water (Dowd 1998) (Dowd 2003). Spores of *E. bieneusi* have been detected in small numbers from the Seine River in France, suggesting that the risk of waterborne transmission to humans is limited (Sparfel 1997) (Fournier 2000).

Fournier *et al.* (Fournier 2002) found a positive sample of unknown species of microsporidia in swimming pool water. Human pathogenic species of microsporidia were detected in irrigation water used for crop production (Thurston-Enriquez 2002).

Reservoirs

Although several animal hosts have been identified, the reservoirs of human microsporidia are still unknown.

E. bieneusi was isolated from farm and companion animals including cattle, pigs, dogs, cats and chickens (Rinder 2000) (Reetz 2002) and also from immunodeficient macaques (Chalifoux 2000). Sulaiman *et al.* (Sulaiman 2003) reported the first isolation of *E. bieneusi* in wild, fur-bearing mammals. However, if *E. bieneusi* may result in zoonotic infection, no direct evidence of transmission from animals to humans has yet been documented.

Survival

Not extensively documented.

In distilled nonchlorinated water, some microsporidia spores can survive extreme temperatures, variations in pH, and multiple freeze-thaw cycles and remains viable for up to 10 years. Spores of *E. intestinalis* stored in water were still infective after 12 months and 3 months at 10°C and 25°C, respectively (Li 2003a).

The survival of *Encephalitozoon* species in sea water at 10 and 20°C lasts between 2 to 12 weeks depending on the species, which suggests that these organisms can potentially remain infectious long enough to become widely dispersed in estuary and coastal waters (Fayer 2004b).

Inactivation/removal

Published data on removal and inactivation are summarized in annex 1.

Data are limited about the response of microsporidia to water treatment processes. The small size of the organisms is likely to make them difficult to remove by filtration process.

Preliminary results showed that conventional drinking water treatment consisting of coagulation, sedimentation, and mixed media filtration allowed the removal of *E. coli* (2.67 log) and *E. intestinalis* (2.47 log) (Gerba 2003).

Encephalitozoon intestinalis spores are more susceptible to chlorine disinfection than *Giardia* cysts, but much more resistant than enteric bacteria and viruses (Gerba 2003). According to Johnson *et al.*

(Johnson C 2003), spores of *Encephalitozoon* are sensitive to chlorination. John *et al.* (John 2005) studied chlorine and ozone disinfection of *Encephalitozoon intestinalis* spores. They found that chlorine Ct values for 99% (2-$\log_{10}$) reduction at 5°C ranged from 12.8 at pH 6 to 68.8 at pH 8 (mg.min/L). Ozone Ct values were approximately an order of magnitude less at 0.59-0.84mg.min/L, depending on initial concentration of *E. intestinalis*. Further studies should be conducted on other species of microsporidia that are infectious to humans.

E. intestinalis seems to be very sensitive to inactivation by UV light. The ultraviolet dose required for a 3-log(10) or 99.9% reduction in the number of infective spores of *Encephalitozoon intestinalis* was determined to be 8.43 mW.s/cm̀ (Huffman 2002) (John 2003). However the sensitivities of other species to UV disinfection have not been evaluated.

Surrogates

Due to the lack of information on sensitivity of infectious species of microsporidia to disinfection, the reliability of *E. coli* (or, alternatively, thermotolerant coliforms) as an index for the presence/absence of these organisms from drinking-water supplies is unknown.

Environmental detection

Standard methods for the sampling, identification, and enumeration of microsporidia in water are not yet available.

A method for the concentration and recovery of microsporidia from tap water was developed (Stine 2005).

Because of the small size of human pathogenic microsporidia, light microscopic methods are not appropriate for water samples and do not allow identification and species determination.

Immunofluorescence assays (IFA) have been developed but the available polyclonal and monoclonal antibodies have shown cross-reactivity among species (Dowd 1999). IFA for routine screening of water did not reveal to be an acceptable approach. A protocol for recovering *Encephalitozoon intestinalis* spores from waters by centrifugation and immunofluorescence microscopy was described (Li 2003b). John *et al.* (John 2003) developed a quantitative cell culture infectivity assay for *E. intestinalis*. Though this approach is very useful for health risk assessment studies, in vitro cultivation of microsporidia is laborious, time consuming and thus not suitable for routine use. In addition, no in vitro cultivation method for the main human pathogen species *E. bieneusi* is available.

Alternative molecular methods have been developed.

A fluorescent *in situ* hybridization assay has been developed for the detection of the human-pathogenic microsporidia, *Encephalitozoon hellem* in water samples (Hester 2000), but this approach does not seem to be sensitive enough for water samples with low contamination levels.

An original approach (Moura 2003) described the use of MALDI-TOF mass spectrometry to characterize four human pathogenic species of microsporidia. This may be a potential alternative to characterize microsporidian isolates.

Several PCR-based detection assays for the main pathogen species *E. bieneusi* and *E. intestinalis* have been described (Sparfel 1997) (Dowd 1999) (Fournier 2000) (Hester 2002). An IMS-PCR assay has been developed for *E. bieneusi* in water samples (Sorel 2003). Recently, a method based on a continuous flow centrifugation for concentration followed by indirect IMS combined with real-time PCR was designed for the detection of *Encephalitozoon intestinalis* in water samples (Hoffman 2007).

The above methods currently available are at the experimental stage and validated protocols do not yet exist. In addition they give qualitative results only.

To date no quantitative molecular method has yet been described for *E. bieneusi* in water samples. No methods which determine the viability of pathogenic microsporidia spores are available.

Typing methods such as Restriction Fragment Length Polymorphism (RFLP) analysis of amplicons can be used to determine genus, species and strain types of various microsporidia.

Research needs

To date, the mode of transmission and environmental occurrence of microsporidia have not been elucidated due to lack of sensitive and specific screening methods.

Little is known about the occurrence of these protozoan parasites in environmental water sources. Further studies are needed to determine whether water supplies may be sources of human pathogenic microsporidia.

Although several animal hosts have been identified, the relevant reservoirs of human microsporidia are still unknown. Also, the routes of spreading are unknown.

Is microsporidiosis a zoonotic disease that will be transmitted through close contact with infected animals or is contaminated surface water responsible for transmission and is it a relevant reservoir?

The effectiveness of drinking water treatment processes including both traditional and novel disinfection technologies needs to be evaluated regarding the removal of microsporidia.

In order to carry out the research work above, reliable and validated methods are needed to detect and quantify human pathogenic species of microsporidia (in particular *Enterocytozoon bieneusi* and *Encephalitozoon intestinalis*) in water. There is also a need to develop methods that determine the viability status of microsporidia detected in drinking water.

Acanthamoeba

Health significance	**High**
Persistence in water supplies	**Long**
Resistance to chlorine	**High**
Relative infectivity	**High**
Important animal source	**No**

Adapted from table 7.1 Waterborne pathogens and their significance in water supplies (WHO 2004).

Microbiology

General description

Acanthamoeba spp. are microscopic free-living amoebae (10–50 μm in diameter) that inhabit a variety of air, soil, and water environments. The organism is characterized by a life cycle of feeding and replicating trophozoite and dormant cyst stages; the trophozoite form, under unfavourable conditions, such as an anaerobic environment, will develop into a dormant cyst that can resist extreme temperatures (−20°C to 56°C), disinfection and desiccation (WHO 2004).

Acanthamoeba was included in the first Contaminant Candidate List (CCL) (U.S. Environmental Protection Agency 1998) but not in the second list (U.S. Environmental Protection Agency 2005).

Classification

Acanthamoeba spp. has been placed in the Family Acanthamoebidae. The genus *Acanthamoeba* spp. has been divided into three morphological groups (I, II, and III) based on cyst size and shape (Marciano-Cabral 2003). The genus contains some 20 species of which *A. castellanii*, *A. polyphaga* and *A. culbertsoni* are known to be human pathogens. However a revision of the taxonomy of the genus through DNA-based approaches is under way.

Human health effects

Clinical features

Acanthamoeba spp. are opportunistic pathogens that produce a multifocal encephalitis called granulomatous amebic encephalitis (GAE), a chronic central nervous system disease of immunocompromised hosts, which includes headaches, stiff neck, nausea, vomiting, malaise, confusion and hallucinations, loss of balance and control of movement and seizures. It is a rare but usually fatal disease. This organism also has been associated with cutaneous lesions and sinusitis in AIDS patients and other immunocompromised individuals.

Acanthamoeba is also a leading cause of destructive keratitis, a chronic ulceration in contact lens wearers. It is a rare disease, which can occur in healthy individuals and may lead to impaired vision, permanent blindness and loss of the eye (Nel 2004a).

A. culbertsoni causes GAE, whereas *A. castellanii* and *A. polyphaga* are associated with acanthamoebic keratitis and acanthamoebic uveitis (WHO 2004).

© 2010 Global Water Research Coalition. Waterborne Pathogens: Review for the Drinking-water Industry. Prepared by Emmanuelle Guillot and Jean-François Loret. ISBN: 9781843391791. Published by IWA Publishing, London, UK.

The prevalence of antibodies to *Acanthamoeba* and the detection of the organism in the upper airways of healthy persons suggest that infection may be common with few apparent symptoms in the vast majority of cases. The pathogenicity of *Acanthamoeba* seems to be related to certain strains and low levels of IgA antibodies in the tears of infected individuals (Nwachuku 2004).

Dose – response data

Dose response experiments in animals suggest that at least 1,000 trophozoites are necessary to cause *Acanthamoeba* eye infection. Risk from *Acanthamoeba* keratitis is complex, depending upon the virulence of the particular strain, exposure, trauma, or other stresses to the eye, and host immune response (Nwachuku 2004). The infectious dose is unknown for GAE.

Target populations

GAE is an opportunistic infection that occurs mainly in immunodeficient individuals such as AIDS sufferers. Keratitis occurs mainly in otherwise healthy people with known risk factors.

Diagnosis

Diagnosis of infection includes direct microscopy of wet mounts of cerebrospinal fluid or stained smears of cerebrospinal fluid sediment, light or electron microscopy of tissues, in vitro cultivation of *Acanthamoeba*, and histological assessment. Immunocytochemistry, chemifluorescent dye staining, PCR, and analysis of DNA sequence variation have also been employed for laboratory diagnosis (Marciano-Cabral 2003).

Treatment

Eye and skin infections are generally treatable, but the resistance of *Acanthamoeba* cysts to most antimicrobial agents makes *Acanthamoeba* keratitis one of the most difficult ocular infections to treat with a mean treatment period of more than 5 months and surgical interventions in 50% of cases.

Reports of successful treatment of *Acanthamoeba* GAE infection are scarce.

A number of health agencies recommend that only sterile water should be used to prepare wash solutions for contact lenses (WHO 2004).

Geographical distribution

The organism is ubiquitous and is probably found worldwide (Marshall 1997).

Epidemiology

Transmission routes

The mode of transmission of GAE has not been well established, but drinking water is not considered to be a source of infection. Inhalation of amoebae through the nasal passages and lungs or introduction through skin lesions have been suggested as possible routes of transmission (Marciano-Cabral 2003).

Acanthamoebic keratitis has been associated with soft contact lenses being washed with contaminated home-made saline solutions or contamination of the contact lens containers. Although the source of the contamination is unknown, tap water is one possibility.

Human-to-human transmission has not yet been reported.

Waterborne outbreaks

Keratitis is the only water-related syndrome caused by *Acanthamoeba* spp. (Karanis 2007a). Most episodes of keratitis occur during warm weather and often follow water exposure or a history of swimming in lakes and ponds while wearing contact lenses (Marshall 1997). Nwachuku *et al.* (Nwachuku 2004) reported a relationship between seasonal distribution of keratitis and abundance of *Acanthamoeba* in surface waters.

Cases of acanthamoebic keratitis have also been associated with drinking water due to use of tap water in preparing solutions for washing contact lenses (Marshall 1997) (Radford 2002) (Kilvington 2004).

Ecology

Regrowth in distribution systems/biofilms

Acanthamoeba may persist and grow in biofilm in domestic water supplies where they may serve as reservoirs for the presence and transmission of other human pathogens.

Reducing the presence of biofilm organisms is likely to reduce food sources and growth of *Acanthamoeba* in distribution systems (WHO 2004).

Occurrence

The wide distribution of *Acanthamoeba* in the natural environment makes soil, airborne dust and water all potential sources. *Acanthamoeba* have been found in surface waters (Tsvetkova 2004) (Declerck 2007), occasionally in marine water and sediments, in tap water (Hoffmann 2001) (Lorenzo-Morales 2005) (Shoff 2008) (Thomas 2008), in bottled water, swimming pools and spas and sewage (Nwachuku 2004). Amoebae concentrations (including *Acanthamoeba*) are particularly high in sediments from distribution systems, up to 10^4 cells/ml of sediment (Loret 2008d).

Acanthamoeba spp. also have been isolated from vegetation, from animals including fish, amphibia, reptiles, and mammals.

Reservoirs

No specific ecological niche has been identified to date. But biofilms and sediments in natural or artificial water systems were reported to be reservoirs for *Acanthamoeba* (Declerck 2007) (Loret 2008a).

Survival

Acanthamoeba feed on bacteria in the environment by trapping them in their cytoplasm. Trophozoites can exist and replicate in water while feeding on bacteria, yeasts and other organisms. Depending on the species, *Acanthamoeba* can grow over a wide temperature range in water, with the optimum temperature for pathogenic species being 30°C (WHO 2004).

Both trophozoites and cysts can retain viable bacteria and may serve as reservoirs for bacteria with human pathogenic potential including *Legionella pneumophila* (Kilvington 1990), *Burkholderia cepacia* (Landers 2000), *Mycobacterium avium* (Steinert 1998), *H. pylori* (Winiecka-Krusnell 2002) and *E. coli* serotype O157 (Barker 1999). Thus, *Acanthamoeba* may serve as a host for a variety of bacteria (Greub 2004), as well as one species of yeast, *Cryptococcus neoformans* (Marciano-Cabral 2003) but it can also protect the bacterial agents against chlorination and other disinfection measures.

Inactivation/removal

The published data on removal and inactivation are summarized in annex 1.

Few studies have documented the efficiency of current drinking water treatment processes on *Acanthamoeba*. Compared with *Giardia* and *Cryptosporidium*, *Acanthamoeba* is relatively large and is amenable to removal from raw water by filtration. Some studies demonstrated the effectiveness of full treatment lines, and especially of the clarification steps, to remove *Acanthamoeba* and other amoebae (Hoffmann 2001) (Loret 2008b). The effectiveness of conventional treatment is however limited by the ability of *Acanthamoeba* and other amoebae to colonize filter media, especially GAC (Loret 2008d).

Acanthamoeba cysts can withstand extreme levels of temperature, desiccation, and disinfection (Marciano-Cabral 2003).

All of the Candidate Contaminant List (CCL) organisms for which data were available were found to be fairly sensitive to chlorine dioxide, except for *Acanthamoeba* cysts, which were found to be the most resistant. The cysts of *Acanthamoeba castellanii* were more resistant to ozone than other protozoa and viruses (Gerba 2003)(Loret 2008c). The trophozoites (active growing form) are much more sensitive to chlorine and other disinfectants used in drinking water treatment. They are however more resistant than bacteria (Nwachuku 2004).

Acanthamoeba cysts are very resistant to chlorine, bromide and iodine inactivation as well as ultraviolet light. For UV, an offset fluence of 30 mW.s/cṁ, before inactivation starts, was observed (Hijnen 2006). A recent study revealed that photocatalytic disinfection of *A. castellini* cysts in water was ineffective (Sökmen 2008).

Surrogates

E. coli (or alternatively, thermotolerant coliforms) cannot be used as an index for the presence/absence of *Acanthamoeba* spp.

Environmental detection

The classical microbiological method relies on *in vitro* culture of amoebae. The basic procedure is the use of non-nutrient agar with *E. coli* as a food source. Once the amoebae are established in a bacterized culture, antibiotics can be added to kill off bacteria and the amoebae can be transferred to the appropriate axenic medium (Schuster 2002). This method is time consuming (2–3 days to 2 weeks), does not allow identification at the species level and cannot distinguish pathogenic from non pathogenic strains.

The use of newer molecular methods such as PCR can provide a more rapid and specific means to identify *Acanthamoeba* in environmental samples. Lorenzo-Morales *et al.* (Lorenzo-Morales 2005) used PCR amplification with a genus-specific primer pair to detect *Acanthamoeba* in water sources. A study (Riviere 2005) described a real-time PCR assay for quantification of *Acanthamoeba* trophozoites and cysts. This promising approach needs to be compared with the conventional microbiological method and then validated.

A variety of molecular methods for genotyping and identification of *Acanthamoeba* strains have been used in clinical samples and environmental water samples such as RFLPs of complete or partial nuclear rRNA18S, variation in complete mitochondrial 16S rRNA, analysis 18S rRNA sequences (Schroeder 2001) (Tsvetkova 2004), and randomly amplified polymorphic DNA analysis of whole-cell DNA (Ortega-Rivas 2005).

Research needs

Acanthamoeba is a relatively newly described pathogen of growing importance because it is responsible for severe pathologies in immunocompetent and immunosuppressed individuals and because it may serve as a potential vector for bacterial infections from water sources. Much remains unknown about the epidemiology and ecology of this organism.

Ecology

Acanthamoeba–bacteria interaction:
Some species of bacteria can grow and reproduce within the cytoplasm and become symbionts. These species include well known pathogens, such as legionellae, but also common environmental bacteria. Very little is known about the fate, resistance, epidemiology of these intracellular bacteria. Is resistance within amoebae a first step towards pathogenicity for environmental bacteria? More information is required on the role of *Acanthamoeba* spp. as potential vectors for bacterial infections from water sources. Bacterial endosymbionts may also play a role in the pathogenicity of *Acanthamoeba*. This topic should be investigated.

Occurrence:
For risk assessment purposes, more information is required on the occurrence and the types of *Acanthamoeba* in tap water and ground water. Information on the type of treatment the water received and/or the level of residual chlorine should be known. Moreover, assessment of the pathogenicity of *Acanthamoeba* strains in tap water by cell culture and molecular methods would be valuable.

Epidemiology

The health effects of *Acanthamoeba* and the significance of water in its transmission should be further investigated in epidemiological studies.

Treatment

Treatment strategies should be developed to better control *Acanthamoeba* and other amoebae in drinking water, especially in granular filtration media. In addition, further research should determine if control of *Acanthamoeba* spp. is needed to better control pathogen transmission through water supplies.

Environmental detection method

A rapid quantitative detection method (based on real-time PCR) should be validated on water samples to be used as an alternative to the conventional culture method.

Conclusions

Over the past decade, new microbiological detection techniques have emerged in order to track specific pathogens both for routine monitoring of water and for the investigation of disease outbreaks. Molecular detection methods especially based on PCR, RT-PCR and recently real-time PCR have been increasingly developed, but still need to be enhanced in the future. There are probably very few or no groups of pathogens for which no molecular detection method has not yet been developed.

Real-time quantitative PCR methods, however, are still lacking for some viruses and bacteria. Standard culture methods are needed for *Helicobacter*, adenovirus serotype 40 and 41, calicivirus, hepatitis E virus and rotavirus.

In the future, evaluation studies, comparison of performance with existing cell culture methods, and standardization of these new methods, as well as development of their capacity to detect viability and virulence, will be necessary to receive their approval from legislative authorities.

Whatever the detection method, sample preparation and recovery of micro-organisms, prior to analysis, will also have to be improved.

Knowledge on waterborne pathogens has significantly been heightened over the past 15 years namely by the development of molecular biology tools but many important questions remain unanswered.

With the purpose of identifying priorities for research, one of these important questions would be the determination of the real burden of waterborne disease. Most of the available statistics are based on outbreak data, which probably reflect only a minor part of the burden. Studies aiming at evaluating endemics and their causes would be necessary.

May water be a significant route for the dissemination of *Helicobacter pylori, Aeromonas, Burkholderia*, non-tuberculosis mycobacteria, *Cyclospora*, microsporidia, as well as adenoviruses, enteroviruses or rotaviruses? The presence of these pathogens in distributed waters should be investigated, traceability studies from source to tap should be conducted and links with epidemiological data should be established.

The removal and inactivation of pathogens by drinking water production processes should be better documented, especially for *Campylobacter*, hepatitis E virus, and *Cyclospora*, for which no specific data are available.

Concerning bacteria, the emergence of antibiotic-resistant strains of enteric bacterial pathogens such as *Salmonella* spp., *Shigella* spp., *E coli O157*, and *Campylobacter* spp. is a serious challenge and many aspects are poorly understood. May drinking water biofilms play a role in the transfer of antibiotic resistance and virulence factors? This question should be an area of future research.

Amongst emerging aquatic pathogens, *Legionella pneumophila* and *non-tuberculous mycobacteria* are unique in terms of their ability to proliferate in hot water systems, their environmental ubiquity, their ability to proliferate under certain circumstances within protozoan cells and their resistance to disinfection, but a great deal remains unknown.

Concerning protozoa, *Cryptosporidium* and *Giardia* arouse significant media attention due to the magnitude of recent outbreaks. Many research studies have essentially focused on *Cryptosporidium*, which constitutes a higher health risk than *Giardia* because no treatment is available against cryptosporidiosis which can be fatal for immunocompromised individuals. In the future, attention should be given to the emerging protozoa *Cyclospora* and *Microsporidia* for which waterborne transmission is plausible but unconfirmed.

The role of *Acanthamoeba* spp. and other amoebae as reservoirs and transmission routes for waterborne pathogens in drinking water systems also requires further research.

A key scientific challenge in the coming years will indisputably be waterborne viruses, about which much remains unknown. Waterborne viruses are increasingly being suspected as major causative agents of gastrointestinal infections. Their incidence is probably dramatically underestimated due in part to the

© 2010 Global Water Research Coalition. Waterborne Pathogens: Review for the Drinking-water Industry. Prepared by Emmanuelle Guillot and Jean-François Loret. ISBN: 9781843391791. Published by IWA Publishing, London, UK.

difficulties inherent to both their specific clinical diagnosis and measurement in drinking water. There are many areas where knowledge is poor regarding the epidemiology of viruses as well as their occurrence and behavior in the environment and the effectiveness of water treatment processes for their removal; this is partly due to the fact that few straightforward and reliable environmental detection methods are available. In addition, many – as yet uncharacterized – pathogenic viruses might emerge in the future.

Most research efforts are currently cast on viruses, bacteria and protozoa in drinking water but other potentially new, little-known infectious agents such as fungi, prions and viroids are likely to prove dangerous, and their presence in water may constitute a risk to human health.

References

Aabo, S., Rasmussen, O.F., Rossen, L., Sorensen, P.D. and Olsen, J.E. (1993). *Salmonella* identification by the polymerase chain reaction. *Mol. Cell. Probes.*, **7**, 171–8.

Abad, F., Pinto, R.M. and Bosch, A. (1998). Flow cytometry detection of infectious rotaviruses in environmental and clinical samples. *Appl. Environ. Microbiol.*, **64**(7), 2392–6.

Abad, F., Pinto, R.M., Diez, J.M. and Bosch, A. (1994). Disinfection of human enteric viruses in water by copper and silver in combination with low levels of chlorine. *Appl. Environ. Microbiol.*, **60**(7), 2377–2383.

Abbaszadegan, M., Huber, M.S., Gerba, C.P. and Pepper, I.L. (1997). Detection of viable *Giardia* cysts by amplification of heat shock-induced mRNA. *Appl. Environ. Microbiol.*, **63**(1), 324–328.

Abbaszadegan, M., Stewart, P. and LeChevallier, M. (1999). A strategy for detection of viruses in groundwater by PCR. *Appl. Environ. Microbiol.*, **65**(2), 444–449.

Abd El Galil, K. and El Sokk, M. *et al.* (2004). Combined immunomagnetic separation-molecular beacon-reverse transcription-PCR assay for detection of hepatitis A virus from environmental samples. *Appl. Environ. Microbiol.*, **70**(7), 4371–4.

Abd el-Galil, K.H., el-Sokkary, M.A., Kheira, S.M., Salazar, A.M., Yates, M.V., Chen W. and Mulchandani A. (2005). Real-time nucleic acid sequence-based amplification assay for detection of hepatitis A virus. *Appl. Environ. Microbiol.*, **71**(11), 7113–6.

Adam, R.D. (2001). Biology of *Giardia* lamblia. *Clin. Microbiol. Rev.*, **14**, 447–475.

Adams, B., Bates, T.C. and Oliver, J.D. (2003). Survival of *Helicobacter pylori* in a natural freshwater environment. *Appl. Environ. Microbiol.*, **69**(12), 7462–6.

Adékambi, T., Ben Salah, S., Khlif, M., Raoult, D. and Drancourt, M. (2006). Survival of environmental mycobacteria in *Acanthamoeba polyphaga*. *Appl. Environl. Microbiol.*, **72**(9), 5974–81.

Agbalika, F., Wullenweber, M. and Prevot, J. (1985). Preliminary evaluation of the ELISA as a tool for the detection of rotaviruses in activated sewage sludge. *Zentralbl Bakteriol Mikrobiol Hyg.*, **180**((5–6)), 534–9.

Aggarval, R. and S. Naik (1994). Hepatitis E: intrafamiliar transmission versus waterborne spread. *J. Hepatol.*, **21**, 718–23.

Alamanos, Y., Maipa, V., Levidiotou, S. and Gessouli, E. (2000). A community waterborne outbreak of gastro-enteritis attributed to *Shigella sonnei*. *Epidemiol. Infect.*, **125**(3), 499–503.

Alary, M. and Nadeau, D. (1990). An outbreak of *Campylobacter* enteritis associated with a community water supply. *Can. J. Public Health*, **81**(4), 268–71.

Albert, M.J., Ansaruzzaman, M., Talukder, K.A., Chopra, A.K., Kuhn, I., Rahman, M., Faruque, A.S., Islam, M.S., Sack, R.B. and Mollby, R. (2000). Prevalence of enterotoxin genes in *Aeromonas* spp. isolated from children with diarrhea, healthy controls, and the environment. *J. Clin. Microbiol.*, **38**(10), 3785–90.

Albinana-Gimenez, N., Clemente-Casares, P., Bofill-Mas, S., Hundesa, A., Galofre, B., Huguet, J.M., Ribas, F. and Girones, R. (2007). Detection and quantification of human and animal viruses as a tool for the control of viral contamination in a drinking water treatment plant. In *European symposium on waterborne pathogens in surface and drinking waters*, p. 19. Centre de Recherche Gabriel Lippmann, Luxembourg.

Albinana-Gimenez, N., Miagostovich, M.P., Calgua, B., Huguet, J.P., Matia, L. and Girones, R. (2009). Analysis of adenoviruses and polyomaviruses quantified by qPCR as indicators of water quality in source and drinking water treatment plants. *Water Research*, **43**, 2011–19.

Aleksic, S. and Bockemuhl, J. (1988). Serological and biochemical characteristics of 416 *Yersinia* strains from well water and drinking water plants in the Federal Republic of Germany: lack of evidence that these strains are of public health importance. *Zentralbl Bakteriol Mikrobiol Hyg [B]*, **185**(6), 527–33.

Alexandrino, M., Grohmann, E. and Szewzyk, U. (2004). Optimization of PCR-based methods for rapid detection of *Campylobacter jejuni*, *Campylobacter coli* and *Yersinia enterocolitica* serovar 0:3 in wastewater samples. *Water Res.*, **38**(5), 1340–6.

Alleron, L., Merlet, N., Lacombe, C. and Frère, J. (2008). Long-term survival of *Legionella pneumophila* in the viable but nonculturable state after monochloramine treatment. *Curr. Microbiol.*, **57**(5), 497–502.

Alvarez, J., Porwollik, S., Laconcha, I., Gisakis, V., Vivanco, A.B., Gonzalez, I., Echenagusia, S., Zabala, N., Blackmer, F., McClelland, M., Rementeria, A. and Garaizar, J. (2003). Detection of a *Salmonella enterica* serovar California strain spreading in spanish feed mills and genetic characterization with DNA microarrays. *Appl. Environ. Microbiol.*, **69**(12), 7531–4.

© 2010 Global Water Research Coalition. Waterborne Pathogens: Review for the Drinking-water Industry. Prepared by Emmanuelle Guillot and Jean-François Loret. ISBN: 9781843391791. Published by IWA Publishing, London, UK.

American Water Works Association (1997). Drinking Water Inspectorate Fact Sheet – Adenoviruses. *Fact Sheet*, **16**(1).

Amvrosieva, T., Titov, L.P., Mulders, M., Hovi, T., Dyakonova, O.V., Votyakov, V.I., Kvacheva, Z.B., Eremin, V.F., Sharko, R.M., Orlova, S.V., Kazinets, O.N. and Bogush, Z.F. (2001). Viral water contamination as the cause of aseptic meningitis outbreak in Belarus. *Cent. Eur. J. Public Health*, **9**(3), 154–7.

Ando, T., Noel, J.S. and Fankhauser, R.L. (2000). Genetic classification of Norwalk-like viruses. *J. Infect. Dis.*, **181**(Suppl. 2), S336-S348.

Ansari, S., Springthorpe, V.S. and Sattar, S.A. (1991). Survival and vehicular spread of human rotaviruses: possible relation to seasonality of outbreaks. *Rev. Infect. Dis.*, **13**(3), 448–61.

APHA, AWWA, *et al.* (1998). Standard Methods for the Examination of Water and Wastewater. 20th ed.

Arando-Rodriguez, R. and Kubwabo, C. *et al.* (2003). Extraction of 15 microcysztins and nodularin using immunoaffinity columns. *Toxicon.*, **42**, 587–599.

Archuleta, R., Mullens, P. and Primm, T.P. (2002). The relationship of temperature to desiccation and starvation tolerance of the *Mycobacterium avium* complex. *Arch. Microbiol.*, **178**(4), 311–4.

Archuleta, R., Yvonne Hoppes, P. and Primm, T.P. (2005). *Mycobacterium avium* enters a state of metabolic dormancy in response to starvation. *Tuberculosis (Edinb).*, **85**(3), 147–58.

Arias, C., Sala, M.R., Domínguez, A., Bartolomé, R., Benavente, A., Veciana, P., Pedrol, A. and Hoyo, G. (2006). Waterborne epidemic outbreak of *Shigella sonnei* gastroenteritis in Santa Maria de Palautordera, Catalonia, Spain. *Epidemiol Infect.*, **134**(3), 598–604.

Armstrong, T.W. and Haas, C.H. (2008). Legionnaire's disease: evaluation of a quantitative microbial risk assessment model. *Journal of Water and Health*, **06**(2), 149–66.

Aronson, T., Holtzman, A., Glover, N., Boian, M., Froman, S., Berlin, O.G., Hill, H. and Stelma, G. Jr. (1999). Comparison of large restriction fragments of Mycobacterium avium isolates recovered from AIDS and non-AIDS patients with those of isolates from potable water. *J. Clin. Microbiol.*, **37**(4), 1008–12.

Artz, R. and Killham, K. (2002). Survival of *Escherichia coli O157:H7* in private drinking water wells: influences of protozoan grazing and elevated copper concentrations. *FEMS Microbiol. Lett.*, **216**(1), 117–22.

Arvanitidou, M., Stathopoulos, G.A., Constantinidis, T.C. and Katsouyannopoulos, V. (1995). The occurrence of *Salmonella*, *Campylobacter* and *Yersinia* spp. in river and lake waters. *Microbiol. Res.*, **150**(2), 153–8.

Arvanitidou, M., Stathopoulos, G.A. and Katsouyannopoulos, VC. (1994). Isolation of *Campylobacter* and *Yersinia* spp. from Drinking Waters. *J. Travel Med.*, **1**(3), 156–159.

Ashdown, L. (1979). Identification of *Pseudomonas pseudomallei* in the clinical laboratory. *J. Clin. Pathol.*, **32**(5), 500–4.

Ashford, R. (1979). Occurrence of an undescribed coccidian in man in Papua New Guinea. *Ann. Trop. Med. Parasitol.*, **73**(5), 497–500.

Assavasilavasukul, P., Lau, B.L., Harrington, G.W., Hoffman, R.M. and Borchardt, M.A. (2008). Effect of pathogen concentrations on removal of *Cryptosporidium* and *Giardia* by conventional drinking water treatment. *Water Res.*, **42**(10–11), 2678–90.

Avery, S. and Undeen, A.H. (1987). The isolation of microsporidia and other pathogens from concentrated ditch water. *J. Am. Mosq. Control Assoc.*, **3**(1), 54–8.

Axelsson-Olsson, D., Ellström, P., Waldenström, J., Haemig, P.D., Brudin, L. and Olsen, B. (2007). *Acanthamoeba-Campylobacter* coculture as a novel method for enrichment of *Campylobacter* species. *Appl. Environ. Microbiol.*, **73**(21), 6864–9.

Axelsson-Olsson, D. and Waldenström, J. *et al.* (2005). Protozoan *Acanthamoeba polyphaga* as a Potential Reservoir for *Campylobacter jejuni*. *Appl. Environ. Microbiol.*, **71**(2), 987–992.

Axon, A. (1996). The transmission of *Helicobacter pylori*: which theory fits the facts? *Eur. J. Gastroenterol. Hepatol.*, **8**(1), 1–2.

Azevedo, N., Pacheco, A.P., Keevil, C.W. and Vieira, M.J. (2004). Nutrient shock and incubation atmosphere influence recovery of culturable *Helicobacter pylori* from water. *Appl. Environ. Microbiol.*, **70**(1), 490–3.

Azevedo, N., Vieira, M.J. and Keevil, C.W. (2003). Establishment of a continuous model system to study *Helicobacter pylori* survival in potable water biofilms. *Water. Sci. Technol.*, **47**(5), 155–60.

Azevedo, N.F., Almeida, C., Fernandes, I., Cerqueira, L., Dias, S., Keevil, C.W. and Vieira, M.J. (2008). Survival of gastric and enterohepatic *Helicobacter* spp. in water: implications for transmission. *Appl. Environ. Microbiol.*, **74**(6), 1805–11.

Baer, G. and J. Walker (1977). Studies of an outbreak of acute hepatitis A: I. Complement level fluctuation. *J. Med. Virol.*, **1**(1), 1–7.

Baeumner, A.J., Humiston, M.C., Montagna, R.A. and Durst, R.A. (2001). Detection of viable oocysts of *Cryptosporidium parvum* following nucleic acid sequence based amplification. *Anal. Chem.*, **73**(6),1176–80.

Baggesen, D.L. and Sandvang, D. *et al.* (2000). Characterization of *Salmonella enterica* serovar Typhimurium DT104 isolated from Denmark and comparison with isolates from Europe and the United States. *J. Clin. Microbiol.*, **38**, 1581–1586.

Baker, J., Entsch, B., Neilan, B.A. and McKay, D.B. (2002). Monitoring changing toxigenicity of a cyanobacterial bloom by molecular methods. *Appl. Environ. Microbiol.*, **68**(12), 6070–6.

Baker, J., Neilan, B.A., Entsch, B. and McKay, D.B. (2001). Identification of cyanobacteria and their toxigenicity in environmental samples by rapid molecular analysis. *Environ. Toxicol.*, **16**(6), 472–82.

Baker, K.H., Hegarty, J.P., Redmond, B., Reed, N.A. and Herson. D.S. (2002). Effect of oxidizing disinfectants (chlorine, monochloramine, and ozone) on *Helicobacter pylori. Appl. Environ. Microbiol.*, **68**, 981–984.

Balayan, M. (1993). Hepatitis E virus infection in Europe: regional situation regarding laboratory diagnosis and epidemiology. *Clin Diagn Virol.*, **1**(1), 1–9.

Balayan, M., Andjaparidze, A.G., Savinskaya, S.S., Ketiladze, E.S., Braginsky, D.M., Savinov, A.P. and Poleschuk, V.F. (1983). Evidence for a virus in non-A, non-B hepatitis transmitted via the fecal-oral route. *Intervirology.*, **20**(1), 23–31.

Balayan, M.S. (1997). Epidemiology of hepatitis E virus infection. *J. Viral Hep.*, **4**(155–65).

Ballard, A.L. and Fry, N.K. *et al.* (2000). Detection of *Legionella pneumophila* using a real-time PCR hybridization assay. *J. Clin. Microbiol.*, **38**(11), 4215–4218.

Bando, S., do Valle, G.R., Martinez, M.B., Trabulsi, L.R. and Moreira-Filho, C.A. (1998). Characterization of enteroinvasive *Escherichia coli* and *Shigella* strains by RAPD analysis. *FEMS Microbiol.*, **165**(1), 159–65.

Banker, D. (2003). Viral hepatitis (Part-I). *Indian J. Med. Sci*, **57**, 363–8.

Barbosa, J., Costa-de-Oliveira, S., Rodrigues, A.G., Hanscheid, T., Shapiro, H. and Pina-Vaz, C. (2008a). A flow cytometric protocol for detection of *Cryptosporidium spp. Cytometry A.*, **73**(1), 44–7.

Barbosa, J., Costa-de-Oliveira, S., Rodrigues, A.G. and Pina-Vaz, C. (2008b). Optimization of a flow cytometry protocol for detection and viability assessment of *Giardia lamblia. Travel Med. Infect Dis.*, **6**(4), 234–9.

Barker, J., Humphrey, T.J. and Brown, M.W. (1999). Survival of *Escherichia coli O157* in a soil protozoan: implications for disease. *FEMS Microbiol. Lett.*, **173**(2), 291–5.

Baron, R., Murphy, F.D., Greenberg, H.B., Davis, C.E., Bregman, D.J., Gary, G.W., Hughes, J.M. and Schonberger, L.B. (1982). Norwalk gastrointestinal illness: an outbreak associated with swimming in a recreational lake and secondary person-to-person transmission. *Am. J. Epidemiol.*, **115**(2), 163–72.

Batigelli, D. and Sobsey, M. *et al.* (1993). The inactivation of hepatitis A and other model viruses by UV irradiation. *Water. Sci. Technol.*, **27**, 339–42.

Baudart, J. and Lemarchand, K. *et al.* (2000). Diversity of Salmonella strains isolated from the aquatic environment as determined by serotyping and amplification of the ribosomal DNA spacer regions. *Appl. Environ. Microbiol.*, **66**, 1544–1552.

Baum, S.G. (1995). Adenovirus. *Principles and practice of infectious diseases*. G. Mandell, R. Douglas and J. Bennett. New York, Churchill Linvingstone: 1382–87.

Beaudeau, P., de Valk, H., Vaillant, V., Mannschott, C., Tillier, C., Mouly, D. and Ledrans, M. (2008). Lessons learned from ten investigations of waterborne gastroenteritis outbreaks, France, 1998–2006. *Journal of Water and Health*, **06**(4), 491–503.

Begue, R., Gonzales, J.L., Correa-Gracian, H. and Tang, S.C. (1998). Dietary risk factors associated with the transmission of *Helicobacter pylori* in Lima, Peru. *Am. J. Trop. Med. Hyg.*, **59**(4), 637–40.

Behets, J., Declerck, P., Delaedt, Y., Creemers, B. and Ollevier, F. (2007). Development and evaluation of a Taqman duplex real-time PCR quantification method for reliable enumeration of *Legionella pneumophila* in water samples. *J. Microbiol. Methods*, **68**(1), 137–44.

Bej, A., DiCesare, J.L., Haff, L., Atlas and R.M. (1991). Detection of *Escherichia coli* and *Shigella* spp. in water by using the polymerase chain reaction and gene probes for uid. *Appl. Environ. Microbiol.*, **57**(4), 1013–7.

Beller, M., Ellis, A., Lee, S.-H., Drebot, M.A., Jenkerson, S.A., Funk, S.A., Sobsey, M.D., Simmons III, O.D., Monroe, S.S., Ando, T., Noel, J., Petric, M., Middaugh, J.P. and Spika, J.S. (1997). Outbreak of viral gastroenteritis due to a contaminated well: international consequences. *JAMA*, **278**, 563–568.

Belliot, G., Lavaux, A., Souihel, D., Agnello, D. and Pothier, P. (2008). Use of murine norovirus as a surrogate to evaluate resistance of human norovirus to disinfectants. *Applied and Environmental Microbiology*, **74**(10), 3315–18.

Belosevic, M., Craik, S.A., Stafford, J.L., Neumann, N.F., Kruithof, J. and Smith, D.W. (2001). Studies on the resistance/reactivation of *Giardia* muris cysts and *Cryptosporidium parvum* oocysts exposed to medium-pressure ultraviolet radiation. *FEMS Microbiol. Lett.*, **204**(1), 197–203.

Belosevic, M., Faubert, G.M., MacLean, J.D., Law, C. and Croll, N.A. (1983). *Giardia lamblia* infections in Mongolian gerbils: an animal model. *J. Infect. Dis.*, **147**(2), 222–6.

Bender, J., Hedberg, C.W. and Besser, J.M. *et al.* (1997). Surveillance for *Escherichia coli O157:H7* infections in Minnesota by molecular subtyping. *N. Engl. J. Med.*, **337**, 388–94.

Benson, J., Fode-Vaughan, K.A. and Collins, M.L. (2004). Detection of *Helicobacter pylori* in water by direct PCR. *Lett. Appl. Microbiol.*, **39**(3), 221–5.

Bergey's Manual of Systematic Bacteriology (2005). Vol. 2, Part B (Krieg, N.R. and Holt, J.G., Eds), Williams and Wilkins, Baltimore, MD.

Berke, T. and D.O. Matson. (2000). Reclassification of the Caliciviridae into distinct genera and exclusion of hepatitis E virus from the family on the basis of comparative phylogenetic analysis. *Arch. Virol.*, **145**, 1421–1436.

Berke, T., Golding, B., Jiang, X., Cubitt, D.W., Wolfaardt, M., Smith, A.W. and Matson, D.O. (1997). Phylogenetic analysis of the Caliciviruses. *J. Med. Virol.*, **52**(4), 419–24.

Beuret, C. (2003). A simple method for isolation of enteric viruses (noroviruses and enteroviruses) in water. *J. Virol. Methods*, **107**(1), 1–8.

Beuret, C., Kohler, D., Baumgartner, A. and Lüthi, T.M. (2002). Norwalk-like virus sequences in mineral waters: one-year monitoring of three brands. *Appl. Environ. Microbiol.*, **68**, 1925–1931.

Beuret, C., Kohler, D. and Luthi, T. (2000). Norwalk-like virus sequences detected by reverse transcription-polymerase chain reaction in mineral waters imported into or bottled in Switzerland. *J. Food Prot.*, **63**, 1576–1582.

Bhagwat, A. (2003). Simultaneous detection of *Escherichia coli O157:H7*, *Listeria monocytogenes* and *Salmonella* strains by real-time PCR. *Int. J. Food Microbiol.*, **25**(2), 217–24.

Bhattacharya, S., Kulka, M., Lampel, K.A., Cebula, T.A. and Goswami, B.B. (2004). Use of reverse transcription and PCR to discriminate between infectious and non-infectious hepatitis A virus. *J. Virol. Methods*, **116**(2), 181–7.

Bigliardi, L., Cesari, C., Zoni, R. and Sansebastiano, G.E. (2004). The concentration of viruses in water using the tangential flow ultrafiltration. Recovery effectiveness in experimental conditions. *Ann. Ig.*, **16**(1–2), 281–9.

Biscardi, D., Castaldo, A., Gualillo, O. and de Fusco, R. (2002). The occurrence of cytotoxic *Aeromonas hydrophila* strains in Italian mineral and thermal waters. *Sci. Total Environ.*, **292**(3), 255–63.

Bishop, R., Davidson, G.P., Holmes, I.H. and Ruck, B.J. (1973). Virus particles in epithelial cells of duodenal mucosa from children with acute non-bacterial gastroenteritis. *Lancet*, **8**(2 (7841)), 1281–3.

Biziagos, E. and Passagot, J. *et al.* (1988). Long-term survival of hepatitis A virus and poliovirus type1 in mineral water. *Appl. Environ. Microbiol.*, **54**, 2705–2710.

Black, R., Levine, M.M., Clements, M.L., Hughes, T.P. and Blaser, M.J. (1988). Experimental *Campylobacter jejuni* infection in humans. *J. Infect. Dis.*, **157**(3), 472–9.

Blackburn, B., Craun, G.F., Yoder, J.S., Hill, V., Calderon, R.L., Chen, N., Lee, S.H., Levy, D.A. and Beach, M.J. (2004). Surveillance for Waterborne-Disease outbreaks associated with drinking water – United states, 2001–2002. *MMWR*, **53**(SS-8): 23–45.

Blackmer, F., Reynolds, K.A., Gerba, C.P. and Pepper, I.L. (2000). Use of integrated cell culture-PCR to evaluate the effectiveness of poliovirus inactivation by chlorine. *Appl. Environ. Microbiol.*, **66**, 2267–2268.

Blanc, D.S., Carrara, P., Zanetti, G. and Francioli, P. (2005). Water disinfection with ozone, copper and silver ions, and temperature increase to control *Legionella*: seven years of experience in a university teaching hospital. *J. Hosp. Infect.*, **60**(1), 69–72.

Blaser, M., Smith, P.F., Wang, W.L. and Hoff, J.C. (1986). Inactivation of *Campylobacter jejuni* by chlorine and monochloramine. *Appl. Environ. Microbiol.*, **51**(2), 307–11.

Blostein, J. (1991). Shigellosis from swimming in a park pond in Michigan. *Public Health Rep.*, **106**(3), 317–22.

Bode, G., Rothenbacher, D., Brenner, H. and Adler, G. (1998). Pets are not a risk factor for *Helicobacter pylori* infection in young children: results of a population-based study in Southern Germany. *Pediatr. Infect. Dis. J.*, **17**(10), 909–12.

Bopp, D., Sauders, B.D., Waring, A.L., Ackelsberg, J., Dumas, N., Braun-Howland, E., Dziewulski, D., Wallace, B.J., Kelly, M., Halse, T., Musser, K.A., Smith, P.F., Morse, D.L. and Limberger, R.J. (2003). Detection, isolation, and molecular subtyping of *Escherichia coli O157:H7* and *Campylobacter jejuni* associated with a large waterborne outbreak. *J. Clin. Microbiol.*, **41**(1), 174–80.

Borchardt, M., Stemper, M.E. and Standridge, J.H. (2003). *Aeromonas* isolates from human diarrheic stool and groundwater compared by pulsed-field gel electrophoresis. *Emerg. Infect. Dis.*, **9**(2), 224–8.

Bosch, A. and Pinto, R. *et al.* (2004). Detection of infectious rotaviruses by flow cytometry. *Methods Mol. Biol.*, **268**, 61–8.

Braganra, S.M., Azevedo, N.F., Simões, L.C., Keevil, C.W. and Vieira, M.J. (2007). Use of fluorescent in situ hybridisation for the visualisation of *Helicobacter pylori* in real drinking water biofilms. *Water. Sci. Technol.*, **55**(8–9), 387–93.

Brandi, G., Sisti, M., Giardini, F., Schiavano, G.F. and Albano, A. (1999). Survival ability of cytotoxic strains of motile *Aeromonas* spp. in different types of water. *Lett. Appl. Microbiol.*, **29**(4), 211–5.

Brandt, C., Kim, H.W., Rodriguez, W.J., Arrobio, J.O., Jeffries, B.C., Stallings, E.P., Lewis, C., Miles, A.J., Chanock, R.M., Kapikian, A.Z. and Parrott, R.H. (1983). Pediatric viral gastroenteritis during eight years of study. *J. Clin. Microbiol.*, **18**(1), 71–8.

Brassard, J. and Seyer, K. *et al.* (2005). Concentration and detection of hepatitis A virus and rotavirus in spring water samples by reverse transcription-PCR. *Virol. Methods*, **123**(2), 163–9.

Brennhovd, O., Kapperud, G. and Langeland, G. (1992). Survey of thermotolerant Campylobacter spp. and Yersinia spp. in three surface water sources in Norway. *Int. J. Food Microbiol.*, **15**(3–4), 327–38.

Brient, L., Lengronne, M., Bertrand, E., Rolland, D., Sipel, A., Steinmann, D., Baudin, I., Legeas, M., Le Rouzic, B. and Bormans, M. (2008). A phycocyanin probe as a tool for monitoring cyanobacteria in freshwater bodies. *J. Environ. Monit.*, **10**(2), 248–55.

Brook, M., Currie, B. and Desmarchelier, P.M. (1997). Isolation and identification of *Burkholderia pseudomallei* from soil using selective culture techniques and the polymerase chain reaction. *J. Appl. Microbiol.*, **82**(5), 589–96.

Brooke, S., Newcombe, G., Nicholson, B. and Klass, G. (2006). Decrease in toxicity of microcystins LA and LR in drinking water by ozonation. *Toxicon.*, **48**(8), 1054–9.

Brooks, H.A., Gersberg, R.M. and Dhar, A.K. (2005). Detection and quantification of hepatitis A virus in seawater via real-time RT-PCR. *J. Virol. Methods*, **127**(2), 109–18.

Brooks, T., Osicki, R., Springthorpe, V., Sattar, S., Filion, L., Abrial, D. and Riffard, S. (2004). Detection and identification of *Legionella* species from groundwaters. *Toxicol. Environ. Health A*, **67**(20–22), 1845–59.

Brown, D. (1999). *The pattern and burden of disease due to human calicivirus infections in the UK.* The International Workshop on Human Caliciviruses, Atlanta, Centers for Disease Control and Prevention.

Bruce, M., Curtis, M.B., Payne, M.M., Gautom, R.K., Thompson, E.C., Bennett, A.L. and Kobayashi, J.M. (2003). Lake-associated outbreak of *Escherichia coli O157:H7* in Clark County, Washington, August 1999. *Arch. Pediatr. Adolesc. Med.*, **157**(10), 1016–21.

Bryan, J., Lehmann, J.D., Setiady, I.F. and Hatch, M.H. (1974). An outbreak of hepatitis-A associated with recreational lake water. *Am. J. Epidemiol.*, **99**(2), 145–54.

Buchbinder, S., Trebesius, K. and Heesemann, J. (2002). Evaluation of detection of *Legionella* spp. in water samples by fluorescence in situ hybridization, PCR amplification and bacterial culture. *Int J. Med. Microbiol.*, **292**(3–4), 241–5.

Bukhari, Z., Weihe, J. and LeChevallier, M. (2004). Development of procedures for rapid detection of *E. coli O157:H7* from source and finished water samples. *Water. Sci. Technol.*, **50**(1), 233–7.

Bunn, J., MacKay, W.G., Thomas, J.E., Reid, D.C. and Weaver, L.T. (2002). Detection of *Helicobacter pylori* DNA in drinking water biofilms: implications for transmission in early life. *Lett. Appl. Microbiol.*, **34**(6), 450–4.

Burgener, M., Candrian, U. and Gilgen, M. (2003). Comparative evaluation of four large-volume RNA extraction kits in the isolation of viral RNA from water samples. *J. Virol. Methods*, **108**, 165–70.

Burke, V., Robinson, J., Gracey, M., Peterson, D., Meyer, N. and Haley, V. (1984). Isolation of *Aeromonas* spp. from an unchlorinated domestic water supply. *Appl. Environ. Microbiol.*, **48**(2), 367–70.

Burke, V., Robinson, J., Gracey, M., Peterson, D. and Partridge, K. (1984). Isolation of *Aeromonas hydrophila* from a metropolitan water supply: seasonal correlation with clinical isolates. *Appl. Environ. Microbiol.*, **48**(2), 361–6.

Burns, B., Saker, M.L., Moffitt, M.C. and Neilan, B.A. (2004). Molecular detection of genes responsible for cyanobacterial toxin production in the genera Microcystis, Nodularia, and Cylindrospermopsis. *Methods Mol. Biol.*, **268**, 213–22.

Buswell, C., Herlihy, Y.M., Lawrence, L.M., McGuiggan, J.T., Marsh, P.D., Keevil, C.W. and Leach, S.A. (1998). Extended survival and persistence of *Campylobacter* spp. in water and aquatic biofilms and their detection by immunofluorescent-antibody and -rRNA staining. *Appl. Environ. Microbiol.*, **64**(2), 733–41.

Buzoleva, L. (2000). The multiplication dynamics of pathogenic bacteria in relation to the trophic and temperature cultivation conditions. *Zh. Mikrobiol. Epidemiol. Immunobiol.*, **2**, 15–8.

Cacciò, S., De Giacomo, M., Aulicino, F.A. and Pozio, E. (2003). *Giardia* Cysts in Wastewater Treatment Plants in Italy. *Appl. Environ. Microbiol.*, **69**(6), 3393–3398.

Campbell, G., Prosser, J., Glover, A. and Killham, K. (2001). Detection of *Escherichia coli O157:H7* in soil and water using multiplex PCR. *J. Appl. Microbiol.*, **91**, 1004–1010.

Cannon, J.L. and Vinjé, J. (2008). Histo-blood group antigen assay for detecting noroviruses in water. *Appl. Environ. Microbiol.*, **74**(21), 6818–9.

Carey, C. and Lee, H.K. *et al.* (2004). Biology, persistence and detection of *Cryptosporidium parvum* and *Cryptosporidium hominis* oocyst. *Water Res.*, **38**, 818–862.

Carmichael, W., Azevedo, S.M., An, J.S., Molica, R.J., Jochimsen, E.M., Lau, S., Rinehart, K.L., Shaw, G.R. and Eaglesham, G.K. (2001). Human fatalities from cyanobacteria: chemical and biological evidence for cyanotoxins. *Environ. Health Perspect.*, **109**(7), 663–8.

Carniel, E. (2008). The Plague *PLoS Med.* Jan 15;5(1)

Caro, A., Got, P., Lesne, J., Binard, S. and Baleux, B. (1999). Viability and virulence of experimentally stressed nonculturable *Salmonella typhimurium. Appl. Environ. Microbiol.*, **65**(7), 3229–32.

Carter, G., Wu, M., Drummond, D.C. and Bermudez, L.E. (2003). Characterization of biofilm formation by clinical isolates of Mycobacterium avium. *J. Med. Microbiol.*, **52**, 747–52.

Carvalho, F.R., Nastasi, F.R., Gamba, R.C., Foronda, A.S. and Pellizari, V.H. (2008). Occurrence and diversity of Legionellaceae in polar lakes of the Antarctic peninsula. *Curr. Microbiol.*, **57**(4), 294–300.

Castignolles, N. and Petit, F. *et al.* (1998). Detection of adenovirus in the waters of the Seine River estuary by nested-PCR. *Mol. Cell. Probes.*, **12**(3), 175–80.

Catalan, V., Garcia, F., Moreno, C., Vila, M.J. and Apraiz, D. (1997). Detection of *Legionella pneumophila* in wastewater by nested polymerase chain reaction. *Res. Microbiol.*, **148**(1), 71–8.

Catalan, V., Moreno, C., Dasi, M.A., Munoz, C. and Apraiz, D. (1994). Nested polymerase chain reaction for detection of *Legionella pneumophila* in water. *Res. Microbiol.*, **145**(8), 603–10.

Catalao Dionisio, L., Joao, M., Ferreiro, V.S., Fidalgo, M.L., Garcia Rosado, M.E. and Borrego, J.J. (2000). Occurrence of *Salmonella* spp in estuarine and coastal waters of Portugal. *Antonie Van Leeuwenhoek.*, **78**(1), 99–106.

CDC. (2000). Shigella surveillance: annual tabulation summary, 1999. Atlanta, Georgia: US Department of Health and Human Services, CDC.

CDC. (2001). From the Centers for Disease Control and Prevention. Shigellosis outbreak associated with an unchlorinated fill-and-drain wading pool–Iowa, 2001. *JAMA*, **286**(16), 1964–5.

Cellini, L., Del Vecchio, A., Di Candia, M., Di Campli, E., Favaro, M. and Donelli, G. (2004). Detection of free and plankton-associated Helicobacter pylori in seawater. *J. Appl. Microbiol.*, **97**(2), 285–92.

Centers for Disease Control. (1991). Outbreks of diarrheal illness associated with cyanobacteria (blue-green algae)-like bodies-Chicago and Nepal, 1989 and 1990. *Morbid. Mortal. Weekly Rep.*, **40**, 325–327.

Centers for Disease Control. (2000). Surveillance for waterborne-disease outbreaks, United States, 1997–1998. *Morb. Mortal. Wkly. Rep.* 49(SS-4).

Chalifoux, L., Carville, A., Pauley, D., Thompson, B., Lackner, A.A. and Mansfield, K.G. (2000). *Enterocytozoon bieneusi* as a cause of proliferative serositis in simian immunodeficiency virus-infected immunodeficient macaques (Macaca mulatta). *Arch. Pathol. Lab. Med.*, **124**(10), 1480–4.

Chalmers, R.M. and Aird, H. *et al.* (2000). Waterborne *Escherichia coli O157. J. Appl. Microbiol.*, **88**, 124S–132S.

Chandran, A. and Mohamed Hatha, A.A. (2005). Relative survival of *Escherichia coli* and *Salmonella typhimurium* in a tropical estuary. *Water Res.*, **39**(7), 1397–403.

Chang, B., Sugiyama, K., Taguri, T., Amemura-Maekawa, J., Kura, F. and Watanabe, H. (2009). Specific detection of viable *Legionella* cells by combined use of photoactivated ethidium monoazide and PCR/real-time PCR. *Appl. Environ. Microbiol.*, **75**(1), 147–53.

Chang, C., Wang, L.Y., Liao, C.Y. and Huang, S.P. (2002). Identification of nontuberculous mycobacteria existing in tap water by PCR-restriction fragment length polymorphism. *Appl. Environ. Microbiol.*, **68**(6), 3159–61.

Chang, H., Loo, L.H., Kuah, B.G. and Heng, B.H. (1995). Comparison of multiplex PCR and culture for detection of Legionellae in cooling tower water samples. *Southeast Asian J. Trop. Med. Public Health*, **26**(2), 258–62.

Chao, W., Ding, R.J. and Chen, R.S. (1988). Survival of *Yersinia enterocolitica* in the environment. *Can. J. Microbiol.*, **34**(6), 753–6.

Chapman, P., Siddons, C.A., Wright, D.J., Norman, P., Fox, J. and Crick, E. (1993). Cattle as a possible source of verocytotoxin-producing *Escherichia coli O157* infections in man. *Epidemiol. Infect.*, **111**(3), 439–47.

Chapron, C.D. and Ballester, N.A. *et al.* (2000). Detection of astroviruses, enteroviruses, and adenovirus types 40 and 41 in surface water collected and evaluated by the information collection rule and an integrated cell culture-nested PCR procedure. *Appl. Environ. Microbiol.*, **66**, 2520–2525.

Chauret, C., Volk, C., Creason, R., Jarosh, J., Robinson, J. and Warnes, C. (2001). Detection of *Aeromonas hydrophila* in a drinking-water distribution system: a field and pilot study. *Can. J. Microbiol.*, **47**(8), 782–6.

Chen, K., Chen, C.J. and Chiu, JP. (2001). A school waterborne outbreak involving both Shigella sonnei and *Entamoeba histolytica. J. Environ. Health*, **64**(4), 9–13.

Chen, Y., Chen, S.C., Kao, C.M. and Chen, Y.L. (2003). Effects of soil pH, temperature and water content on the growth of *Burkholderia pseudomallei. Folia Microbiol. (Praha).*, **48**(2), 253–6.

Cho, H. and Lee, S. *et al.* (2000). Detection of adenoviruses and enteroviruses in tap water and river water by reverse transcription multiplex PCR. *Can. J. Microbiol.*, **46**(5), 417–24.

Cho, J. and Kim, S.J. (1999). Viable, but non-culturable, state of a green fluorescence protein-tagged environmental isolate of *Salmonella typhi* in groundwater and pond water. *FEMS Microbiol. Lett.*, **170**(1), 257–64.

Choi, S. and Jiang, S.C. (2005). Real-time PCR quantification of human adenoviruses in urban rivers indicates genome prevalence but low infectivity. *Appl. Environ. Microbiol.*, **71**(11), 7426–33.

Clayson, E.T., Snitbhan, R. and Ngarmpochana, M. (1996). Evidence that the hepatitis E virus (HEV) is a zoonotic virus: detection of natural infection among swine, rats, and chickens in an area endemic for human disease. *Enterically transmitted hepatitis viruses.* Y. Buisson, P. Coursaget, and M. Kane (ed.). Joue-les-Tours, France. p. 329–335.

Clemente-Casares, P. and Pina, S. *et al.* (2003). Hepatitis E virus epidemiology in industrialized countries. *Emerg. Infect. Dis.*, **9**(4), 448–54.

Codd, G., Morrison, L.F. and Metcalf, J.S. (2005). Cyanobacterial toxins: risk management for health protection. *Toxicol. Appl. Pharmacol.*, **203**(3), 264–72.

Colwell, R. and Brayton, P. *et al.* (1985). Viable but non-culturable Vibrio cholerae and related pathogens in the environment: implications for release of genetically engineered microorganisms. *Bio/Technology*, **3**, 817–20.

Cools, I., Uyttendaele, M., Caro, C., D'Haese, E., Nelis, H.J. and Debevere, J. (2003). Survival of *Campylobacter jejuni* strains of different origin in drinking water. *J. Appl. Microbiol.*, **94**(5), 886–92.

Cooper, A. and Barnes, H. *et al.* (2004). Assessing risk of *Legionella. ASHRAE J.*, **46**(4), 22–27.

Cooper, I.R., White, J., Mahenthiralingam, E. and Hanlon, G.W. (2008). Long-term persistence of a single *Legionella pneumophila* strain possessing the mip gene in a municipal shower despite repeated cycles of chlorination. *J. Hosp. Infect.*, **70**(2), 154–9.

Costa, J., Tiago, I., da Costa, M.S. and Verissimo, A. (2005). Presence and persistence of *Legionella* spp. in groundwater. *Appl. Environ. Microbiol.*, **71**(2), 663–71.

Cotte, L., Rabodonirina, M., Chapuis, F., Bailly, F., Bissuel, F., Raynal, C., Gelas, P., Persat, F., Piens, M. and Trepo, C. (1999). Waterborne outbreak of intestinal microsporidiosis in persons with and without human immunodeficiency virus infection. *J. Infect. Dis.*, **180**, 2003–2008.

Coursaget, P. and Buisson, Y. *et al.* (1998). Outbreak of enterically-transmitted hepatitis due to hepatitis A and hepatitis E viruses. *J. Hepatol.*, **28**(5), 745–50.

Cox, P.A., Banack, S.A. and Murch, S.J. (2003). Biomagnification of cyanobacterial neurotoxins and neurodegenerative disease among the Chamorro people of Guam. *Proc. Natl. Acad. Sci. USA*, **100**(23), 13380–3.

Crabtree, K.D. (1996). Waterborne Rotavirus And Coxsackievirus: A Risk Assessment Approach. *Drinking Water & Health Newsletter* 1 July.

Crabtree, K.D., Gerba, C.P., Rose, J.B. and Haas, C.N. (1997). Waterborne adenovirus: a risk assessment. *Water Science & Technology*, **35**(11–12), 1–6.

Crance, J. and Gantzer, C. *et al.* (1998). Effect of temperature on the survival of hepatits A virus and its capsidal antigen in synthetic seawater. *Environ. Toxicol. Water Qual.*, **13**, 89–92.

Craun, G.F. (1986). *Waterborne diseases in the United States.* Boca Raton, F.L., CRC Press.

Craun, G.F. (1991). Causes of waterborne outbreaks in the United States. *Water Sci. Tech.*, **24**, 17–20.

Craun, G.F. (ed.) (1996). *Waterborne diseases in the United States.* Boca Raton, F.L., CRC Press.

Crockett, C.S., Haas, C.N., Fazil, A., Rose, J.B. and Gerba, C.P. (1996). Prevalence of shigellosis in the U.S.: consistency with dose-response information. *Int. J. Food Microbiol.*, **30**, 87–99.

Cromeans, T.L., Lu, X., Erdman, D.D., Humphrey, C.D. and Hill, V.R. (2008). Development of plaque assays for adenoviruses 40 and 41. *J. Virol. Methods*, **151**(1), 140–5.

Cukor, G. and Blacklow, N.R. (1984). Human viral gastroenteritis. *Microbiol. Rev.*, **48**(2), 157–79.

Currie, B., Mayo, M., Anstey, N.M., Donohoe, P., Haase, A. and Kemp, D.J. (2001). A cluster of melioidosis cases from an endemic region is clonal and is linked to the water supply using molecular typing of *Burkholderia pseudomallei* isolates. *Am. J. Trop. Med. Hyg.*, **65**(3), 177–9.

da Silva, A., Schwartz, D.A., Visvesvara, G.S., de Moura, H., Slemenda, S.B. and Pieniazek, N.J. (1996). Sensitive PCR diagnosis of Infections by *Enterocytozoon bieneusi* (microsporidia) using primers based on the region coding for small-subunit rRNA. *J. Clin. Microbiol.*, **34**(4), 986–7.

Dagan, R. (1996). Nonpolio enteroviruses and the febrile young infant: epidemiologic, clinical and diagnostic aspects. *Pediatr. Infect. Dis. J.*, **15**, 67–71.

Dahling, D. (2002). An improved filter elution and cell culture assay procedure for evaluating public groundwater systems for culturable enteroviruses. *Water Environ. Res.*, **74**(6), 564–8.

Dailloux, M. and Laurin, C. *et al.* (1999). Water and nontuberculous mycobacteria. *Wat. Res.*, **33**(10), 2219–28.

Dalton, C., Goater, A.D., Pethig, R. and Smith, H.V. (2001). Viability of *Giardia intestinalis* cysts and viability and sporulation state of *Cyclospora cayetanensis* oocysts determined by electrorotation. *Appl. Environ. Microbiol.*, **67**(2), 586–590.

Daly, R.I., Ho, L. and Brookes, J.D. (2007). Effect of chlorination on *Microcystis aeruginosa* cell integrity and subsequent microcystin release and degradation. *Environ. Sci. Technol.*, **41**(12), 4447–53.

Dance, D. (1999). Melioidosis. *Tropical Infectious Diseases: Principles, Pathogens and Practice.* R. Guerrant, Walker, D.H., Weller, P.F. Philadelphia., Churchill Livingstone., **1**, 430–7.

Dance, D. (2000). Ecology of *Burkholderia pseudomallei* and the interactions between environmental Burkholderia spp. and human-animal hosts. *Acta Trop.*, **74**(2–3), 159–68.

Dance, D. (2002). Melioidosis. *Curr. Opin. Infect. Dis.*, **15**(2), 127–32.

D'Antonio, R., Winn, R.E., Taylor, J.P., Gustafson, TL, Current, W.L., Rhodes, M.M., Gary, G.W, Jr. and Zajac R.A. (1985). A waterborne outbreak of cryptosporidiosis in normal hosts. *Ann. Intern. Med.*, **103**, 886–8.

De Serres, G. and Cromeans, T. *et al.* (1999). Molecular confirmation of hepatitis A virus from well water: epidemiology and public health implications. *J. Infect. Dis.*, **179**(1), 37–43.

Declerck, P., Verelst, L., Duvivier, L., Van Damme, A. and Ollevier, F. (2003a). A detection method for *Legionella* spp in (cooling) water: fluorescent in situ hybridization (FISH) on whole bacteria. *Water. Sci. Technol.*, **47**(3), 143–6.

Declerck, P., Verelst, L., Duvivier,. L, Van Damme, A. and Ollevier, F. (2003b). PCR as a test for the presence or absence of *Legionella* in (cooling) water. *Water. Sci. Technol.*, **47**(3), 103–7.

Declerck, P., Behets, J., Lammertyn, E., Lebeau, I., Anné, J. and Ollevier, F. (2006). Detection and quantification of *Legionella pneumophila* in water samples using competitive PCR. *Can. J. Microbiol.*, **52**(6), 584–90.

Declerck, P., Behets, J., Van Hoef, V. and Ollevier, F. (2007). Detection of *Legionella* spp. and some of their amoeba hosts in floating biofilms from anthropogenic and natural aquatic environments. *Water Research*, **41**, 3159–67.

Dedman, D., Laurichesse, H., Caul, E.O. and Wall, P.G. (1998). Surveillance of small round structured virus (SRSV) infection in England and Wales, 1990–5. *Epidemiol. Infect.*, **121**(1), 139–49.

Degnan, A., Sonzogni, W.C. and Standridge, J.H. (2003). Development of a plating medium for selection of *Helicobacter* pylori from water samples. *Appl. Environ. Microbiol.*, **69**(5), 2914–8.

Delaedt, Y., Daneels, A., Declerck, P., Behets, J., Ryckeboer, J., Peters, E. and Ollevier, F. (2008). The impact of electrochemical disinfection on *Escherichia coli* and *Legionella pneumophila* in tap water. *Microbiol. Res.*, **163**(2), 192–9.

Den Boer, J., Yzerman, E.P., Schellekens, J., Lettinga, K.D., Boshuizen, H.C., Van Steenbergen, J.E., Bosman, A., Van den Hof, S., Van Vliet, H.A., Peeters, M.F., Van Ketel, R.J., Speelman, P., Kool, J.L. and Conyn-Van Spaendonck, M.A. (2002). A large outbreak of Legionnaires' disease at a flower show, the Netherlands, 1999. *Emerg. Infect. Dis.*, **8**(1), 37–43.

Denis-Mize, K. and Fout, G. *et al.* (2004). Detection of human enteric viruses in stream water with RT-PCR and cell culture. *J. Water Health*, **2**(1), 37–47.

deRegnier, D., Cole, L., Schupp, D.G. and Erlandsen, S.L. (1989). Viability of *Giardia* cysts suspended in lake, river, and tap water. *Appl. Environ. Microbiol.*, **55**(5), 1223–9.

De Roda Husman, A.M., Bijkerk, P., Lodder, W., Van Den Berg, H., Pribil, W., Cabaj, A., Gehringer, P., Sommer, R. and Duizer, E. (2004). Calicivirus inactivation by nonionizing (253.7-nanometer-wavelength [UV]) and ionizing (gamma) radiation *Appl. Environ. Microbiol.*, **70**(9), 5089–93.

Dev, V., Main, M. and Gould, I. (1991). Waterborne outbreak of Escherichia coli O157. *Lancet*, **337**, 412.

Diederen, B.M.W., De Jong, C.M.A., Aarts, I., Peeters, M.F. and Van der Zee, A. (2007). Molecular evidence for the ubiquitous presence of *Legionella* species in Dutch tap water installations. *Journal of water and health*, **05**(3), 375–83.

Di Giovanni, G. and LeChevallier, M.W. (2005). Quantitative-PCR assessment of *Cryptosporidium parvum* cell culture infection. *Appl. Environ. Microbiol.*, **71**(3), 1495–500.

Didier, E., Stovall, M.E., Green, L.C., Brindley, P.J., Sestak, K. and Didier, P.J. (2004). Epidemiology of microsporidiosis: sources and modes of transmission. *Vet. Parasitol.*, **126**(1–2), 145–66.

Didier, E., Varner, P.W., Didier, P.J., Aldras, A.M., Millichamp, N.J., Murphey-Corb, M., Bohm, R. and Shadduck, J.A. (1994). Experimental microsporidiosis in immunocompetent and immunodeficient mice and monkeys. *Folia Parasitol (Praha).*, **41**(1), 1–11.

Diergaardt, S., Venter, S.N., Spreeth, A., Theron, J. and Brozel, V.S. (2004). The occurrence of campylobacters in water sources in South Africa. *Water Res.*, **38**(10), 2589–95.

Divizia, M., Gabrieli, R., Donia D., Macaluso, A., Bosch, A., Guix, S., Sanchez, G., Villena, C., Pinto, R.M., Palombi, L., Buonuomo, E., Cenko, F., Leno, L., Bebeci, D. and Bino, S. (2004). Waterborne gastroenteritis outbreak in Albania. *Water. Sci. Technol.*, **50**(1), 57–6.

Divizia, M. and Ruscio, V. *et al.* (1998). Hepatitis A virus detection in wastewater by PCR and hybridization. *New Microbiol.*, **21**(2), 161–7.

Doern, G., Herrmann, J.E., Henderson, P., Stobbs-Walro, D., Perron, D.M. and Blacklow, N.R. (1986). Detection of rotavirus with a new polyclonal antibody enzyme immunoassay (Rotazyme II) and a commercial latex agglutination test (Rotalex): comparison with a monoclonal antibody enzyme immunoassay. *J. Clin. Microbiol.*, **23**(2), 226–9.

Donaldson, K.A. and Griffin, D.W. *et al.* (2002). Detection, quantitation and identification of enteroviruses from surface waters and sponge tissue from the Florida Keys using real-time RT-PCR. *Water Research*, **36**, 2505–2514.

Donohue, M.J., Best, J.M., Smallwood, A.W., Kostich, M., Rodgers, M. and Shoemaker, J.A. (2007). Differentiation of *Aeromonas* isolated from drinking water distribution systems using matrix-assisted laser desorption/ionization-mass spectrometry. *Anal. Chem.*, **79**(5), 1939–46.

Dorsch, M., Ashbolt, N.J., Cox, P.T. and Goodman, A.E. (1994). Rapid identification of *Aeromonas* species using 16S rDNA targeted oligonucleotide primers: a molecular approach based on screening of environmental isolates. *J. Appl. Bacteriol.*, **77**(6), 722–6.

Doultree, J.C., Druce, J.D., Birch, C.J., Bowden, D.S. and Marshall, J.A. (1999). Inactivation of feline calicivirus, a Norwalk virus surrogate. *Journal of Hospital Infection*, **41**, 5157.

Dowd, S., Gerba, C.P., Kamper, M. and Pepper, I.L. (1999). Evaluation of methodologies including immunofluorescent assay (IFA) and the polymerase chain reaction (PCR) for detection of human pathogenic microsporidia in water. *Microbiol. Methods*, **35**(1), 43–52.

Dowd, S., Gerba, C.P. and Pepper, I.L. (1998). Confirmation of the human-pathogenic microsporidia *Enterocytozoon bieneusi*, *Encephalitozoon intestinalis*, and *Vittaforma corneae* in water. *Appl. Environ. Microbiol.*, **64**(9), 3332–5.

Dowd, S., John, D., Eliopolus, J., Gerba, C.P., Naranjo, J., Klein, R., Lopez, B., de Mejia, M., Mendoza, C.E. and Pepper, I.L. (2003). Confirmed detection of *Cyclospora cayetanesis*, *Encephalitozoon intestinalis* and *Cryptosporidium parvum* in water used for drinking. *J. Water Health*, **1**(3), 117–23.

Dubois E., Hennechart, C., Merle, G., Burger, C., Hmila, N., Ruelle, S., Perelle, S. and Ferré, V. (2007). Detection and quantification by real-time RT-PCR of hepatitis A virus from inoculated tap waters, salad vegetables, and soft fruits: characterization of the method performances. *Int. J. Food Microbiol.*, **117**(2), 141–9.

Dubois, E., Le Guyader, F., Haugarreau, L., Kopecka, H., Cormier, M. and Pommepuy, M. (1997). Molecular epidemiological survey of rotaviruses in sewage by reverse transcriptase seminested PCR and restriction fragment length polymorphism assay. *Appl. Environ. Microbiol.*, **63**(5), 1794–800.

Duizer, E., Bijkerk, P., Rockx, B., De Groot, A., Twisk, F. and Koopmans, M. (2004). Inactivation of caliciviruses. *Appl. Environ. Microbiol.*, **70**(8), 4538–43.

DuPont, H., Chappell, C.L., Sterling, C.R., Okhuysen, P.C., Rose, J.B. and Jakubowski, W. (1995). The infectivity of *Cryptosporidium parvum* in healthy volunteers. *N. Engl. J. Med.*, **332**(13), 855–9.

DuPont, H., Levine, M.M., Hornick, R.B. and Formal, S.B. (1989). Inoculum size in shigellosis and implications for expected mode of transmission. *J. Infect. Dis.*, **159**(6), 1126–8.

Dusserre, E., Ginevra, C., Hallier-Soulier, S., Vandenesch, F., Festoc, G., Etienne, J., Jarraud, S. and Molmeret, M. (2008). A PCR-based method for monitoring *Legionella pneumophila* in water samples detects viable but noncultivable legionellae that can recover their cultivability. *Appl. Environ. Microbiol.*, **74**(15), 4817–24.

DWI (1998). Standard operating protocol for the monitoring of *Cryptosporidium* oocysts in treated water supplies to satisfy water supply (water quality) (amendments) regulations. SI N° 1524. Drinking Water Inspectorate. UK.

Eaton, K., Catrenich, C.E., Makin, K.M. and Krakowka, S. (1995). Virulence of coccoid and bacillary forms of Helicobacter pylori in gnotobiotic piglets. *J. Infect. Dis.*, **171**(2), 459–62.

Eden, K., Rosenberg, M.L., Stoopler, M., Wood, B.T., Highsmith, A.K., Skaliy, P., Wells, J.G. and Feeley, J.C. (1977). Waterborne gastrointestinal illness at a ski resort. Isolation of *Yersinia enterocolitica* from drinking water. *Public Health Rep.*, **92**(3), 245–50.

Egger, D., Pasamontes, L., Ostermayer, M. and Bienz, K. (1995). Reverse transcription multiplex PCR for differentiation between polio- and enteroviruses from clinical and environmental samples. *J. Clin. Microbiol.*, **33**(6), 1442–7.

Egoz, N., Shmilovitz, M., Kretzer, B., Lucian, M., Porat, V. and Raz, R. (1991). An outbreak of *Shigella sonnei* infection due to contamination of a municipal water supply in northern Israel. *J. Infect.*, **22**(1), 87–93.

Eischeid, A.C., Meyer, J.N. and Linden, K.G. (2009). UV disinfection of adenoviruses: molecular indications of DNA damage efficiency. *Appl. Environ. Microbiol.*, **75**(1), 23–8.

Eisenstein, L., Bodager, D. and Ginzl, D. (2008). Outbreak of giardiasis and cryptosporidiosis associated with a neighborhood interactive water fountain–Florida, 2006. *J. Environ. Health*, **71**(3), 18–22; quiz 49–50.

El-sharkwai, F. and El-attar, L. *et al.* (1989). Some environmental factors affecting suvival of faecal pathogens and indicator. *Water. Sci. Technol.*, **21**, 115–120.

Emerson, M.A., Sproul, O.J. and Buck, C.E. (1982). Ozone inactivation of cell-associated viruses. *Appl. Environ. Microbiol.*, **43**(3), 603–08.

Engstrand, L. (2001). *Helicobacter* in water and waterborne routes of transmission. *J. Appl. Microbiol.*, **90**, 80S-4S.

Enriquez, C. and Hurst, C.J. *et al.* (1995). Survival of the Enteric Adenovirus 40 and 41 in Tap, Sea and Waste Water. *Water Res.*, **29**(11), 2548–2553.

Enroth, H. and Engstrand, L. (1995). Immunomagnetic separation and PCR for detection of *Helicobacter pylori* in water and stool specimens. *J. Clin. Microbiol.*, **33**(8), 2162–5.

Environment Agency. (1998). The determination of microcystin algal toxins in raw and treated waters by high performance chromatography. *Methods for the examination of waters and associated materials.* Bristol, pp. 13.

Erova, T.E., Pillai, L., Fadl, A.A., Sha, J., Wang, S., Galindo C.L. and Chopra A.K. (2006). DNA adenine methyltransferase influences the virulence of *Aeromonas hydrophila. Infect Immun.*, **74**(1), 410–24.

Escudero, M., Velazquez, L., De Cortinez, Y.M., Di Genaro, M.S., de Guzman, A.M. (1994). *Yersinia* ssp. in surface water in San Luis, Argentina. *Folia Microbiol. (Praha).*, **39**(6), 459–62.

Espigares, M. and Garcia, F. *et al.* (1999). Detection of Hepatitis A virus in wastewater. *Environ. Toxicol.*, **14**, 391–96.

Estes, M.K., Graham, D.Y. Smith, E.M. and Gerba, C.P. (1979). Rotavirus stability and inactivation. *J. Gen. Virol.*, **43**, 403–409.

Falcao, J., Brocchi, M., Proenca-Modena, J.L., Acrani, G.O., Correa, E.F. and Falcao, D.P. (2004). Virulence characteristics and epidemiology of *Yersinia enterocolitica* and Yersiniae other than *Y. pseudotuberculosis* and *Y. pestis* isolated from water and sewage. *J. Appl. Microbiol.*, **96**(6), 1230–6.

Falconer, I., Beresford, A.M. and Runnegar, M.T. (1983). Evidence of liver damage by toxin from a bloom of the blue-green alga, *Microcystis aeruginosa. Med. J. Aust.*, **1**(11), 511–4.

Falconer, I. and Buckley, T.H. (1989). Tumour promotion by Microcystis sp., a blue-green alga occurring in water supplies. *Med. J.*, **150**(6), 351.

Falkinham, Jr. (2002). Nontuberculous mycobacteria in the environment. *Clin. Chest. Med.*, **23**, 529–551.

Falkinham, Jr. (2003). Factors influencing the chlorine susceptibility of *Mycobacterium avium*, *Mycobacterium intracellulare*, and *Mycobacterium scrofulaceum*. *Appl. Environ. Microbiol.*, **69**(9), 5685–9.

Falkinham, Jr., Norton, C.D. and LeChevallier, M.W. (2001). Factors influencing numbers of *Mycobacterium avium, Mycobacterium intracellulare*, and other Mycobacteria in drinking water distribution systems. *Appl. Environ. Microbiol.*, **67**(3), 1225–31.

Fan, X., Chua, A., Li, T.G. and Zeng, Q.S. (1998). Survival of *Helicobacter pylori* in milk and tap water. *J. Gastroenterol Hepatol.*, **13**(11), 1096–8.

FAO/WHO. (2002). Microbial Risk Assessment Series 2. Risk assessments of *Salmonella* in eggs and broiler chickens. (http://whqlibdoc.who.int/publications/2002/9291562293.pdf)

Faruque, S., Chowdhury, N., Khan, R., Hasan, M.R., Nahar, J., Islam, M.J., Yamasaki, S., Ghosh, A.N., Nair, G.B. and Sack, D.A. (2003). *Shigella dysenteriae* type 1-specific bacteriophage from environmental waters in Bangladesh. *Appl. Environ. Microbiol.*, **69**(12), 7028–31.

Faruque, S.M., Khan, R., Kamruzzaman, M., Yamasaki, S., Ahmad, Q.S., Azim, T., Nair, G.B., Takeda, Y. and Sack, D.A. (2002). Isolation of *Shigella dysenteriae* type 1 and *S. flexneri* strains from surface waters in Bangladesh: comparative molecular analysis of environmental Shigella isolates versus clinical strains. *Appl. Environ. Microbiol.*, **68**, 3908–3913.

Fayer, R. (2004a). *Cryptosporidium*: a water-borne zoonotic parasite. *Vet. Parasitol.*, **126**(1–2), 37–56.

Fayer, R. (2004b). Infectivity of microsporidia spores stored in seawater at environmental temperatures. *J. Parasitol.*, **90**(3), 654–7.

Fergusson, K.M. and Saint, C.P. (2000). Molecular phylogeny of *Anabaena circinalis* and its identification in environmental samples by PCR. *Appl. Environ. Microbiol.*, **66**(9), 4145–8.

Fergusson, K.M. and Saint, C.P. (2003). Multiplex PCR assay for *Cylindrospermopsis raciborskii* and cylindrospermopsin-producing cyanobacteria. *Environ. Toxicol.*, **18**(2), 120–5.

Fernandez, M., Giampaolo, B.N., Ibanez, S.B., Guagliardo, M.V., Esnaola, M.M., Conca, L., Valdivia, P., Stagnaro, S.M., Chiale, C. and Frade, H. (2000). Aeromonas hydrophila and its relation with drinking water indicators of microbiological quality in Argentine. *Genetica.*, **108**(1), 35–40.

Ferrero Poschetto, L., Ike, A., Papp, T., Mohn, U., Böhm, R. and Marschang, R.E. (2007). Comparison of the sensitivities of noroviruses and feline calicivirus to chemical disinfection under field-like conditions. *Applied and Environmental Microbiology*, **73**(17), 5494–500.

Fewtrell, L. and Godfree, A. *et al.* (1994). *Pathogenic Microorganisms in temperate environmental waters.* Cardigan, Dyfed, Samara Publishing.

Fey, A., Eichler, S., Flavier, S., Christen, R., Hofle, M.G. and Guzman, C.A. (2004). Establishment of a real-time PCR-based approach for accurate quantification of bacterial RNA targets in water, using *Salmonella* as a model organism. *Appl. Environ. Microbiol.*, **70**(6), 3618–23.

Figueras, M.J., Soler, L., Chacon, M.R., Guarro, J. and Martinez-Murcia, A.J. (2000). Extended method for discrimination of *Aeromonas* spp. by 16S rDNA RFLP analysis. *Int. J. Syst. Evol. Microbiol.* **50**Pt 6, 2069–73.

Figueras, M., Suarez-Franquet, A., Chacon, M.R., Soler, L., Navarro, M., Alejandre, C., Grasa, B., Martinez-Murcia, A.J. and Guarro, J. (2005). First record of the rare species *Aeromonas culicicola* from a drinking water supply. *Appl. Environ. Microbiol.*, **71**(1), 538–41.

Finch, G., Black, E.K., Labatiuk, C.W., Gyurek, L. and Belosevic, M. (1993). Comparison of *Giardia lamblia* and Giardia muris cyst inactivation by ozone. *Appl. Environ. Microbiol.*, **59**(11), 3674–80.

Fiore, A. (2004). Hepatitis A transmitted by food. *Food Safety*, **38**, 705–15.

Fischer, T., Steinsland, H. and Valentiner-Branth, P. (2002). Rotavirus particles can survive storage in ambient tropical temperatures for more than 2 months. *J. Clin. Microbiol.*, **40**(12), 4763–4.

Fiume, L., Bucca Sabattini, M.A. and Poda, G. (2005). Detection of *Legionella pneumophila* in water samples by species-specific real-time and nested PCR assays. *Lett Appl Microbiol.*, **41**(6), 470–5.

Fleming, C., Caron, D., Gunn, J.E., Horine, M.S., Matyas, B.T. and Barry, M.A. (2000). An outbreak of *Shigella sonnei* associated with a recreational spray fountain. *Am. J. Public Health*, **90**(10), 1641–2.

Fluit, A., Widjojoatmodjo, M.N. and Verhoef, J. (1991). Detection of Salmonella species in fecal samples by immunomagnetic separation and PCR. *J. Clin. Microbiol.*, **33**(4), 1046–7.

Fode-Vaughan, K., Maki, J.S., Benson, J.A. and Collins, M.L. (2003). Direct PCR detection of *Escherichia coli O157:H7*. *Lett. Appl. Microbiol.*, **37**(3), 239–43.

Fontaine, M. and Guillot, E. (2002). Development of a TaqMan Quantitative PCR assay specific for *Cryptosporidium parvum*. *FEMS Microbiol.ogy Letters*, **214**, 13–17.

Fontaine, M. and Guillot, E. (2003). An immunomagnetic separation-real time PCR method for quantification of *Cryptosporidium parvum* in water samples. *J. Microbiol. Methods*, **54**, 29–36.

Formiga-Cruz, M., Hundesa, A., Clemente-Casares, P., Albinana-Gimenez, N., Allard, A. and Girones, R. (2005). Nested multiplex PCR assay for detection of human enteric viruses in shellfish and sewage. *J. Virol. Methods*, **125**(2), 111–8.

Foulds, I., Granacki, A., Xiao, C., Krull, U.J., Castle, A. and Horgen, P.A. (2002). Quantification of microcystin-producing cyanobacteria and E. coli in water by 5-nuclease PCR. *J. Appl. Microbiol.*, **93**(5), 825–34.

Fournier, S., Dubrou, S., Liguory, O., Gaussin, F., Santillana-Hayat, M., Sarfati, C., Molina, J.M. and Derouin, F. (2002). Detection of Microsporidia, cryptosporidia and *Giardia* in swimming pools: a one-year prospective study. *FEMS Immunol Med. Microbiol.*, **33**(3), 209–13.

Fournier, S., Liguory, O., Santillana-Hayat, M., Guillot, E., Sarfati, C., Dumoutier, N., Molina, J. and Derouin, F. (2000). Detection of microsporidia in surface water: a one-year follow-up study. *FEMS Immunol Med. Microbiol.*, **2**, 95–100.

Fout, G., Martinson, B.C., Moyer, M.W. and Dahling, D.R. (2003). A multiplex reverse transcription-PCR method for detection of human enteric viruses in groundwater. *Appl. Environ. Microbiol.*, **69**(6), 3158–64.

Fout, G.S., Schaefer III, F.W., Messer, J.W., Dahling, D.R. and Stetler, R.E. (1996). *ICR Microbial Laboratory Manual*. Washington, D.C, U.S. Environmental Protection Agency.

Franklin, L.J., Fielding, J.E., Gregory, J., Gullan, L., Lightfoot, D., Poznanski, S.Y. and Vally, H. (2008). An outbreak of *Salmonella Typhimurium* 9 at a school camp linked to contamination of rainwater tanks. *Epidemiol. Infect.*, **8**, 1–7.

Franzen, C. and Muller, A. (1999). Molecular techniques for detection, species differentiation, and phylogenetic analysis of microsporidia. *Clin. Microbiol. Rev.*, **12**, 243–285.

Fraser, D., Tsai, T.R., Orenstein, W., Parkin, W.E., Beecham, H.J., Sharrar, R.G., Harris, J., Mallison, G.F., Martin, S.M., McDade, J.E., Shepard, C.C. and Brachman, P.S. (1977). Legionnaires' disease: description of an epidemic of pneumonia. *N. Engl. J. Med.*, **297**(22), 1189–97.

Fricker, C. (1987). The isolation of salmonellas and campylobacters: a review. *J. Appl. Bacteriol.*, **63**(2), 99–116.

Fritsche, T., Sobek, D. and Gautom, R.K. (1998). Enhancement of in vitro cytopathogenicity by *Acanthamoeba* spp. following acquisition of bacterial endosymbionts. *FEMS Microbiol. Lett.*, **166**(2), 231–6.

Fuhrman, J.A., Liang, X. and Noble, R.T. (2005). Rapid detection of enteroviruses in small volumes of natural waters by real-time quantitative reverse transcriptase PCR. *Appl. Environ. Microbiol.*, **71**(8), 4523–30.

Fukushima, H., Gomyoda, M., Tsubokura, M. and Aleksic, S. (1995). Isolation of *Yersinia pseudotuberculosis* from river waters in Japan and Germany using direct KOH and HeLa cell treatments. *Zentralbl. Bakteriol.*, **282**(1), 40–9.

Fukushima, H., Tsunomori, Y. and Seki, R. (2003). Duplex real-time SYBR green PCR assays for detection of 17 species of food- or waterborne pathogens in stools. *J. Clin. Microbiol.*, **41**(11), 5134–46.

Furukawa, K., Noda, N., Tsuneda, S., Saito, T., Itayama, T. and Inamori, Y. (2006). Highly sensitive real-time PCR assay for quantification of toxic cyanobacteria based on microcystin synthetase A gene. *J. Biosci. Bioeng.*, **102**(2), 90–6.

Gaafar, M.R. (2007). Effect of solar disinfection on viability of intestinal protozoa in drinking water. *J. Egypt Soc. Parasitol.*, **37**(1), 65–86.

Gannon, V., Graham, T.A., Read, S., Ziebell, K., Muckle, A., Mori, J., Thomas, J., Selinger, B., Townshend, I. and Byrne, J. (2004). Bacterial pathogens in rural water supplies in Southern Alberta, Canada. *J. Toxicol. Environ. Health*, **67**(20–22), 1643.

Garcia-Aljaro, C., Bonjoch, X. and Blanch, A.R. (2005). Combined use of an immunomagnetic separation method and immunoblotting for the enumeration and isolation of *Escherichia coli O157* in wastewaters. *J. Appl. Microbiol.*, **98**(3), 589–97.

Garcia-Fulgueiras, A., Navarro, C., Fenoll, D., Garcia, J., Gonzales-Diego, P., Jimenez-Bunuelas, T., Rodriguez, M., Lopez, R., Pacheco, F., Ruiz, J., Segovia, M., Balandron, B. and Pelaz, C. (2003). Legionnaires' disease outbreak in Murcia, Spain. *Emerg. Infect. Dis.*, **9**(8), 915–21.

Gassilloud, B., Duval, M., Schwartzbrod, L. and Gantzer, C. (2003a). Recovery of feline calicivirus infectious particles and genome from water: comparison of two concentration techniques. *Water Sci. Technol.*, **47**(3), 97–101.

Gassilloud, B., Schwartzbrod, L. and Gantzer, C. (2003b). Presence of Viral Genomes in Mineral Water: a Sufficient Condition To Assume Infectious Risk? *Appl. Environ. Microbiol.*, **69**(7), 3965–3969.

Gerba, C., Nwachuku, N. and Riley, K.R. (2003). Disinfection resistance of waterborne pathogens on the United States Environmental Protection Agency's Contaminant Candidate List (CCL). *J. Wat. Supply Res. Technol-Aqua*, **52**(2), 81–94.

Gerba, C., Riley, K.R., Nwachuku, N., Ryu, H. and Abbaszadegan, M. (2003). Removal of *Encephalitozoon intestinalis*, calicivirus, and coliphages by conventional drinking water treatment. *J. Environ. Sci. Health A Tox Hazard Subst. Environ. Eng.*, **38**(7), 1259.

Gerrity, D., Ryu, H., Crittenden, J. and Abbaszadegan, M. (2008). UV inactivation of adenovirus type 4 measured by integrated cell culture qPCR. *J. Environ. Sci. Health Part A*. **43**(14): 1628–38.

Gersberg, R.M., Rose, M.A., Robles-Sikisaka, R. and Dhar, A.K. (2006). Quantitative detection of hepatitis a virus and enteroviruses near the United States-Mexico border and correlation with levels of fecal indicator bacteria. *Appl. Environ. Microbiol.*, **72**(12), 7438–44.

Ghosh, A. and Sehgal, S.C. (2001). Survivability of *Shigella dysenteriae* type 1 & S. flexneri 2a in laboratory conditions simulating aquatic environment. *Indian J. Med. Res.*, **114**, 199–206.

Gião, M.S., Azevedo, N.F., Wilks, S.A., Vieira, M.J. and Keevil, C.W. (2008a). Persistence of *Helicobacter pylori* in heterotrophic drinking-water biofilms. *Appl. Environ. Microbiol.*, **74**(19), 5898–904.

Gião, M.S., Wilks, S.A., Azevedo, N.F., Vieira, M.J. and Keevil, C.W. (2008b). Validation of SYTO 9/Propidium Iodide Uptake for Rapid Detection of Viable but Noncultivable *Legionella pneumophila. Microb. Ecol.*

Giglio, S., Monis, P.T. and Saint, C.P. (2005). *Legionella* confirmation using real-time PCR and SYTO9 is an alternative to current methodology. *Appl. Environ. Microbiol.*, **71**(12), 8944–8.

Gilgen, M. and Germann, D. *et al.* (1997). Three-step isolation method for sensitive detection of enterovirus, rotavirus, hepatitis A virus, and small round structured viruses in water samples. *Int. J. Food Microbiol.*, **37**(2–3), 189–99.

Glass, R., Bresee, J.S., Parashar, U.D., Miller, M.A. and Gentsch, J.R. (1997). Rotavirus vaccines at the threshold. *Nat. Med.*, **3**, (10–11).

Godoy, D. and Randle, A. *et al.* (2003). Multilocus sequence typing and evolutionary relationships among the causative agents of melioidosis and glanders, *Burkholderia pseudomallei* and *Burkholderia mallei. J. Clin. Microbiol.*, **41**(5), 2068–2079.

Godoy, P., Nuín, C., Alsedà, M., Llovet, T., Mazana, R. and Domínguez, A. (2006). Waterborne outbreak of gastroenteritis caused by norovirus transmitted through drinking water. *Rev. Clín. Esp.*, **2006**(9), 435–7.

Gofti-Laroche, L., Gratacap-Cavallier, B., Genoulaz, O., Joret, J.C., Hartemann, P., Seigneurin, J.M. and Zmirou, D. (2001). A new analytical tool to assess health risks associated with the virological quality of drinking water (EMIRA study). *Water. Sci. Technol.*, **43**(12), 39–48.

Goh, S., Reacher, M., Casemore, D.P., Verlander, N.Q., Charlett, A., Chalmers, R.M., Knowles, M., Pennington, A., Williams, J., Osborn, K. and Richards, S. (2004). Sporadic cryptosporidiosis decline after membrane filtration of public water supplies, England, 1996–2002. *Emerg. Infect. Dis.*, **11**(2), 251–9.

Goodwin, C., Mendall, M.M. and Northfield, T.C. (1997). *Helicobacter pylori* infection. *Lancet.*, **349**(9047), 265–9.

Grabow, W. (2002). Enteric Hepatitis viruses. *Guidelines for drinking water quality, 2nd edition, Addendum: Microbiological agents in drinking water*. Geneva, World Health Organization, pp. 18–39.

Grabow, W., Gauss-Muller, V., Prozesky, O.W. and Deinhardt, F. (1983). Inactivation of hepatitis A virus and indicator organisms in water by free chlorine residuals. *Appl. Environ. Microbiol.*, **46**(3), 619–24.

Graczyk, T.K., Fayer, R. and Cranfield, M.R. (1997a) Zoonotic transmission of *Cryptosporidium parvum*: Implications for water-borne cryptosporidiosis. *Parasitol. Today*, **9**, 348–51.

Graczyk, T.K., Fayer, R., Cranfield, M.R. and Owens, R. (1997b). *Cryptosporidium parvum* oocysts recovered from water by the membrane filter dissolution method retain their infectivity. *J. Parasitol.*, **83**(1), 111–4.

Graham, D. (1987). Minimal infectious dose of rotavirus in volunteers: safety issues. *J. Infect. Dis.*, **156**(2), 416–8.

Gratacap-Cavallier and Genoulaz, B. *et al.* (2000). Detection of human and animal rotavirus sequences in drinking water. *Appl. Environ. Microbiol.*, **66**(6), 2690–2692.

Green, K. (1997). The role of human caliciviruses in epidemic gastroenteritidis. *Arch. Virol. Suppl.*, **13**, 153–165.

Greening, G. and Hewitt, J. *et al.* (2002). Evaluation of integrated cell culture-PCR (C-PCR) for virological analysis of environmental samples. *J. Appl. Microbiol.*, **93**(5), 745–50.

Gregersen, P., Grunnet, K., Uldum, S.A., Andersen, B.H. and Madsen, H. (1999). Pontiac fever at a sewage treatment plant in the food industry. *Scand. J. Work, Environ. & Health*, **25**(3), 291–95.

Gregor, J., Marsálek, B. and Sípková, H. (2007). Detection and estimation of potentially toxic cyanobacteria in raw water at the drinking water treatment plant by in vivo fluorescence method. *Water Res.*, **41**(1), 228–34.

Gregory, J.B., Litaker, R.W. and Noble, R.T. (2006). Rapid one-step quantitative reverse transcriptase PCR assay with competitive internal positive control for detection of enteroviruses in environmental samples. *Appl. Environ. Microbiol.*, **72**(6), 3960–7.

Greub, G. and Raoult, D. (2004). Microorganisms resistant to free-living amoebae. *Clinical Microbiology Reviews*, **17**(2), 413–33.

Griffin, P. and Tauxe, R.V. (1991). The epidemiology of infections caused by *Escherichia coli O157:H7*, other enterohemorrhagic E. coli, and the associated hemolytic uremic syndrome. *Epidemiol Rev.*, **13**, 60–98.

Grimm, A.C. and Fout, G.S. (2002). Development of a molecular method to identify hepatitis E virus in water. *J. Virol. Methods*, **101**, 175–188.

Guillot, E. and Courtois, S. *et al.* (2003). *A New Breakthrough Process Using DNA Chip Technology for Controlling the Microbiological Quality of Water.* AWWA Water Quality Technology Conference, Philadelphia, PA.

Guntupalli, R., Lakshmanan, R.S., Hu, J., Huang, T.S., Barbaree, J.M., Vodyanoy, V. and Chin, B.A. (2007). Rapid and sensitive magnetoelastic biosensors for the detection of *Salmonella typhimurium* in a mixed microbial population. *J. Microbiol. Methods*, **70**(1), 112–8.

Gutiérrez-Aguirre, I., Banjac, M., Steyer, A., Poljšak-Prijatelj, M., Peterka, M., Strancar, A. and Ravnikar, M. (2008a). Concentrating rotaviruses from water samples using monolithic chromatographic supports. *J. Chromatogr. A*

Gutiérrez-Aguirre, I., Steyer, A., Boben, J., Gruden, K., Poljsak-Prijatelj, M. and Ravnikar, M. (2008b). Sensitive detection of multiple rotavirus genotypes with a single reverse transcription-real-time quantitative PCR assay. *J. Clin. Microbiol.*, **46**(8), 2547–54.

Guthmann, J.P., Klovstad, H., Boccia, D., Hamid, N., Pinoges, L., Nizou, J.Y., Tatay, M., Diaz, F., Moren, A., Grais, R.F., Ciglenecki, I., Nicand, E. and Guerin, P.J. (2006). A large outbreak of hepatitis E among a displaced population in Darfur, Sudan, 2004: the role of water treatment methods. *Clin. Infect. Dis.*, **42**(12), 1685–91.

Guttman-Bass, N., Tchorsh, Y. and Marva, E. (1987). Comparison of methods for rotavirus detection in water and results of a survey of Jerusalem wastewater. *Appl. Environ. Microbiol.*, **53**(4), 761–7.

Guttman-Bass, N.a.A., R. (1983). Concentration of Simian Rotavirus SA-11 from Tap Water by Membrane Filtration and Organic Flocculation. *Appl. Environ. Microbiol.*, **45**(3), 850–855.

Guy, R., Payment, P., Krull, U.J. and Horgen, P.A. (2003). Real-time PCR for quantification of *Giardia* and *Cryptosporidium* in environmental water samples and sewage. *Appl. Environ. Microbiol.*, **69**(9), 5178–85.

Haas, C.N. (1983). Estimation of risk due to low doses of microorganisms: a comparison of alternative methodologies. *American Journal of Epidemiology*, **118**, 576–582.

Haas, C. and Rose, J. (1994). *Reconciliation of microbial risk models and outbreaks epidemiology: the case of the Milwaukee outbreak*. American Water Works Association, Denver, CO, pp. 517–523.

Haas, C.N. and Crockett, C.S. *et al.* (1996). Assessing the risk posed by oocysts in drinking water. *Journal AWWA*, **88**(9), 131–136.

Haas, C.N., Rose, J.B. and Gerba, C.P. (1999). Quantitative microbial risk assessment. New York: John Wiley & Sons.

Haas, C.N., Thayyar-Madabusi, A., Rose, J.B. and Gerba, C.P. (2000). Development of a dose-response relationship for Escherichia coli O157:H7. *Int. J. Food Microbiol.*, **56**, 153–159.

Hafliger, D., Hubner, P. and Luthy, J. (2000). Outbreak of viral gastroenteritis due to sewage-contaminated drinking water. *Int. J. Food Microbiol.*, **10**(54), 123–6.

Hall, R.M. and M.D. Sobsey (1993). Inactivation of Hepatitis A Virus and MS2 by Ozone and Ozone-Hydrogen peroxide in Buffered Water. *Wat. Sci. Tech.*, **27**(3/4), 371–378.

Halliday, M. and Kang, L. *et al.* (1991). An epidemic of hepatitis A attributable to the ingestion of raw clams in Shanghai, China. *J. Infect. Dis.*, **164**(5), 852–9.

Hallier-Soulier, S. and Guillot, E. (2000). Detection of *Cryptosporidia* and *Cryptosporidium parvum* oocysts in environmental water samples by immunomagnetc separation-polymerase chain reaction. *J. Appl. Microbiol.*, **89**, 5–10.

Hallier-Soulier, S. and Guillot, E. (2003). An immunomagnetic separation-reverse transcription polymerase chain reaction (IMS-RT-PCR) test for sensitive and rapid detection of viable waterborne *Cryptosporidium parvum*. *Environ. Microbiol.*, **5**(7), 592–598.

Hamamoto, A., Mori, M., Takahashi, A., Nakano, M., Wakikawa, N., Akutagawa, M., Ikehara, T., Nakaya, Y. and Kinouchi, Y. (2007). New water disinfection system using UVA light-emitting diodes. *J. Appl. Microbiol.*, **103**(6), 2291–8.

Han, T., Paik, I.K. and Kim, S.J. (2003). Molecular relatedness between isolates *Yersinia pseudotuberculosis* from a patient and an isolate from mountain spring water. *J. Korean Med. Sci.*, **18**(3), 425–8.

Handfield, M., Simard, P., Couillard, M. and Letarte, R. (1996a). *Aeromonas hydrophila* isolated from food and drinking water: hemagglutination, hemolysis, and cytotoxicity for a human intestinal cell line (HT-29). *Appl. Environ. Microbiol.*, **62**, 3459–61.

Handfield, M., Simard, P. and Letarte, R. (1996b). Differential media for quantitative recovery of waterborne *Aeromonas hydrophila*. *Appl. Environ. Microbiol.*, **62**(9), 3544–7.

Hanninen, M., Niskanen, M. and Korhonen, L. (1998). Water as a reservoir for *Campylobacter jejuni* infection in cows studied by serotyping and pulsed-field gel electrophoresis (PFGE). *Zentralbl. Veterinarmed*, **45**(1), 37–42.

Harakeh, M.S. and Butler, M. (1985). Factors influenting the ozone inactivation of enteric viruses in effluent. *Ozone Sci. & Eng.*, **6**, 235–243.

Haramoto, E., Katayama, H. and Ohgaki, S. (2004). Detection of noroviruses in tap water in Japan by means of a new method for concentrating enteric viruses in large volumes of freshwater. *Appl. Environ. Microbiol.*, **70**(4), 2154–2160.

Harley, D., Harrower, B., Lyon, M. and Dick, A. (2001). A primary school outbreak of pharyngoconjunctival fever caused by adenovirus type 3. *Commun. Dis. Intell.*, **25**(1), 9–12.

Havelaar, A., Berwald, L.G., Groothuis, D.G. and Baas, J.G. (1985). Mycobacteria in semi-public swimming-pools and whirlpools. *Zentralbl. Bakteriol. Mikrobiol. Hyg [B]*, **180**(5–6), 505–14.

Havelaar, A., During, M. and Versteegh, J.F. (1987). Ampicillin-dextrin agar medium for the enumeration of Aeromonas species in water by membrane filtration. *J. Appl. Bacteriol.*, **62**(3), 279–87.

Havelaar, A., Schets, F.M., van Silfhout, A., Jansen, W.H., Wieten, G. and van der Kooij, D. (1992). Typing of *Aeromonas* strains from patients with diarrhoea and from drinking water. *J. Appl. Bacteriol.*, **72**(5), 435–44.

Hayes, S.L., White, K.M. and Rodgers, M.R. (2006). Assessment of the effectiveness of low-pressure UV light for inactivation of Helicobacter pylori. *Appl. Environ. Microbiol.*, **72**(5), 3763–65.

He, J.W. and Jiang, S. (2005). Quantification of enterococci and human adenoviruses in environmental samples by real-time PCR. *Appl. Environ. Microbiol.*, **71**(5), 2250–5.

Hedberg, C.W. and Osterholm, M.T. (1993). Outbreaks of food-borne and waterborne viral gastroenteritis. *Clin. Microbiol. Rev.*, **6**, 199–210.

Hedlund, K.O., Rubilar-Abreu, E. and Svensson, L. (2000). Epidemiology of Calicivirus Infections in Sweden, 1994–1998. *The Journal of Infectious Diseases*, **181**, S275– S280.

Hegarty, J., Dowd, M.T. and Baker, K.H. (1999). Occurrence of *Helicobacter pylori* in surface water in the United States. *J. Appl. Microbiol.*, **87**(5), 697–701.

Hejkal, T., Smith, E.M. and Gerba, C.P. (1984). Seasonal occurrence of rotavirus in sewage. *Appl. Environ. Microbiol.*, **47**(3), 588–90.

Helmi, K., Skraber, S., Gantzer, C., Willame, R., Hoffmann, L. and Cauchie, H.M. (2008). Interactions of *Cryptosporidium parvum*, *Giardia lamblia*, vaccinal poliovirus type 1, and bacteriophages phiX174 and MS2 with a drinking water biofilm and a wastewater biofilm. *Appl. Environ. Microbiol.*, **74**(7), 2079–88.

Heresztyn, T. and Nicholson, B.C. (2001). Determination of cyanobacterial hepatotoxins directly in water using a protein phosphatase inhibition assay. *Water Res.*, **35**(13), 3049–56.

Hermon-Taylor, J., Bull, T.J., Sheridan, J.M., Cheng, J., Stellakis, M.L. and Sumar, N. (2000). Causation of Crohn's disease by Mycobacterium avium subspecies paratuberculosis. *Can. J. Gastroenterol.*, **14**(6), 521–39.

Hernandez, J., Alonso, J.L., Fayos, A., Amoros, I. and Owen, R.J. (1995a). Development of a PCR assay combined with a short enrichment culture for detection of *Campylobacter jejuni* in estuarine surface waters. *FEMS Microbiol. Lett.*, **127**(3), 201–6.

Hernandez, J., Fayos, A., Alonso, J.L. and Owen, R.J. (1996). Ribotypes and AP-PCR fingerprints of thermophilic campylobacters from marine recreational waters. *J. Appl. Bacteriol.*, **80**(2), 157–64.

Hernandez, J., Fayos, A., Ferrus, M.A. and Owen, R.J. (1995b). Random amplified polymorphic DNA fingerprinting of *Campylobacter jejuni* and *C. coli* isolated from human faeces, seawater and poultry products. *Res. Microbiol.*, **146**(8), 685–96.

Herwaldt, B. (2000). *Cyclospora cayetanensis*: a review, focusing on the outbreaks of cyclosporiasis in the 1990s. *Clin. Infect. Dis.*, **31**(4), 1040–57.

Hester, J., Lindquist, H.D., Bobst, A.M. and Schaefer, F.W. 3rd. (2000). Fluorescent in situ detection of *Encephalitozoon hellem* spores with a 6-carboxyfluorescein-labeled ribosomal RNA-targeted oligonucleotide probe. *J. Eukaryot. Microbiol.*, **47**(3), 299–308.

Hester, J.D., Varma, M., Bobst, A.M., Ware, M.W., Lindquist, H.D. and Schaefe, F.W. 3rd. (2002). Species-specific detection of three human-pathogenic microsporidial species from the genus *Encephalitozoon* via fluorogenic 5′ nuclease PCR assays. *Mol. Cell. Probes.*, **16**(6), 435–44.

Higgins, J., Fayer, R., Trout, J.M., Xiao, L., Lal, A.A., Kerby, S. and Jenkins, M.C. (2001). Real-time PCR for the detection of *Cryptosporidium parvum*. *J. Microbiol. Methods*, **47**(3), 323–37.

Higgins, J., Trout, J.M., Fayer, R., Shelton, D. and Jenkins, M.C. (2003). Recovery and detection of *Cryptosporidium parvum* oocysts from water samples using continuous flow centrifugation. *Water Res.*, **37**(15), 3551–60.

Hijnen, W.A.M., Beerendonk, E.F. and Medema, G.J. (2006). Inactivation credit of UV radiation for viruses, bacteria and protozoan (oo)cysts in water: a review. *Water Res.*, **40**, 3–22.

Hilborn, E.D., Covert, T.C., Yakrus, M.A., Harris, S.I., Donnelly, S.F., Rice, E.W., Toney, S., Bailey, S.A. and Stelma, G.N. Jr. (2006). Persistence of nontuberculous mycobacteria in a drinking water system after addition of filtration treatment. *Appl. Environ. Microbiol.*, **72**(9), 5864–9.

Hilborn, E.D., Yakrus, M.A., Covert, T.C., Harris, S.I., Donnelly, S.F., Schmitt, M.T., Toney, S., Bailey, S.A. and Stelma, G.N. (2008). Molecular comparison of Mycobacterium avium isolates from clinical and environmental sources. *Appl. Environ. Microbiol.*, **74**(15), 4966–68.

Hill, V.R., Polaczyk, A.L., Hahn, D., Narayanan, J., Cromeans, T.L., Roberts, J.M. and Amburgey, J.E. (2005). Development of a rapid method for simultaneous recovery of diverse microbes in drinking water by ultrafiltration with sodium polyphosphate and surfactants. *Appl. Environ. Microbiol.*, **71**(11), 6878–84.

Hisbergues, M., Christiansen, G., Rouhiainen, L., Sivonen, K. and Borner, T. (2003). PCR-based identification of microcystin-producing genotypes of different cyanobacterial genera. *Arch. Microbiol.*, **180**(6), 402–10.

Hlavsa, M., Watson, J.C. and Beach, M.J. (2005). Cryptosporidiosis surveillance–United States 1999–2002. *MMWR Surveill Summ.*, **54**(1), 1–8.

HMSO. (1994). *The Microbiology of Water 1994, Part 1, Drinking Water.* London, HMSO.

Ho, B. and Tam, T. (2000). Rapid enumeration of *Salmonella* in environmental waters and wastewater. *Wat. Res.*, **34**(8), 2397–99.

Ho, M., Chen, E.R., Hsu, K.H., Twu, S.J., Chen, K.T., Tsai, S.F., Wang, J.R. and Shih, S.R. (1999). An epidemic of enterovirus 71 infection in Taiwan. Taiwan Enterovirus Epidemic Working Group. *N. Engl. J. Med.*, **341**(13), 929–35.

Hoeger, S.J., Dietrich, D.R. and Hitzfeld, B.C. (2002). Effect of ozonation on the removal of cyanobacterial toxins during drinking water treatment. *Environ. Health Perspect.*, (11), 1127–32

Hoeger, S., Hitzfeld, B.C. and Dietrich, D.R. (2005). Occurrence and elimination of cyanobacterial toxins in drinking water treatment plants. *Toxicol. Appl. Pharmacol.*, **203**(3), 231–42.

Hoff, J., Rice, E.W. and Schaefer, F.W. 3rd. (1985). Comparison of animal infectivity and excystation as measures of *Giardia muris* cyst inactivation by chlorine. *Appl. Environ. Microbiol.*, **50**(4), 1115–7.

Hoffmann, R. and Michel, R. (2001). Distribution of free-living amoeba (FLA) during preparation and supply of drinking water. *Int. J. Hyg. Environ. Health*, **203**, 215–219.

Hoffman, R.M., Wolk, D.M., Spencer, S.K. and Borchardt, M.A. (2007). Development of a method for the detection of waterborne microsporidia. *J. Microbiol. Methods*, **70**(2), 312–8.

Holme, R. (2003). Drinking water contamination in Walkerton, Ontario: positive resolutions from a tragic event. *Water. Sci. Technol.*, **47**(3), 1–6.

Homan, W. and Gilsing, M. *et al.* (1998). Characterization of *Giardia duodenalis* by polymerase-chain-reaction fingerprinting. *Parasitol. Res.*, **84**, 707–714.

Hongmanee, P., Stender, H. and Rasmussen, O.F. (2001). Evaluation of a fluorescence in situ hybridization assay for differentiation between tuberculous and nontuberculous Mycobacterium species in smears of Lowenstein-Jensen and Mycobacteria Growth Indicator Tube cultures using peptide nucleic acid probes. *J. Clin. Microbiol.*, **39**(3), 1032–5.

Hopkins, R., Gaspard, B., Williams, F.P., Karlin, R.J., Cukor, G. and Blacklow, N.R. (1984). A community waterborne gastroenteritis outbreak: evidence for rotavirus as the agent. *Am. J. Public Health*, **74**(3), 263–5.

Hopkins, R., Vial, P.A., Ferreccio, C., Ovalle, J., Prado, P., Sotomayor, V., Russell, R.G., Wasserman, S.S. and Morris, J.G. Jr. (1993). Seroprevalence of *Helicobacter pylori* in Chile: vegetables may serve as one route of transmission. *J. Infect. Dis.*, **168**(1), 222–6.

Horiuchi, T., Ohkusa, T., Watanabe, M., Kobayashi, D., Miwa, H. and Eishi, Y. (2001). Helicobacter pylori DNA in drinking water in Japan. *Microbiol. Immunol.*, **45**, 515–519.

Hörman, A. and Rimhanen-Finne, R. *et al.* (2004). *Campylobacter* spp., *Giardia* spp., *Cryptosporidium* spp., Noroviruses, and Indicator Organisms in Surface Water in Southwestern Finland, 2000–2001. *Appl. Environ. Microbiol.*, **70**(1), 87–95.

Horng, Y.T., Soo, P.C., Shen, B.J., Hung, Y.L., Lo, K.Y., Su, H.P., Wei, J.R., Hsieh, S.C., Hsueh, P.R. and Lai, H.C. (2006). Development of an improved PCR-ICT hybrid assay for direct detection of Legionellae and *Legionella pneumophila* from cooling tower water specimens. *Water Res.*, **40**(11), 2221–9.

Horsburgh, C.J. (1996). Epidemiology of disease caused by nontuberculous mycobacteria. *Semin Respir. Infect.*, **11**(4), 244–51.

Hot, D., Legeay, O., Jacques, J., Gantzer, C., Caudrelier, Y., Guyard, K., Lange, M. and Andreoletti, L. (2003). Detection of somatic phages, infectious enteroviruses and enterovirus genomes as indicators of human enteric viral pollution in surface water. *Water Res.*, **19**, 4703–10.

House, J., Ho, L., Newcombe, G. and Burch, M. (2004). Management strategies for toxic blue-green algae: Literature review. GWRC report.

Howard, K. and Inglis, T.J. (2003). The effect of free chlorine on *Burkholderia pseudomallei* in potable water. *Water Res.*, **37**(18), 4425–32.

Howard, K. and Inglis, T.J. (2005). Disinfection of *Burkholderia pseudomallei* in potable water. *Water Res.*, **39**(6), 1085–92.

Hsu, B., Chen, C.H., Kung, C.M., Wan, M.T. and Shen, S.M. (2007). Evaluation of enterovirus recovery in surface water by different adsorption and elution procedures. *Chemosphere.*, **66**(5), 964–9.

Hsu, B., Wu, N.M., Jang, H.D., Shih, F.C., Wan, M.T. and Kung, C.M. (2005). Using the flow cytometry to quantify the *Giardia* cysts and *Cryptosporidium* oocysts in water samples. *Environ. Monit Assess.*, **104**(1–3), 155–62.

Hsu, W.B., Wang, J.H., Chen, P.C., Lu, Y.S. and Chen, J.H. (2007). Detecting low concentrations of *Shigella sonnei* in environmental water samples by PCR. *FEMS Microbiol. Lett.*, **270**(2), 291–8.

Huffman, D., Gennaccaro, A., Rose, J.B. and Dussert, B.W. (2002). Low- and medium-pressure UV inactivation of microsporidia *Encephalitozoon intestinalis*. *Water Res.*, **36**(12), 3161–4.

Huffman, D., Nelson, K.L. and Rose, J.B. (2003). CalicivirusAn Emerging Contaminant in Water: State of the Art. *Environmental Engineering Science*, **20**(5), 503–515.

Hulten, K., Enroth, H., Nystrom, T. and Engstrand, L. (1998). Presence of *Helicobacter* species DNA in Swedish water. *J. Appl. Microbiol.*, **85**(2), 282–6.

Hulten, K., Han, S.W., Enroth, H., Klein, P.D., Opekun, A.R., Gilman, R.H., Evans, D.G., Engstrand, L., Graham, D.Y. and El-Zaatari, F.A. (1996). *Helicobacter pylori* in the drinking water in Peru. *Gastroenterology*, **110**(4), 1031–5.

Hung, T., Chen, G.M., Wang, C.G., Yao, H.L., Fang, Z.Y., Chao, T.X., Chou, Z.Y., Ye, W., Chang, X.J. and Den, S.S. *et al.* (1984). Waterborne outbreak of rotavirus diarrhoea in adults in China caused by a novel rotavirus. *Lancet*, **26**(1), 1139–42.

Hunter, P. (1991). Human illness associated with freshwater cyanobacteria (blue-green algae). *PHLS Microbiology Digest*, **8**, 96–100.

Hunter, P. (1993). The microbiology of bottled natural mineral waters. *J. Appl. Bacteriol.*, **74**(4), 345–52.

Hunter, P. (1997). *Waterborne disease; Epidemiology and ecology*. Chichester, John Wiley & Sons Ltd.

Hurst, C. (1991). Presence of enteric viruses in freshwater and their removal by the conventional drinking water treatment process. *Bull World Health Organ.*, **69**(1), 113–9.

Ibekwe, A. and Grieve, C.M. (2003). Detection and quantification of *Escherichia coli O157:H7* in environmental samples by real-time PCR. *J. Appl. Microbiol.*, **94**(3), 421–31.

Ibekwe, A., Watt, P.M., Grieve, C.M., Sharma, V.K. and Lyons, S.R. (2002). Multiplex fluorogenic real-time PCR for detection and quantification of *Escherichia coli O157:H7* in dairy wastewater wetlands. *Appl. Environ. Microbiol.*, **68**(10), 4853–62.

Ikeda, R., Kondracki, S.F., Drabkin, P.D., Birkhead, G.S. and Morse, D.L. (1993). Pleurodynia among football players at a high school. An outbreak associated with coxsackievirus B1. *JAMA*, **270**(18), 2205–6.

Inglis, T., Garrow, S.C., Henderson, M., Clair, A., Sampson, J., O'Reilly, L. and Cameron, B. (2000). *Burkholderia pseudomallei* traced to water treatment plant in Australia. *Emerg. Infect. Dis.*, **6**(1), 56–9.

Inoue, M., Nakashima, H., Ishida, T. and Tsubokura, M. (1988a). Three outbreaks of *yersinia pseudotuberculosis* infection. *Zentralbl Bakteriol Mikrobiol Hyg [B]*, **186**(5–6), 504–11.

Inoue, M., Nakashima, H., Ishida, T., Tsubokura, M. and Sakazaki, R. (1988b). Isolation of Yersinia pseudotuberculosis from water. *Zentralbl Bakteriol Mikrobiol Hyg [B]*, **186**(4), 338–43.

Inoue, H., Takama, T., Agawa, Y., Onodera, J., Ishima, T., Agata, K., Saitoh, K. and Huruhata, K. (2007). Evaluation of Duopath *Legionella* kit for the rapid identification of *Legionella* strains isolated from water samples. *Biocontrol Sci.*, **12**(4), 155–8.

InVS. (2004). Epidémie communautaire de légionellose Pas-de-Calais, France, novembre 2003 – janvier 2004, rapport d'investigation. In. Institut National de Veille Sanitaire.

Isaacson, M., Canter, P.H., Effler, P., Arntzen, L., Bomans, P. and Heenan, R. (1993). Haemorrhagic colitis epidemic in Africa. *Lancet.*, **341**(8850), 961.

Islam, M., Doyle, M.P., Phatak, S.C., Millner, P. and Jiang, X. (2004). Persistence of enterohemorrhagic *Escherichia coli O157:H7* in soil and on leaf lettuce and parsley grown in fields treated with contaminated manure composts or irrigation water. *J. Food Prot.*, **67**(7), 1365–70.

Islam, M., Hasan, M.K., Miah, M.A., Sur, G.C., Felsenstein, A., Venkatesan, M., Sack, R.B. and Albert, M.J. (1993). Use of the polymerase chain reaction and fluorescent-antibody methods for detecting viable but nonculturable *Shigella dysenteriae* type 1 in laboratory microcosms. *Appl. Environ. Microbiol.*, **59**(2), 536–40.

Islam, D. and Lindberg, A.A. (1992). Detection of *Shigella dysenteriae* type 1 and *Shigella flexneri* in feces by immunomagnetic isolation and polymerase chain reaction. *J. Clin. Microbiol.*, **30**(11), 2801–6.

Ito, E., Kondo F., Terao, K. and Harada, K. (1997). Neoplastic nodular formation in mouse liver induced by repeated intraperitoneal injections of microcystin-LR. *Toxicon.*, **35**(9), 1453–7.

Jakopanec, I., Borgen, K., Vold, L., Lund, H., Forseth, T., Hannula, R. and Nygård, K. (2008). A large waterborne outbreak of campylobacteriosis in Norway: the need to focus on distribution system safety. *BMC Infect Dis.*, **24**(8), 128.

Jarmey-Swan, C. and Gibbs, R. *et al.* (2000). A novel method for detection of viable *Giardia* cysts in water samples. *Wat. Res.*, **34**(6), 1948–1951.

Jenkins, M.C., Trout, J., Abrahamsen, M.S., Lancto, C.A., Higgins, J. and Fayer, R. (2000). Estimating viability of *Cryptosporidium parvum* oocysts using reverse transcriptase-polymerase chain reaction (RT-PCR) directed at mRNA encoding amyloglucosidase. *J. Microbiol. Methods*, **43**(2), 97–106.

Jephcott, A. and Begg, N. *et al.* (1986). Outbreak of giardiasis associated with main water in the United Kingdom. *Lancet*, 730–732.

Jiang, J., Alderisio, K.A., Singh, A. and Xiao, L. (2005). Development of procedures for direct extraction of *Cryptosporidium* DNA from water concentrates and for relief of PCR inhibitors. *Appl. Environ. Microbiol.*, **71**(3), 1135–41.

Jiang, S. (2006). Human adenoviruses in water: occurrence and health implications: a critical review. *Environ. Sci. Technol.*, **40**(23), 7132–40.

Jiang, S. and Chu, W. (2004). PCR detection of pathogenic viruses in southern California urban rivers. *J. Appl. Microbiol.*, **97**(1), 17–28.

Jiang, X. and Pickering, L.K. (2002). Update on caliciviruses and human acute gastroenteritis. *Pediatr. Infect. Dis. J.*, **21**(11), 1069–70.

Johansson, P., Torven, M., Hammarlund, A.C., Bjorne, U., Hedlund, K.O. and Svensson, L. (2002). Food-borne outbreak of gastroenteritis associated with genogroup I calicivirus. *J. Clin. Microbiol.*, **40**(3), 794–8.

John, D., Haas, C.N., Nwachuku, N. and Gerba, C.P. (2005). Chlorine and ozone disinfection of *Encephalitozoon intestinalis* spores. *Water Res.* in press.

John, D., Nwachuku, N., Pepper, I.L. and Gerba, C.P. (2003). Development and optimization of a quantitative cell culture infectivity assay for the microsporidium *Encephalitozoon intestinalis* and application to ultraviolet light inactivation. *J. Microbiol. Methods*, **52**(2), 183–96.

Johnson, A., Linden, K., Ciociola, K.M., De Leon, R., Widmer, G. and Rochelle, P.A. (2005). UV inactivation of *Cryptosporidium hominis* as measured in cell culture. *Appl. Environ. Microbiol.*, **71**(5), 2800–2.

Johnson, C., Marshall, M.M., DeMaria, L.A., Moffet, J.M. and Korich, D.G. (2003). Chlorine inactivation of spores of *Encephalitozoon* spp. *Appl. Environ. Microbiol.*, **69**(2), 1325–6.

Johnson, C.H., Rice, E.W. and Reasoner, D.J. (1997). Inactivation of *Helicobacter pylori* by chlorination. *Appl. Environ. Microbiol.*, **63**, 4969–4970.

Johnson, D., Pieniazek, N.J., Griffin, D.W., Misener, L. and Rose, J.B. (1995). Development of a PCR protocol for sensitive detection of *Cryptosporidium* oocysts in water samples. *Appl. Environ. Microbiol.*, **61**(11), 3849–55.

Johnson, J., Thomas, J.E., Graham, T.A., Townshend, I., Byrne, J., Selinger, L.B. and Gannon, V.P. (2003). Prevalence of *Escherichia coli O157:H7* and *Salmonella* spp. in surface waters of southern Alberta and its relation to manure sources. *Can. J. Microbiol.*, **49**(5), 326–35.

Joly, P., Falconnet, P.A., André, J., Weill, N., Reyrolle, M., Vandenesch, F., Maurin, M., Etienne, J. and Jarraud, S. (2006). Quantitative real-time *Legionella* PCR for environmental water samples: data interpretation. *Appl. Environ. Microbiol.*, **72**(4), 2801–8.

Jones, D., Sutcliffe, E.M. and Curry, A. (1991). Recovery of viable but non-culturable *Campylobacter* jejuni. *J. Gen. Microbiol.*, **137**(10), 2477–82.

Jones, I. and Roworth, M. (1996). An outbreak of *Escherichia coli O157* and campylobacteriosis associated with contamination of a drinking water supply. *Public Health*, **110**(5), 277–82.

Jones, K. (2001). Campylobacters in water, sewage and environment. *J. Appl. Microbiol.*, **90**, 68S–79S.

Jones, K., Betaieb, M. and Telford, D.R. (1990). Seasonal variation of thermophilic campylobacters in sewage sludge. *J. Appl. Bacteriol.*, **69**(2), 185–9.

Jothikumar, N. and Cliver, D. *et al.* (1998). Immunomagnetic capture PCR for rapid concentration and detection of hepatitis A virus from environmental samples. *Appl. Environ. Microbiol.*, **64**(2), 504–8.

Jothikumar, N., Cromeans, T.L., Robertson, B.H., Meng, X.J. and Hill, V.R. (2006). A broadly reactive one-step real-time RT-PCR assay for rapid and sensitive detection of hepatitis E virus. *J. Virol. Methods*, **131**(1), 65–71.

Jothikumar, N. and Griffiths, M.W. (2002). Rapid detection of *Escherichia coli O157:H7* with multiplex real-time PCR assays. *Appl. Environ. Microbiol.*, **68**(6), 3169–3171.

Jothikumar, N.A.K., Kamatchiammal, S., Paulmurugan, R., Saravanadevi, S. and Khanna, P. (1993). Detection of hepatitis E virus in raw and treated wastewater with the polymerase chain reaction. *Appl. Environ. Microbiol.*, **59**(8), 2558–62.

Jothikumar, N., Kang, G. and Hill, V.R. (2009). Broadly reactive TaqMan(R) assay for real-time RT-PCR detection of rotavirus in clinical and environmental samples. *J. Virol. Methods*, **155**(2), 126–31.

Jothikumar, N.P.R., Padmanabhan, P., Sundari, R.B., Kamatchiammal, S. and Rao, K.S. (2000). Duplex RT-PCR for simultaneous detection of hepatitis A and hepatitis E virus isolated from drinking water samples. *J. Environ. Monit.*, **2**(6), 587–90.

Jothikumar, N.K.P., Paulmurugan, R., Kamatchiammal, S. and Padmanabhan, P. (1995). A simple device for the concentration and detection of enterovirus, hepatitis E virus and rotavirus from water samples by reverse transcription-polymerase chain reaction. *J. Virol. Methods*, **55**(3), 401–15.

June, G., Sherrod, P.S., Amaguana, R.M., Andrews, W.H. and Hammack, T.S. (1993). Effectiveness of the Bacteriological Analytical Manual culture method for the recovery of Shigella sonnei from selected foods. *J. AOAC*, **76**(6), 1240–8.

Kadoi, K. and Kadoi, B.K. (2001). Stability of feline caliciviruses in marine water maintained at different temperatures. *New Microbiol.*, **24**(1), 17–21.

Kaestli, M., Mayo, M., Harrington, G., Watt, F., Hill, J., Gal, D. and Currie, B.J. (2007). Sensitive and specific molecular detection of *Burkholderia pseudomallei*, the causative agent of melioidosis, in the soil of tropical northern Australia. *Appl. Environ. Microbiol.*, **73**(21), 6891–7.

Kamma, S., Tang, L., Leung, K., Ashton, E., Newman, N. and Suresh, M.R. (2008). A rapid two dot filter assay for the detection of *E. coli O157* in water samples. *J. Immunol. Methods*, **336**(2), 159–65.

Kang, C.D., Cao, C., Lee, J., Choi, I.S., Kim, B.W. and Sim, S.J. (2008). Surface plasmon resonance-based inhibition assay for real-time detection of *Cryptosporidium parvum* oocyst. *Water Res.*, **42**(6–7), 1693–9.

Kapikian, A.Z., Wyatt, R.G., Dolin, R., Thornhill, T.S., Kalica, A.R. and Channock, R.M. (1972). Visualization by immune electron microscopy of a 27 nm particle associated with acute infectious nonbacterial gastroenteritis. *Journal of Virology*, **10**, 1075–1081.

Kaplan, J., Feldman, R., Campbell, D.S., Lookabaugh, C. and Gary, G.W. (1982a). The frequency of a Norwalk-like pattern of illness in outbreaks of acute gastroenteritis. *Am. J. Public Health*, **72**(12), 1329–32.

Kaplan, J., Goodman, R.A., Schonberger, L.B., Lippy, E.C. and Gary, G.W. (1982b). Gastroenteritis due to Norwalk virus: an outbreak associated with a municipal water system. *J. Infect. Dis.*, **146**(2), 190–7.

Kapperud, G., Nesbakken, T. and Dommarsnes, K. (1991). Application of restriction endonuclease analysis and genetic probes in the epidemiology of Yersinia enterocolitica infection. *Contrib Microbiol. Immunol.*, **12**, 68–74.

Kapperud, G. and Vardund, T. (1995). Detection of pathogenic *Yersinia enterocolitica* in food, water, and feces by nested polymerase chain reactions and immunomagnetic separation. *Contrib Microbiol. Immunol.*, **13**(130–3).

Kappus, K., Marks, J.S., Holman, R.C., Bryant, J.K., Baker, C., Gary, G.W. and Greenberg, H.B. (1982). An outbreak of Norwalk gastroenteritis associated with swimming in a pool and secondary person-to-person transmission. *Am. J. Epidemiol.*, **116**(5), 834–9.

Karanis, P., Kourenti, C. and Smith, H. (2007a). Waterborne transmission of protozoan parasites: a worldwide review of outbreaks and lessons learnt. *J. Water Health*, **5**(1), 1–38.

Karanis, P., Thekisoe, O., Kiouptsi, K., Ongerth, J., Igarashi, I. and Inoue, N. (2007b). Development and preliminary evaluation of a loop-mediated isothermal amplification procedure for sensitive detection of cryptosporidium oocysts in fecal and water samples. *Appl. Environ. Microbiol.*, **73**(17), 5660–2.

Karapinar, M. and Gonul, S.A. (1991). Survival of *Yersinia enterocolitica* and *Escherichia coli* in spring water. *Int. J. Food Microbiol.*, **18**(4), 315–9.

Karita, M., Teramukai, S. and Matsumoto, S. (2003). Risk of *Helicobacter pylori* transmission from drinking well water is higher than that from infected intrafamilial members in Japan. *Dig. Dis. Sci.*, **48**(6), 1062–7.

Katayama, H., Shimasaki,A. and Ohgaki, S. (2002). Development of a virus concentration method and its application to detection of enterovirus and Norwalk virus from coastal seawater. *Appl. Environ. Microbiol.*, **68**, 1033–1039.

Kaucner, C. and Stinear, T. (1998). Sensitive and rapid detection of viable Giardia cysts and Cryptosporidium parvum oocysts in large-volume water samples with wound fiberglass cartridge filters and reverse transcription-PCR. *Appl. Environ. Microbiol.*, **64**(5), 1743–9.

Kauffmann, F. (1966). The bacteriology of *Enterobacteriaceae*. Munksgaard, Copenhagen, Denmark

Keegan, A., Fanok, S., Monis, P.T. and Saint, C.P. (2003). Cell culture-Taqman PCR assay for evaluation of *Cryptosporidium parvum* disinfection. *Appl. Environ. Microbiol.*, **69**(5), 2505–11.

Keegan, A., Daminato, D., Saint, C.P. and Monis, P.T. (2008). Effect of water treatment processes on *Cryptosporidium* infectivity. *Water Res.*, **42**(6–7), 1805–11.

Keene, W., McAnulty, J.M., Hoesly, F.C., Williams, L.P. Jr., Hedberg, K., Oxman, G.L., Barrett, T.J., Pfaller, M.A. and Fleming, D.W. (1994). A swimming-associated outbreak of hemorrhagic colitis caused by *Escherichia coli O157:H7* and Shigella sonnei. *N. Engl. J. Med.*, **331**(9), 579–84.

Kerr, M., Fitzgerald, M., Sheridan, J.J., McDowell, D.A. and Blair, I.S. (1999). Survival of *Escherichia coli* O157:H7 in bottled natural mineral water. *J. Appl. Microbiol.*, **87**, 833–841.

Keswick, B., Gerba, C.P., DuPont, H.L. and Rose, J.B. (1984). Detection of enteric viruses in treated drinking water. *Appl. Environ. Microbiol.*, **47**(6), 1290–4.

Keswick, B., Gerba, C.P. and Goyal, S.M. (1981). Occurrence of enteroviruses in community swimming pools. *Am. J. Public Health*, **71**(9), 1026–30.

Keswick, B.H., Satterwhite, T.K., Johnson, P.C., Du Point, H.L., Secor, S.L.,Bitsura, J.A., Gary, G.W. and Hoff, J.C. (1985). Inactivation of Norwalk virus in drinking water by chlorine. *Appl. Environ. Microbiol.*, **50**, 261–264.

Khalil, K., Lindblom, G.B., Mazhar, K. and Kaijser, B. (1994). Flies and water as reservoirs for bacterial enteropathogens in urban and rural areas in and around Lahore, Pakistan. *Epidemiol. Infect.*, **113**(3), 435–44.

Khan, A. and Cerniglia, C.E. (1997). Rapid and sensitive method for the detection of *Aeromonas* caviae and Aeromonas trota by polymerase chain reaction. *Lett. Appl. Microbiol.*, **24**(4), 233–9.

Khan, I.U. and Edge, T.A. (2007). Development of a novel triplex PCR assay for the detection and differentiation of thermophilic species of *Campylobacter* using 16S-23S rDNA internal transcribed spacer (ITS) region. *J. Appl. Microbiol.*, **103**(6), 2561–9.

Kilvington, S., Gray, T., Dart, J., Morlet, N., Beeching, J.R., Frazer, D.G. and Matheson, M. (2004). *Acanthamoeba keratitis*: the role of domestic tap water contamination in the United Kingdom. *Invest Ophthalmol. Vis. Sci.*, **45**(1), 165–9.

Kilvington, S. and Price, J. (1990). Survival of *Legionella pneumophila* within cysts of Acanthamoeba *polyphaga* following chlorine exposure. *J. Appl. Bacteriol.*, **68**, 519–525.

Kim-Farley, R., Rutherford, G., Lichfield, P., Hsu, S.T., Orenstein, W.A., Schonberger, L.B., Bart, K.J., Lui, K.J. and Lin, C.C. (1984). Outbreak of paralytic poliomyelitis, Taiwan. *Lancet*, **8**(2), 1322–4.

King, B.J., Keegan, A.R., Monis, P.T. and Saint, C.P. (2005). Environmental temperature controls *Cryptosporidium* oocyst metabolic rate and associated retention of infectivity. *Appl. Environ. Microbiol.*, **71**(7), 3848–57.

King, C., Shotts, E.B. Jr., Wooley, R.E. and Porter, K.G. (1988). Survival of coliforms and bacterial pathogens within protozoa during chlorination. *Appl. Environ. Microbiol.*, **54**(12), 3023–33.

King, B.J., Hoefel, D., Daminato, D.P., Fanok, S. and Monis, P.T. (2008). Solar UV reduces *Cryptosporidium parvum* oocyst infectivity in environmental waters. *J. Appl. Microbiol.*, **104**(5), 1311–23.

Kirk, R. and Rowe, M.T. (1994). A PCR assay for the detection of *Campylobacter jejuni* and *Campylobacter coli* in water. *Lett. Appl. Microbiol.*, **19**(5), 301–3.

Kirschke, D., Jones, T.F., Buckingham, S.C., Craig, A.S. and Schaffner, W. (2002). Outbreak of aseptic meningitis associated with echovirus 13. *Pediatr. Infect. Dis. J.*, **21**(11), 1034–8.

Kittigul, L. and Ekchaloemkiet, S. *et al.* (2005). An efficient virus concentration method and RT-nested PCR for detection of rotaviruses in environmental water samples. *J. Virol. Methods*, **124**(1–2), 117–22.

Knochel, S. (1991). Chlorine resistance of motile *Aeromonas* spp. *Water. Sci. Technol.*, **24**, 327–330.

Ko, G. and Cromeans, T. *et al.* (2003). Detection of infectious adenovirus in cell culture by mRNA reverse transcription-PCR. *Appl. Environ. Microbiol.*, **69**(12), 7377–84.

Ko, G., Jothikumar, N., Hill, V.R. and Sobsey, M.D. (2005). Rapid detection of infectious adenoviruses by mRNA real-time RT-PCR. *J. Virol. Methods*, **127**(2), 148–53.

Koenraad, P.M. F.J. and Rombouts, F.M. *et al.* (1997). Epidemiological aspects of thermophilic *Campylobacter* in water-related environments: a review. *Water Environ. Res.*, **69**, 52–63.

Koff, R. (1992). Clinical manifestations and diagnosis of hepatitis A virus infection. *Vaccine*, **10**(Suppl. 1), S15–7.

Kogure, K. and Ikemoto, E. (1997). Wide occurrence of enterohemorragic *Eschrichia coli O157* in natural freshwater environment. *Nippon Saikingaku Zasshi.*, **52**(3), 601–7.

Koide, M., Haranaga, S., Higa, F., Tateyama, M., Yamane, N. and Fujita, J. (2007). Comparative evaluation of Duopath *Legionella* lateral flow assay against the conventional culture method using *Legionella pneumophila* and *Legionella anisa* strains. *Jpn. J. Infect. Dis.*, **60**(4), 214–6.

Koide, M., Saito, A., Kusano, N. and Higa, F. (1993). Detection of *Legionella* spp. in cooling tower water by the polymerase chain reaction method. *Appl. Environ. Microbiol.*, **59**(6), 1943–6.

Kolling, G. and Matthews, K. (2001). Examination of recovery in vitro and in vivo of nonculturable *Escherichia coli O157:H7*. *Appl. Environ. Microbiol.*, **67**(9), 3928–3933.

Kong, R.Y.C. and Lee, S.K.Y. *et al.* (2002). Rapid detection of six types of bacterial pathogens in marine waters by multiplex PCR. *Water Research*, **36**, 2802–2812.

Konishi, K., Saito, N., Shoji, E., Takeda, H., Kato, M., Asaka, M. and Ooi, H.K. (2007). *Helicobacter pylori*: longer survival in deep ground water and sea water than in a nutrient-rich environment. *APMIS*, **115**(11), 1285–91.

Korhonen, L. and Martikainen, P.J. (1991a). Comparison of the survival of *Campylobacter jejuni* and *Campylobacter coli* in culturable form in surface water. *Can. J. Microbiol.*, **37**(7), 530–3.

Korhonen, L. and Martikainen, P.J. (1991b). Survival of *Escherichia coli* and *Campylobacter jejuni* in untreated and filtered lake water. *J. Appl. Bacteriol.*, **71**(4), 379–82.

Korol, S., Fortunato, M.S., Paz, M., Sanahuja, M.C., Lazaro, E., Santini, P. and D'Aquino, M. (1995). Water disinfection: comparative activities of ozone and chlorine on a wide spectrum of bacteria. *Rev. Argent. Microbiol.*, **27**(4), 175–83.

Kothary, M. and Babu, U. (2001). Infective dose of foodborne pathogens in volunteers: a review. *Journal of Food Science*, **21**(1).

Koskenniemi, K., Lyra, C., Rajaniemi-Wacklin, P., Jokela, J. and Sivonen, K. (2007). Quantitative real-time PCR detection of toxic *Nodularia* cyanobacteria in the Baltic Sea. *Appl. Environ. Microbiol.*, **73**(7), 2173–9.

Koujitani, E., Horisaka, T., Nomura, Y., Hara-Kudo, Y., Okatani, A.T., Iwata, T., Kumagai, S. and Hayashidani, H. (2006). Immuno-magnetic separation and agar layer methods for the isolation of freeze-injured *Yersinia enterocolitica O:8* from water. *J. Vet. Med. Sci.*, **68**(3), 195–9.

Krishnan, P., Fernandes, S., Savio, J., Ross, C.R., Pradeep, R., Choudhary, R., Shet, A.S. and Pais, P. (2008). Melioidosis. *J. Assoc. Physicians India*, **56**, 636–9.

Kuan, M. (1997). Detection and rapid differentiation of human enteroviruses following genomic amplification. *J. Clin. Microbiol.*, **35**(10), 2598–601.

Kühn, I., Allestam, G., Huys, G., Janssen, P., Kersters, K. and Krovacek, K. *et al.* (1997). Diversity, persistence, and virulence of *Aeromonas* strains isolated from drinking water distribution systems in Sweden. *Appl. Environ. Microbiol.*, **63**, 2708–15.

Kuiper, M., Wullings, B.A., Akkermans, A.D., Beumer, R.R. and van der Kooij, D. (2004). Intracellular proliferation of *Legionella pneumophila* in Hartmannella vermiformis in aquatic biofilms grown on plasticized polyvinyl chloride. *Appl. Environ. Microbiol.*, **70**(11), 6826–33.

Kuiper-Goodman, T. and Falconer, I. *et al.* (1999). Human health aspects. *Toxic cyanobacteria in water*. I. Chorus, Bartram J. London, E and F.N. Spon, pp. 113–153.

Kukkula, M. and Arstila, P. *et al.* (1997). Waterborne outbreak of viral gastroenteritis. *Scand J. Infect. Dis.*, **29**(4), 415–8.

Kukkula, M., Maunula, L., Silvennoinen, E. and von Bonsdorff, C.-H. (1999). Outbreak of viral gastroenteritis due to drinking water contaminated by Norwalk-like viruses. *J. Infect. Dis.*, **180**, 1771–1776.

Kumar, S., Balakrishna, K. and Batra, H.V. (2006). Detection of *Salmonella enterica* serovar Typhi (S. Typhi) by selective amplification of invA, viaB, fliC-d and prt genes by polymerase chain reaction in mutiplex format. *Lett. Appl. Microbiol.*, **42**(2), 149–54.

Kurmayer, R. and Kutzenberger, T. (2003). Application of real-time PCR for quantification of microcystin genotypes in a population of the toxic cyanobacterium Microcystis sp. *Appl. Environ. Microbiol.*, **69**(11), 6723–30.

Kurokawa, M., Ono, K., Ishiyama, S. and Rai, S.K. (2004). Evaluation of rapid diagnostic methods for pediatric viral diarrhea using samples collected in Nepal and Japan. *Nepal Med. Coll. J.*, **6**(2), 78–82.

Kusnetsov, J., Jousimies-Somer, H.R., Nevalainen, A.I. and Martikainen, P.J. (1994). Isolation of *Legionella* from water samples using various culture methods. *J. Appl. Bacteriol.*, **76**(2), 155–62.

Kusnetsov, J., Torvinen, E., Lehtola, M. and Miettinen, I.T. (2006). High concentrations of legionellae and mycobacteria in waste water systems of Finnish forest industry. In *EWGLI Lisbon 2006* (eds EWGLI). EWGLI, Lisbon.

Kusnetsov, J.M., Ruotsalainen, E., Seppälä, I., Kanervo-Nordström, A., Virolainen-Julkunen, A. and Lyytikäinen, O. (2008). A cluster of Pontiac fever among employees in the biological waste-water treatment plant of a chemical mill. In *European Working Group for Legionella Infections, 23rd meeting*, p. 21, Madrid.

Kuznetsov, V. and Timchenko, N.F. (1998). Interactions between bacteria of the genus *Yersinia* in a water environment. *Zh. Mikrobiol. Epidemiol. Immunobiol.*, **6**, 26–9.

Kuusi, M., Klemets, P., Miettinen, I., Laaksonen, I., Sarkkinen, H., Hanninen, M.L., Rautelin, H., Kela, E. and Nuorti, J.P. (2004). An outbreak of gastroenteritis from a non-chlorinated community water supply. *J. Epidemiol. Community Health*, **58**(4), 273–7.

Kwang, J., Littledike, E.T. and Keen, J.E. (1996). Use of the polymerase chain reaction for *Salmonella* detection. *Lett. Appl. Microbiol.*, **22**(1), 46–51.

Kwo, P., Schlauder, G.G., Carpenter, H.A., Murphy, P.J., Rosenblatt, J.E., Dawson, G.J., Mast, E.E., Krawczynski, K. and Balan, V. (1997). Acute hepatitis E by a new isolate acquired in the United States. *Mayo Clin. Proc.*, **72**(12), 1133–6.

Laconcha, I., Lopez-Molina, N., Rementeria, A., Audicana, A., Perales, I. and Garaizar, J. (1998). Phage typing combined with pulsed-field gel electrophoresis and random amplified polymorphic DNA increases discrimination in the epidemiological analysis of *Salmonella* enteritidis strains. *Int. J. Food Microbiol.*, **40**(1–2), 27–34.

Lagatolla, C., Dolzani, L., Tonin, E., Lavenia, A., Di Michele, M., Tommasini, T., Monti-Bragadin, C. (1996). PCR ribotyping for characterizing *Salmonella* isolates of different serotypes. *J. Clin. Microbiol.*, **34**(10), 2440–3.

Lalonde, L.F. and Gajadhar, A.A. (2008). Highly sensitive and specific PCR assay for reliable detection of *Cyclospora cayetanensis* oocysts. *Appl. Environ. Microbiol.*, **74**(14), 4354–8.

Lambert, J.R., Lin, S.K. and Aranda-Michel, J. (1995). Helicobacter pylori. *Scand. J. Gastroenterol.* **30**(Suppl. 208), 33–46.

Lambertini, E., Spencer, S.K., Bertz, P.D., Loge, F.J., Kieke, B.A. and Borchardt, M.A. (2008). Concentration of enteroviruses, adenoviruses, and noroviruses from drinking water by use of glass wool filters. *Appl. Environ. Microbiol.*, **74**(10), 2990–6.

Lampel, K., Keasler, S.P. and Hanes, D.E. (1996). Specific detection of *Salmonella enterica* serotype Enteritidis using the polymerase chain reaction. *Epidemiol. Infect.*, **116**(2), 137–45.

Landers, P., Kerr, K.G. and Rowbotham, T.J. *et al.* (2000). Survival and growth of *Burkholderia cepacia* within the free-living amoeba *Acanthamoeba polyphaga. Eur J. Clin. Microbiol. Infect. Dis.*, **19**, 121–123.

Langer, B. and Frosner, G.G. (1996). Relative importance of the enterically transmitted human hepatitis viruses type A and E as a cause of foreign travel associated hepatitis. *Arch. Virol. Suppl.*, **11**, 171–9.

La Rosa, G., Fontana, S., Di Grazia, A., Iaconelli, M., Pourshaban, M. and Muscillo, M. (2007). Molecular identification and genetic analysis of Norovirus genogroups I and II in water environments: comparative analysis of different reverse transcription-PCR assays. *Appl. Environ. Microbiol.*, **73**(13), 4152–61.

Lawrence, C., Reyrolle, M., Dubrou, S., Forey, F., Decludt, B., Goulvestre, C., Matsiota-Bernard, P., Etienne, J. and Nauciel, C. (1999). Single clonal origin of a high proportion of *Legionella pneumophila* serogroup 1 isolates from patients and the environment in the area of Paris, France, over a 10-year period. *J. Clin. Microbiol.*, **37**(8), 2652–5.

Lawton, L., Campbell, D.L., Beattie, K.A. and Codd, G.A. (1990). Use of a rapid bioluminescence assay for detecting cyanobacterial microcystin toxicity. *Lett. Appl. Microbiol.*, **11**(4), 205–7.

Laxer, M., Timblin, B.K. and Patel, R.J. (1991). DNA sequences for the specific detection of *Cryptosporidium parvum* by the polymerase chain reaction. *Am. J. Trop. Med. Hyg.*, **45**(6), 688–94.

LeChevallier, M.W., Di Giovanni, G.D., Clancy, J.L., Bukhari, Z., Bukhari, S., Rosen, J.S., Sobrinho, J. and Frey, M.M. (2003). Comparison of method 1623 and cell culture-PCR for detection of *Cryptosporidium* spp. in source waters. *Appl. Environ. Microbiol.*, **69**(2), 971–9.

LeChevallier, M., Evans, T.M., Seidler, R.J., Daily, O.P., Merrell, B.R. and Rollins, D.M. *et al.* (1982). *Aeromonas sobria* in chlorinated drinking water supplies. *Microbial. Ecology 1*, **982**(8), 325–3.

LeChevallier, M., Norton, C.D., Falkinha, J.O., Willimas, M.D., Taylor, R.H. and Cowan, H.E. (2001). Occurrence and control of *Mycobacterium avium* complex, AWWA Research Foundation and American Water Works Association, Denver, CO, pp. 1–115.

Leclerc, H. and Schwartzbrod, L. *et al.* (2002). Microbial agents associated with waterborne diseases. *Critical Reviews in Microbiology*, **28**(4), 371–409.

Leclerc, H., Schwartzbrod, L. and Dei-Cas, E. (2004). Microbial agents associated with waterborne diseases. *Microbial Waterborne Pathogens*. T. Cloete, J.B. Rose, L. Nel and T. Ford. London, UK, IWA Publishing.

Le Dantec, C., Duguet, J.P., Montiel, A., Dumoutier, N., Dubrou, S. and Vincent, V. (2002). Occurrence of mycobacteria in water treatment lines and in water distribution systems. *Appl. Environ. Microbiol.*, **68**(11), 5318–5325.

Lee, C. and Lee, S. *et al.* (2004). Use of cell culture-PCR assay based on combination of A549 and BGMK cell lines and molecular identification as a tool to monitor infectious adenoviruses and enteroviruses in river water. *Appl. Environ. Microbiol.*, **70**(11), 6695–705.

Lee, D.Y., Lauder, H., Cruwys, H., Falletta, P. and Beaudette, L.A. (2008). Development and application of an oligonucleotide microarray and real-time quantitative PCR for detection of wastewater bacterial pathogens. *Sci. Total Environ.*, **398**(1–3), 203–11.

Lee, H.K. and Jeong, Y.S. (2004). Comparison of total culturable virus assay and multiplex integrated cell culture-PCR for reliability of waterborne virus detection. *Appl. Environ. Microbiol.*, **70**(6), 3632–3636.

Lee, J., Zoh, K. and Ko, G. (2008). Inactivation and UV disinfection of murine norovirus with TiO2 under various environmental conditions. *Appl. Environ. Microbiol.*, **74**(7), 2111–7.

Lee, S. and Kim, S. (2002). Detection of infectious enteroviruses and adenoviruses in tap water in urban areas in Korea. *Water Res.*, **36**(1), 248–56.

Lee, S.H., Levy, D.A., Craun, G.F., Beach, M.J. and Calderon R.L. (2002). Surveillance for waterborne-disease outbreaks – United States, 1999–2000. *MMWR*, **51**(SS08), 1–28.

Lee, Y., Gomez, L.L., McAuliffe, I.T. and Tsang, V.C. (2004). Evaluation of *Cryptosporidium parvum* oocyst recovery efficiencies from various filtration cartridges by electrochemiluminescence assays. *Lett. Appl. Microbiol.*, **39**(2), 156–62

Lee, S.U., Joung, M., Yang, D.J., Park, S.H., Huh, S., Park, W.Y. and Yu, J.R. (2008). Pulsed-UV light inactivation of *Cryptosporidium parvum. Parasitol. Res.*, **102**(6), 1293–9.

Le Guyader, F. and Dubois, E. *et al.* (1994). Detection of hepatitis A virus, rotavirus, and enterovirus in naturally contaminated shellfish and sediment by reverse transcription-seminested PCR. *Appl. Environ. Microbiol.*, **60**(10), 3665–71.

Legnani, P., Leoni, E., Soppelsa, F. and Burigo, R. (1998). The occurrence of *Aeromonas* species in drinking water supplies of an area of the Dolomite Mountains, Italy. *J. Appl. Microbiol.*, **85**(2), 271–6.

Lehtola, M.J., Loades, C.J. and Keevil, C.W. (2005). Advantages of peptide nucleic acid oligonucleotides for sensitive site directed 16S rRNA fluorescence in situ hybridization (FISH) detection of *Campylobacter jejuni*, *Campylobacter coli* and *Campylobacter lari*. *J. Microbiol. Methods*, **62**(2), 211–9.

Lehtola, M.J., Pitkänen, T., Miebach, L. and Miettinen, I.T. (2006a). Survival of *Campylobacter jejuni* in potable water biofilms: a comparative study with different detection methods. *Water. Sci. Technol.*, **54**(3), 57–61.

Lehtola, M.J., Torvinen, E., Miettinen, I.T. and Keevil, C.W. (2006b). Fluorescence in situ hybridization using peptide nucleic acid probes for rapid detection of *Mycobacterium avium* subsp. *avium* and *Mycobacterium avium* subsp. *paratuberculosis* in potable-water biofilms. *Appl. Environ. Microbiol.*, **72**(1), 848–53.

Le Minor, L. and Popoff, M. (1987). Request for an Opinion. Designation of Salmonella enterica sp. nov., nom. rev., as the type and only species of the genus Salmonella. *Int. J. Syst. Bacteriol.*, **37**, 465–468.

Le Minor, L., Veron, M. and Popoff, M. (1982). The taxonomy of Salmonella. *Ann Microbiol. (Paris)*. **133**(2), 223–43.

Lemos, V, Graczyk, T.K., Alves, M., Lobo, M.L., Sousa, M.C., Antunes, F. and Matos, O. (2005). Identification and determination of the viability of Giardia lamblia cysts and Cryptosporidium parvum and Cryptosporidium hominis oocysts in human fecal and water supply samples by fluorescent in situ hybridization (FISH) and monoclonal antibodies. *Parasitol. Res.*, **98**(1), 48–53.

Lepeuple, A., Jovic, M. and de Roubin, M.R. (2004). Molecular typing of the *Legionella pneumophila* population isolated from several locations in a contaminated water network. *Water. Sci. Technol.*, **50**(1), 281–5.

Levi, K., Smedley, J. and Towner, K.J. (2003). Evaluation of a real-time PCR hybridization assay for rapid detection of *Legionella pneumophila* in hospital and environmental water samples. *Clin. Microbiol. Infect.*, **9**(7), 754–8.

Levy, D., Bens, M.S., Craun, G.F., Calderon, R.L. and Herwaldt, B.L. (1998). Surveillance for waterborne-disease outbreaks–United States, 1995–1996. *MMWR CDC Surveill Summ.*, **47**(5), 1–34.

Lew, A. and Desmarchelier, P.M. (1994). Detection of *Pseudomonas pseudomallei* by PCR and hybridization. *J. Clin. Microbiol.*, **32**(5), 1326–32.

Li, D., Craik, S.A., Smith, D.W. and Belosevic, M. (2008). Survival of *Giardia lamblia* trophozoites after exposure to UV light. *FEMS Microbiol. Lett.*, **278**(1), 56–61.

Li, J., Shi, X.Q., Chao, F.H., Wang, X.W., Zheng, J.L. and Song, N. (2004a). A study on detecting and identifying enteric pathogens with PCR. *Biomed. Environ. Sci.*, **17**(1), 109–20.

Li, J. and Wang, X. *et al.* (2002). Detection of enteroviruses and hepatitis A virus in water by consensus primer multiplex RT-PCR. *World J. Gastroenterol.*, **8**(4), 699–702.

Li, J. and Xin, Z. *et al.* (2004b). Mechanisms of inactivation of hepatitis A virus in water by chlorine dioxide. *Water Res.*, **38**(6), 1514–9.

Li, X., Palmer, R., Trout, J.M. and Fayer, R. (2003a). Infectivity of microsporidia spores stored in water at environmental temperatures. *J. Parasitol.*, **89**(1), 185–8.

Li, X., Tate, K.W., Dunbar, L.A., Huang, B. and Atwill, E.R. (2003b). Efficiency for recovering *Encephalitozoon intestinalis* spores from waters by centrifugation and immunofluorescence microscopy. *J. Eukaryot. Microbiol.* **50**(Suppl.), 579–80.

Liao, C. and Shollenberger, L.M. (2003). Survivability and long-term preservation of bacteria in water and in phosphate-buffered saline. *Lett. Appl. Microbiol.*, **37**(1), 45–50.

Lin, C., Hung, C.L., Hsu, S.C., Tsai, C.C. and Tsen, H.Y. (2004). An improved PCR primer pair based on 16S rDNA for the specific detection of *Salmonella* serovars in food samples. *J. Food Prot.*, **67**(7), 1335–43.

Linden, K.G., Thurston, J., Schaefer, R. and Malley, J.P. Jr. (2007). Enhanced UV inactivation of adenoviruses under polychromatic UV lamps. *Appl. Environ. Microbiol.*, **73**(23), 7571–4.

Lippy, E.C. and Erb, J. (1976). Gastrointestinal illness at Sewickley, Pa. *Journal AWWA*, **68**, 606–610.

Lisle, J.T., Broadway, S.C., Prescott, A.M., Pyle, B.H., Fricker, C. and McFeters, G.A. (1998). Effects of starvation on physiological activity and chlorine disinfection resistance in *Escherichia coli O157:H7. Appl. Environ. Microbiol.*, **64**, 4658–4662.

Liu, Y., Gilchrist, A., Zhang, J. and Li, X.F. (2008a). Detection of viable but nonculturable *Escherichia coli O157:H7* bacteria in drinking water and river water. *Appl. Environ. Microbiol.*, **74**(5), 1502–7.

Liu, Y.J., Zhang, C.M. and Wang, X.C. (2008b). Simultaneous detection of enteric bacteria from surface waters by QPCR in comparison with conventional bacterial indicators. *Environ. Monit Assess.*, **18**.

Lodder, W.J. and Vinjé, J. *et al.* (1999). Molecular detection of norwalk-like caliciviruses in sewage. *Appl. Environ. Microbiol.*, **65**(12), 5624–5627.

Lopman, B., Reacher, M.H., Van Duijnhoven, Y., Hanon, F.X., Brown, D. and Koopmans, M. (2003). Viral gastroenteritis outbreaks in Europe, 1995–2000. *Emerg. Infect. Dis.*, **9**(1), 90–6.

Lorenzo-Morales, J., Coronado-Alvarez, N., Martínez-Carretero, E., Maciver, S.K. and Valladares, B. (2007). Detection of four adenovirus serotypes within water-isolated strains of Acanthamoeba in the Canary Islands, Spain. *Am. J. Trop. Med. Hyg.*, **77**(4), 753–6.

Lorenzo-Morales, J., Ortega-Rivas, A., Foronda, P., Martinez, E. and Valladares B. (2005). Isolation and identification of pathogenic *Acanthamoeba* strains in Tenerife, Canary Islands, Spain from water sources. *Parasitol. Res.*, **95**(4), 273–7.

Loret, J.F., Robert, S., Saucedo, G., Catalan, V., Corsaro, D. and Greub, G. (2008a). Characterization of amoebae and intra-amoebal bacteria in drinking water, and identification of control strategies. In *Water Quality and Technology Conference*. American Water Works Association, Cincinnati.

Loret, J.F., Jousset, M., Robert, S., Anselme, C., Saucedo, G., Ribas, F., Martinez, L. and Catalan, V. (2008b). Elimination of free-living amoebae by drinking water treatment processes. *European Journal of Water Quality*, **39**(1), 37–50.

Loret, J.F., Jousset, M., Robert, S., Saucedo, G., Ribas, F., Thomas, V. and Greub, G. (2008c). Amoebae-resisting bacteria in drinking water: risk assessment and management. *Wat. Sci. Tech.*, **58**(3), 571–77.

Loret, J.F., Robert, S., Saucedo, G., Catalan, V., Corsaro, D. and Greub, G. (2008d). Characterization of amoebae and intra-amoebal bacteria in drinking water, and identification of control strategies. In *Water Quality and Technology Conference*. American Water Works Association, Cincinnati.

Lu, Y., Redlinger, T.E., Avitia, R., Galindo, A. and Goodman, K. (2002). Isolation and genotyping of Helicobacter pylori from untreated municipal wastewater. *Appl. Environ. Microbiol.*, **68**(3), 1436–9.

Lumb, R., Stapledon, R., Scroop, A., Bond, P., Cunliffe, D., Goodwin, A., Doyle, R. and Bastian, I. (2004). Investigation of spa pools associated with lung disorders caused by Mycobacterium avium complex in immunocompetent adults. *Appl. Environ. Microbiol.*, **70**(8), 4906–10.

Lyautey, E., Lacoste, B., Ten-Hage, L., Rols, J.L. and Garabetian, F. (2005). Analysis of bacterial diversity in river biofilms using 16S rDNA PCR-DGGE: methodological settings and fingerprints interpretation. *Water Res.*, **39**(2–3), 380–8.

MacGregor, R., Dreyer, K., Herman, S., Hocknell, P.K., Nghiem, L., Tevere, V.J. and Williams, A.L. (1999). Use of PCR in detection of *Mycobacterium avium* complex (MAC) bacteremia: sensitivity of the assay and effect of treatment for MAC infection on concentrations of human immunodeficiency virus in plasma. *J. Clin. Microbiol.*, **37**(1), 90–4.

MacKenzie, W., Schell, W.L., Blair, K.A., Addiss, D.G., Peterson, D.E., Hoxie, N.J., Kazmierczak, J.J. and Davis, J.P. (1995). Massive outbreak of waterborne *cryptosporidium* infection in Milwaukee, Wisconsin: recurrence of illness and risk of secondary transmission. *Clin. Infect. Dis.*, **21**(1), 57–62.

Mahbubani, M., Bej, A.K., Perlin, M.H., Schaefer, F.W. 3rd., Jakubowski, W. and Atlas, R.M. (1991). Detection of *Giardia* cysts by using the polymerase chain reaction and distinguishing live from dead cysts. *Appl. Environ. Microbiol.*, **57**(12), 3456–61.

Mahbubani, M., Bej, A.K., Perlin, M.H., Schaefer, F.W. 3rd., Jakubowski, W. and Atlas, R.M. (1992). Differentiation of *Giardia duodenalis* from other *Giardia* spp. by using polymerase chain reaction and gene probes. *J. Clin. Microbiol.*, **30**(1), 74–8.

Mahbubani, M., Schaefer, F.W. 3rd., Jones, D.D. and Bej, A.K. (1998). Detection of *Giardia* in environmental waters by immuno-PCR amplification methods. *Curr. Microbiol.*, **36**(2), 107–13.

Mahoney, F., Farley, T.A., Kelso, K.Y., Wilson, S.A., Horan, J.M. and McFarland, L.M. (1992). An outbreak of hepatitis A associated with swimming in a public pool. *J. Infect. Dis.*, **165**(4), 613–8.

Maiwald, M., Kissel, K., Srimuang, S., von Knebel Doeberitz, M. and Sonntag, H.G. (1994). Comparison of polymerase chain reaction and conventional culture for the detection of *legionellas* in hospital water samples. *J. Appl. Bacteriol.*, **73**(3), 216–25.

Majid, S., Reza, N.M., Mohammad, B., Jamileh, G. and Mahmood, J.T. (2007). Detection of *Legionella pneumophila* by PCR-ELISA method in industrial cooling tower water. *Pak. J. Biol. Sci.*, **10**(22), 4015–21.

Makary, P., Maunula, L., Niskanen, T., Kuusi, M., Virtanen, M., Pajunen, S., Ollgren, J. and Tran Minh, N.N. (2008). Multiple norovirus outbreaks among workplace canteen users in Finland, July 2006. *Epidemiol. Infect.*, **4**, 1–6.

Malorny, B., Paccassoni, E., Fach, P., Bunge, C., Martin, A. and Helmuth, R. (2004). Diagnostic real-time PCR for detection of *Salmonella* in food. *Appl. Environ. Microbiol.*, **70**(12), 7046–52.

Mansfield, L. and Gajadhar, A.A. (2004). Cyclospora cayetanensis, a food- and waterborne coccidian parasite. *Vet. Parasitol.*, **126**(1–2), 73–90.

Manz, W., Amann, R., Szewzyk, R., Szewzyk, U., Stenstrom, T.A., Hutzler, P. and Schleifer, K.H. (1995). In situ identification of Legionellaceae using 16S rRNA-targeted oligonucleotide probes and confocal laser scanning microscopy. *Microbiology*, **141**, 29–39.

Marciano-Cabral, F. and Cabral, G. (2003). *Acanthamoeba* spp. as agents of disease in humans. *Clin. Microbiol. Rev.*, **16**, 273–307.

Marranzano, M., Coniglio, M.A. and Mauro, L. (2003). Evaluation of pulsed-field gel electrophoresis (PFGE) for *Yersinia enterocolitica* molecular epidemiology investigations. *Adv. Exp. Med. Biol.*, **529**, 375–8.

Marshall, B. and Warren, J.R. (1984). Unidentified curved bacilli in the stomach of patients with gastritis and peptic ulceration. *Lancet.*, **16**(1), 1311–5.

Marshall, M., Naumovitz, D., Ortega, Y. and Sterling, C.R. (1997). Waterborne protozoan pathogens. *Clin. Microbiol. Rev.*, **10**(1), 67–85.

Martin S., Penttinen, P., Hedin, G., Ljungström, M., Allestam, G., Andersson, Y. and Giesecke, J. (2006). A case-cohort study to investigate concomitant waterborne outbreaks of Campylobacter and gastroenteritis in Söderhamn, Sweden, 2002–3. *J. Water Health*, **4**(4), 417–24.

Massa, S., Cesaroni, D., Poda, G. and Trovatelli, L.D. (1988). Isolation of Yersinia enterocolitica and related species from river water. *Zentralbl. Mikrobiol.*, **143**(8), 575–81.

Mast, E.E. and Krawczynski, K. (1996). Hepatitis E: an overview. *Annu. Rev. Med.*, **47**, 257–66.

Maule, A. (2000). Survival of verocytotoxigenic *Escherichia coli O157* in soil, water and on surfaces. *J. Appl. Microbiol.*, **88**, 71S–78S.

Maunula, L., Miettinen, I.T. and Von Bonsdorff, C.H. (2005). Norovirus outbreaks from drinking water. *Emerging Infectious Diseases*, **11**(11), 1716–21.

Maurer, A. and Sturchler, D. (2000). A waterborne outbreak of small round structured virus, *campylobacter* and *shigella* co-infections in La Neuveville, Switzerland, 1998. *Epidemiol. Infect.*, **125**(2), 325–32.

Mbithi, J. and Springthorpe, V. *et al.* (1990). Chemical disinfection of hepatitis A virus on environmental surfaces. *Appl. Environ. Microbiol.*, **56**(11), 3601–4.

Mbithi, J. and Springthorpe, V. *et al.* (1991). Effect of relative humidity and air temperature on survival of hepatitis A virus on environmental surfaces. *Appl. Environ. Microbiol.*, **57**(5), 1394–9.

McCoy, W. (2004). *Legionella. Microbial waterborne pathogens*. T. Cloete, J.B. Rose, L. Nel and T. Ford. London, UK, IWA Publishing, pp. 100–142.

McDaniels, A.E., Wymer, L., Rankin, C. and Haugland, R. (2005). Evaluation of quantitative real time PCR for the measurement of *Helicobacter pylori* at low concentrations in drinking water. *Water Res.*, **39**(19), 4808–16.

McElhiney, J., Drever, M., Lawton, L.A. and Porter, A.J. (2002). Rapid isolation of a single-chain antibody against the cyanobacterial toxin microcystin-LR by phage display and its use in the immunoaffinity concentration of microcystins from water. *Appl. Environ. Microbiol.*, **68**(11), 5288–95.

McElhiney, J. and Lawton, L.A. (2005). Detection of the cyanobacterial hepatotoxins microcystins. *Toxicol. Appl. Pharmacol.*, **203**(3), 219–30.

McKeown, I., Orr, P., Macdonald, S., Kabani, A., Brown, R., Coghlan, G., Dawood, M., Embil, J., Sargent, M., Smart, G. and Bernstein, C.N. (1999). *Helicobacter pylori* in the Canadian arctic: seroprevalence and detection in community water samples. *Am. J. Gastroenterol.*, **94**(7), 1823–9.

Mead, P. and Griffin, P.M. (1998). *Escherichia coli O157:H7. Lancet*, **352**, 1207–12.

Medema, G.J., Teunis, P.F.M., Havelaar, A.H. and Haas, C.N. (1996). Assessment of the dose-response relationship of *Campylobacter jejuni*. *International Journal of Food Microbiology*, **30**, 101–111.

Medema, G. and Schets, F. *et al.* (1998). Improved detection and vital staining of *Cryptosporidium* and *Giardia* with flow cytometry. *Wat. Res.*, **38**(12), 61–65.

Medema, G.E.A. (1991). Effectivity of chlorine dioxide to control *Aeromonas* in drinking-water distribution systems. *Water. Sci. Technol.*, **24**, 325–326.

Medema, G., Wullings, B., Roeleveld, P. and Van der Kooij, D. (2002). Risk assessment of Legionella and enteric pathogens in sewage treatment works. In *IWA-AWWA International Symposium on Waterborne Pathogens* (eds AWWA), Cascais/Lisbon, Portugal.

Mehnert, D. and Stewien, K.E. (1993a). Detection and distribution of rotavirus in raw sewage and creeks in São Paulo, Brazil. *Appl. Environ. Microbiol.*, **59**(1), 140–3.

Mehnert, D., Stewien, K.E., Harsi, C.M., Queiroz, A.P., Candeias, J.M. and Candeias, J.A. (1993b). Detection of rotavirus in sewage and creek water: efficiency of the concentration method. *Mem Inst Oswaldo Cruz.*, **92**(1), 97–100.

Melby, K., Gondrosen, B., Gregusson, S., Ribe, H. and Dahl, O.P. (1991). Waterborne campylobacteriosis in northern Norway. *Int. J. Food Microbiol.*, **12**(2–3), 151–6.

Melnick, J. (1990). Enteroviruses: plioviruses, coxsackieviruses, echoviruses and newer enteroviruses. *Virology*. B.N. F.a. D.M. Knipe. New York, Raven Press, pp. 549–605.

Mena, K.D., Gerba, C.P., Haas, C.N. and Rose, J.B. (2003). Risk assessment of waterborne coxsackievirus. *Journal AWWA*, **95**(7), 122–31.

Meng, X. (2000). Novel strains of hepatitis E virus identified from humans and other animal species:is hepatitis E a zoonosis? *J. Hepatol.*, **33**, 842–845.

Mercanoglu, B. and Griffiths, M.W. (2005). Combination of immunomagnetic separation with real-time PCR for rapid detection of *Salmonella* in milk, ground beef, and alfalfa sprouts. *J. Food Prot.*, **68**(3), 557–61.

Merino, S., Rubires, X., Knochel, S. and Tomas, J.M. (1995). Emerging pathogens: *Aeromonas* spp. *Int. J. Food Microbiol.*, **28**(2), 157–68.

Metcalf, J. and Codd, G.A. (2003). Analysis of cyanobacterial toxins by immunological methods. *Chem Res. Toxicol.*, **16**(2), 103–12.

Meyer, E. and Jarroll, E.L. (1980). Giardiasis. *Am. J. Epidemiol.*, **111**(1), 1–12.

MicroRisk. (2007). Quantitative Microbial Risk Assessment in the Water Safety Plan. In *Report for the European Commission under the Fifth Framework Programme, Theme 4 Energy, environment and sustainable development (contract EVK1CT200200123)* (eds G. Medema, J.F. Loret, T.A. Stenström and N.J. Ashbolt). Kiwa Water Research, Nieuwegein, The Netherlands.

Miettinen, I.T., Zacheus O., von Bonsdorff C.-H. and Vartiainen, T. (2001). Waterborne epidemics in Finland 1998–1999. *Water Sci. Technol.*, **43**, 67–71.

Minenoa, T. and Avery, M.A. (2003). Giardiasis: recent progress in chemotherapy and drug development. *Curr Pharm Des.*, **9**(11), 841–55.

Miyamoto, M., Yamaguchi, Y. and Sasatsu, M. (2000). Disinfectant effects of hot water, ultraviolet light, silver ions and chlorine on strains of *Legionella* and nontuberculous mycobacteria. *Microbios.*, **101**(398), 7–13.

Mizunoe, Y., Wai, S.N., Takade, A. and Yoshida, S. (1999). Restoration of culturability of starvation-stressed and low-temperature-stressed *Escherichia coli O157:H7* cells by using H2O2-degrading compounds. *Arch. Microbiol.*, **172**, 63–67.

Monceyron, C. and Grinde B. (1994). Detection of hepatitis A virus in clinical and environmental samples by immunomagnetic separation and PCR. *J. Virol. Methods*, **46**, 157–166.

Montemayor, M., Galofré, B., Ribas, F. and Lucena, F. (2007). Comparative study between two laser scanning cytometers and epifluorescence microscopy for the detection of *Cryptosporidium* oocysts in water. *Cytometry A.*, **71**(3), 163–9.

Moore, J., Caldwell, P. and Millar, B. (2001). Molecular detection of *Campylobacter* spp. in drinking, recreational and environmental water supplies. *Int. J. Hyg. Environ. Health 2001 Nov;204(2–3)*, **204**(2–3), 185–9.

Morace, G., Pisani, G., Divizia, M. and Pana, A. (1993). Detection of hepatitis A virus in concentrated river water by polymerase chain reaction. *Zentralbl. Hyg. Umweltmed.*, **193**(6), 521–7.

Moreno, Y., Botella, S., Alonso, J.L., Ferrus, M.A., Hernandez, M. and Hernandez, J. (2003a). Specific detection of *Arcobacter* and *Campylobacter* strains in water and sewage by PCR and fluorescent in situ hybridization. *Appl. Environ. Microbiol.*, **69**(2), 1181–6.

Moreno, Y. and Ferrus, M. *et al.* (2002). Comparison of 23S polymerase chain reaction-restriction fragment length polymorphism and amplified fragment length polymorphism techniques as typing systems for thermophilic campylobacters. *FEMS Microbiol. Lett.*, **211**(1), 97–103.

Moreno, Y., Ferrus, M.A., Alonso, J.L., Jimenez, A. and Hernandez, J. (2003b). Use of fluorescent in situ hybridization to evidence the presence of *Helicobacter pylori* in water. *Water Res.*, **37**(9), 2251–6.

Moreno, Y., Piqueres, P., Alonso, J.L., Jiménez, A., Gonzalez, A. and Ferrus, M.A. (2007). Survival and viability of *Helicobacter pylori* after inoculation into chlorinated drinking water. *Water Research*, **41**, 3490–96.

Morgan-Ryan, U., Fall, A., Ward, L.A., Hijjawi, N., Sulaiman, I., Fayer, R., Thompson, R.C., Olson, M., Lal, A. and Xiao, L. (2002). *Cryptosporidium hominis* n. sp. (Apicomplexa: Cryptosporidiidae) from Homo sapiens. *J. Eukaryot. Microbiol.*, **49**(6), 433–40.

Morin, N., Gong, Z. and Li, X.F. (2004). Reverse transcription-multiplex PCR assay for simultaneous detection of *Escherichia coli O157:H7*, *Vibrio cholerae O1*, and Salmonella Typhi. *Clin. Chem.*, **50**(11), 2037–44.

Morinigo, M., Borrego, J.J. and Romero, P. (1986). Comparative study of different methods for detection and enumeration of *Salmonella* spp. in natural waters. *J. Appl. Bacteriol.*, **61**(2), 169–76.

Moura, H., Ospina, M., Woolfitt, A.R., Barr, J.R. and Visvesvara, G.S. (2003). Analysis of four human microsporidian isolates by MALDI-TOF mass spectrometry. *J. Eukaryot. Microbiol.*, **50**(3), 156–63.

Moyer, N., Luccini, G.M., Holcomb, L.A., Hall, N.H. and Altwegg, M. (1992). Application of ribotyping for differentiating aeromonads isolated from clinical and environmental sources. *Appl. Environ. Microbiol.*, **58**(6), 1940–4.

Muder, R. and Yu, V.L. (2002). Infection due to *Legionella* species other than *L. pneumophila*. *Clin. Infect. Dis.*, **35**(8), 990–8.

Murga, R., Forster, T.S., Brown, E., Pruckler, J.M., Fields, B.S. and Donlan, R.M. (2001). Role of biofilms in the survival of *Legionella pneumophila* in a model potable-water system. *Microbiology*, **147**, 3121–6.

Myrmel, M., Rimstad, E. and Wasteson, Y. (2000). Immunomagnetic separation of a Norwalk-like virus (genogroup I) in artificially contaminated environmental water samples. *Int. J. Food Microbiol.*, **62**(1–2), 17–26.

Myrmel, M., Rimstad, E., Wasteson, Y. and Skjerve, E. (1999). Detection of small round structured viruses in artificially contaminated water using filter adsorption and reverse transcription polymerase chain reaction. *Int. J. Food Microbiol.*, **49**(1–2), 85–94.

Nail, S. and Aggarwal, R. *et al.* (1992). A large waterborne viral hepatitis E epidemic in Kanpur, India. *Bull World Health Organ.*, **70**, 597–604.

Nasser, A. and Metcalf, T.G. (1987). An A-ELISA to detect hepatitis A virus in estuarine samples. *Appl. Environ. Microbiol.*, **53**(5), 1192–5.

Nayak, A.K. and Rose, J.B. (2007). Detection of *Helicobacter pylori* in sewage and water using a new quantitative PCR method with SYBR green. *J. Appl. Microbiol.*, **103**(5), 1931–41.

Neilan, B. (1995). Identification and Phylogenetic Analysis of Toxigenic Cyanobacteria by Multiplex Randomly Amplified Polymorphic DNA PCR. *Appl. Environ. Microbiol.*, **61**(6), 2286–2291.

Neilan, B. and Jacobs, D. *et al.* (1995). Genetic diversity and phylogeny of toxic cyanobacteria determined by DNA polymorphisms within the phycocyanin locus. *Appl. Environ. Microbiol.*, **61**(11), 3875–3883.

Nel, L. and Markotter, W. *et al.* (2004a). Emerging infectious waterborne diseases: protozoan agents. *Microbial waterborne pathogens*. T. Cloete, J.B. Rose, L. Nel and T. Ford. London, UK, IWA Publishing.

Nel, L. and J. Weyer (2004b). Emerging infectious waterborne diseases: viral agents. *Microbial waterborne pathogens*. T. Cloete, J.B. Rose, L. Nel and T. Ford. London, UK, IWA Publishing.

Neubauer, H., Sauer, T., Becker, H., Aleksic, S. and Meyer, H. (1998). Comparison of systems for identification and differentiation of species within the genus *Yersinia*. *J. Clin. Microbiol.*, **36**(11), 3366–8.

Newcombe G. and Nicholson, B. (2002). Treatment options for the saxitoxin class of cyanotoxins *Water Supply*, **2**(5–6), 271–275.

Ngazoa, E.S., Fliss, I. and Jean, J. (2008). Quantitative study of persistence of human norovirus genome in water using TaqMan real-time RT-PCR. *J. Appl. Microbiol.*, **104**(3), 707–15.

Nichols, G.E.A. (1996). Health significance of bacteria in distribution systems-review of *Aeromonas*. London, UK Water Industry Research Ltd.

Nicholson, B.C., Rosinato, J. and Burch, MD (1994). Destruction of cyanobacterial peptide hepatotoxins by chlorine and chloramines. *Wat. Res.*, **28**, 1297–1303.

Nishiwaki-Matsushima, R., Ohta, T., Nishiwaki, S., Suganuma, M., Kohyama, K., Ishikawa, T., Carmichael, W.W. and Fujiki, H. (1992). Liver tumor promotion by the cyanobacterial cyclic peptide toxin microcystin-LR. *J. Cancer Res. Clin. Oncol.*, **118**(6), 420–4.

Niyogi, S. (2005). Shigellosis. *J. Microbiol.*, **43**(2), 133–43.

Norton, C., LeChevallier, M.W. and Falkinham, J.O. 3rd. (2004). Survival of *Mycobacterium avium* in a model distribution system. *Water Res.*, **38**(6), 1457–66.

Nuanualsuwan, S., Mariam, T., Himathongkham, S. and Cliver, D.O. (2002). Ultraviolet inactivation of feline calicivirus, human enteric viruses and coliphages. *Photochem. Photobiol.*, **76**(4), 406–10.

Numanovic, F., Hukic, M., Nurkic, M., Delibegovic, Z., Gegic M. and Tihic, N. (2004). Viruses in drinking water: HAV and enteroviruses. *Med. Arh.*, **58**(2), 105–8.

Nwachuku, N. and Gerba, C.P. (2004). Health effects of *Acanthamoeba* spp. and its potential for waterborne transmission. *Rev Environ. Contam Toxicol.*, **180**, 93–131.

Nwachuku, N., Gerba, C.P., Oswald, A. and Mashadi, F.D. (2005). Comparative inactivation of adenovirus serotypes by UV light disinfection. *Appl. Environ. Microbiol.*, **71**(9), 5633–6.

Nygard, K., Gondrosen, B. and Lund, V. (2003a). Water-borne disease outbreaks in Norway. *Tidsskr Nor Laegeforen.*, **123**(23), 3410–3.

Nygard, K., Torven, M., Ancker, C., Knauth, S.B., Hedlund, K.O., Giesecke, J., Andersson, Y. and Svensson, L. (2003b). Emerging genotype (GGIIb) of norovirus in drinking water, Sweden. *Emerg. Infect. Dis.*, **9**(12), 1548–52.

Oberste, M., Maher, K., Kilpatrick, D.R., Flemister, M.R., Brown, B.A. and Pallansch, M.A. (1999). Typing of human enteroviruses by partial sequencing of VP1. *J. Clin. Microbiol.*, **37**(5), 1288–93.

Ochiai, Y., Takada, C. and Hosaka, M. (2005). Detection and discrimination of *Cryptosporidium parvum* and *C. hominis* in water samples by immunomagnetic separation-PCR. *Appl. Environ. Microbiol.*, **71**(2), 898–903.

Olvera, M., Eguía, A., Rodríguez, O., Chong, E., Pillai, S.D. and Ilangovan, K. (2008). Inactivation of *Cryptosporidium parvum* oocysts in water using ultrasonic treatment. *Bioresour Technol.*, **99**(6), 2046–9.

Orlandi, P.A. and Lampel, K.A. (2000). Extraction-free, filter-based template preparation for rapid and sensitive PCR detection of pathogenic parasitic protozoa. *J. Clin. Microbiol.*, **38**, 2271–2277.

Orlandi, P., Carter, L., Brinker, A.M., da Silva, A.J., Chu, D.M., Lampel, K.A. and Monday, S.R. (2003). Targeting single-nucleotide polymorphisms in the 18S rRNA gene to differentiate *Cyclospora* species from Eimeria species by multiplex PCR. *Appl. Environ. Microbiol.*, **69**(8), 4806–13.

Ormen, O. and Ostensvik, O. (2001). The occurrence of aerolysin-positive *Aeromonas* spp. and their cytotoxicity in Norwegian water sources. *J. Appl. Microbiol.*, **90**(5), 797–802.

Ormen, O., Granum, P.E., Lassen, J. and Figueras, M.J. (2005). Lack of agreement between biochemical and genetic identification of *Aeromonas* spp. *APMIS*, **113**(3), 203–7.

Ortega-Rivas, A., Lorenzo-Morales, J., Martinez, E., Villa, M., Clavel, A., Valladares, B. and del Castillo, A. (2005). A specific primer pair for the diagnosis and identification of *Acanthamoeba astronyxis* by random amplified polymorphic DNA-polymerase chain reaction. *J. Parasitol.*, **91**(1), 122–6.

Orru, G., Masia, G., Orru, G., Romano, L., Piras, V. and Coppola, R.C. (2004). Detection and quantitation of hepatitis E virus in human faeces by real-time quantitative PCR. *J. Virol. Methods*, **118**(2), 77–82

Ouahid, Y., Perez-Silva, G. and Campo, F.F. (2005). Identification of potentially toxic environmental Microcystis by individual and multiple PCR amplification of specific microcystin synthetase gene regions. *Environ. Toxicol.*, **20**(3), 235–42.

Oyofo, B. and Rollins, D.M. (1993). Efficacy of filter types for detecting *Campylobacter jejuni* and *Campylobacter coli* in environmental water samples by polymerase chain reaction. *Appl. Environ. Microbiol.*, **59**(12), 4090–5.

Paintlia, A., Mahajan, R.C., Chakraborti, A., Sehgal, R. and Ganguly, N.K. (1999). Characterization of Giardia lamblia groups A and B from North India by isoenzyme and random amplified polymorphic DNA analysis. *Parasitol. Res.*, **85**(6), 510–2.

Palmer, C., Tsai, Y.L., Paszko-Kolva, C., Mayer, C. and Sangermano, L.R. (1993). Detection of *Legionella* species in sewage and ocean water by polymerase chain reaction, direct fluorescent-antibody, and plate culture methods. *Appl. Environ. Microbiol.*, **59**(11), 3618–24.

Pan, H., Song, L., Liu, Y. and Borner, T. (2002). Detection of hepatotoxic Microcystis strains by PCR with intact cells from both culture and environmental samples. *Arch. Microbiol.*, **178**(6), 421–7.

Papapetropoulou, M., Tsintzou, A. and Vantarakis, A. (1997). Environmental mycobacteria in bottled table waters in Greece. *Can. J. Microbiol.*, **43**(5), 499–502.

Papapetropoulou, M. and A. Vantarakis (1998). Detection of adenovirus outbreak at a municipal swimming pool by nested PCR amplification. *J. Infect.*, **36**(1), 101–3.

Parashar, U., Hummelman, E.G., Bresee, J.S., Miller, M.A. and Glass, R.I. (2003). Global illness and deaths caused by rotavirus disease in children. *Emerg. Infect. Dis.*, **9**(5), 565–72.

Park, S., Mackay, W.G. and Reid, D.C. (2001). *Helicobacter* sp. recovered from drinking water biofilm sampled from a water distribution system. *Water Res.*, **35**(6), 1624–6.

Parshionikar, S., Willian-True, S., Fout, G.S., Robbins, D.E., Seys, S.A., Cassady, J.D. and Harris, R. (2003). Waterborne outbreak of gastroenteritis associated with a norovirus. *Appl. Environ. Microbiol.*, **69**(9), 5263–8.

Patti, A.M., Santi, A.L., Fiore, L., Vellucci, L., De Stefano, D., Bellelli, E., Barbuti, S. and Fara, G.M. (2003). Environmental surveillance of poliovirus in Italy: pilot study. *Ann. Ig.*, **15**(2), 97–105.

Payment, P. (1981). Isolation of viruses from drinking water at the Point-Viau water treatment plant. *Can. J. Microbiol.*, **27**(4), 417–20.

Payment, P. and Tremblay, M. *et al.* (1985). Relative resistance to chlorine of poliovirus and Coxsackie virus isolates from environmental sources and drinking water. *Environ. Microbiol.*, **49**, 981.

Pearson, L.A. and Neilan, B.A. (2008). The molecular genetics of cyanobacterial toxicity as a basis for monitoring water quality and public health risk. *Curr. Opin. Biotechnol.*, **19**(3), 281–8.

Pelayo, L., Fraga, J., Nunez, F.A., Mendoza, D., Torres, D.R. and Finlay, C.M. (2003). Genetic characterization by random amplified polymorphic DNA analysis (RAPD) of 18 isolates of *Giardia lamblia* obtained from day care children. *Exp. Parasitol.*, **104**(3–4), 162–6.

Peng, X., Murphy, T. and Holden, N.M. (2008). Evaluation of the effect of temperature on the die-off rate for *Cryptosporidium parvum* oocysts in water, soils, and feces., *Appl. Environ. Microbiol.*, **74**(23), 7101–07.

Pereira, J.T., Costa, A.O., de Oliveira, Silva, M.B., Schuchard, W., Osaki, S.C., de Castro, E.A., Paulino, R.C. and Soccol, V.T. (2008). Comparing the efficacy of chlorine, chlorine dioxide, and ozone in the inactivation of *Cryptosporidium parvum* in water from Parana State, Southern Brazil. *Appl. Biochem. Biotechnol.*, **151**(2–3), 464–73.

Perelle, S., Cavellini, L., Burger, C., Blaise-Boisseau, S., Hennechart-Collette, C., Merle, G. and Fach, P. (2008). Use of a robotic RNA purification protocol based on the NucliSens((R)) easyMAGtrade mark for real-time RT-PCR detection of hepatitis A virus in bottled water. *J. Virol. Methods.*

Perola, O., Kauppinen, J., Kusnetsov, J., Karkkainen, U.M., Luck, P.C. and Katila, M.L. (2005). Persistent *Legionella pneumophila* colonization of a hospital water supply: efficacy of control methods and a molecular epidemiological analysis. *APMIS*, **113**(1), 45–53.

Peterson, D., Hurley, T.R., Hoff, J.C. and Wolfe, L.G. (1983). Effect of chlorine treatment on infectivity of hepatitis A virus. *Appl. Environ. Microbiol.*, **45**(1), 223–7.

Pianetti, A., Falcioni, T., Bruscolini, F., Sabatini, L., Sisti, E. and Papa, S. (2005). Determination of the viability of *Aeromonas hydrophila* in different types of water by flow cytometry, and comparison with classical methods. *Appl. Environ. Microbiol.*, **71**(12), 7948–54.

Pianetti, A., Manti, A., Boi, P., Citterio, B., Sabatini, L., Papa, S., Rocchi, M.B. and Bruscolini, F. (2008a). Determination of viability of *Aeromonas hydrophila* in increasing concentrations of sodium chloride at different temperatures by flow cytometry and plate count technique. *Int. J. Food Microbiol.*, **127**(3), 252–60.

Pianetti, A., Sabatini, L., Citterio, B., Sisti, E., Pierfelici, L. and Bruscolini, F. (2008b). Inactivation of *Legionella pneumophila* by combined systems of copper and silver ions and free chlorine. *Ig Sanit Pubbl.*, **64**(1), 27–40.

Pina, S., Buti, M., Jardi, R., Clemente-Casares, P., Jofre, J. and Girones, R. (2001). Genetic analysis of hepatitis A virus strains recovered from the environment and from patients with acute hepatitis. *J. Gen. Virol.*, **82**, 2955–63.

Pina, S. and Jofre, J. *et al.* (1998a). Characterization of a strain of infectious hepatitis E virus isolated from sewage in an area where hepatitis E is not endemic. *Appl. Environ. Microbiol.*, **64**(11), 4485–8.

Pina, S. and Puig, M. *et al.* (1998b). Viral pollution in the environment and in shellfish: human adenovirus detection by PCR as an index of human viruses. *Appl. Environ. Microbiol.*, **64**(9), 3376–82.

Piqueres, P., Moreno, Y., Alonso, J.L. and Ferrús, M.A. (2006). A combination of direct viable count and fluorescent in situ hybridization for estimating *Helicobacter pylori* cell viability. *Res. Microbiol.*, **157**(4), 345–9.

Piriou, P., Clinquart, S. and Laine, J.M. (2003). Persistence of Cryptosporidium parvum in drinking water distribution systems: pipe loop studies. Water Quality Technology Conference, Philadelphia, AWWA.

Polaczyk, A.L., Narayanan, J., Cromeans, T.L., Hahn, D., Roberts, J.M., Amburgey, J.E. and Hill, V.R. (2008). Ultrafiltration-based techniques for rapid and simultaneous concentration of multiple microbe classes from 100-L tap water samples. *J. Microbiol. Methods*, **73**(2), 92–9.

Pontrelli, G., Boccia, D., DI Renzi, M., Massari, M., Giugliano, F., Celentano, L.P., Taffon, S., Genovese D., DI Pasquale, S., Scalise, F., Rapicetta, M., Croci, L. and Salmaso, S. (2008). Epidemiological and virological characterization of a large community-wide outbreak of hepatitis A in southern Italy. *Epidemiol. Infect.*, **136**(8), 1027–34.

Popovitch, G. and Bondarenko, V. (1982). *Shigella* and *Salmonella* viability in river water at varying temperatures. *Biol. Abstr.*, **74**.

Porter, J., Mobbs, K., Hart, C.A., Saunders, J.R., Pickup, R.W. and Edwards, C. (1997). Detection, distribution and probable fate of Escherichia coli O157 from asymptomatic cattle on a dairy farm. *J. Appl. Microbiol.*, **83**(3), 297–306.

Powell, M.R. (2000). Dose-Response Envelope for *Escherichia coli O157:H7*. *Quantitative Microbiology*, **2**, 141–163.

Pozio, E., Rezza, G., Boschini, A., Pezzotti, P., Tamburrini, A., Rossi, P., Di Fine, M., Smacchia, C., Schiesari, A., Gattei, E., Zucconi, R. and Ballarini, P. (1997). Clinical cryptosporidiosis and human immunodeficiency virus (HIV)-induced immunosuppression: findings from a longitudinal study of HIV-positive and HIV-negative former injection drug users. *J. Infect. Dis.*, **176**(4), 969–75.

Preston, M., Johnson, W., Khakhria, R. and Borczyk, A. (2000). Epidemiologic subtyping of *Escherichia coli serogroup O157* strains isolated in Ontario by phage typing and pulsed-field gel electrophoresis. *J. Clin. Microbiol.*, **38**(6), 2366–8.

Puig, M. and Jofre, J. *et al.* (1994). Detection of adenoviruses and enteroviruses in polluted waters by nested PCR amplification. *Appl. Environ. Microbiol.*, **60**(8), 2963–70.

Pusch, D., Oh, D.Y., Wolf, S., Dumke, R., Schroter-Bobsin, U., Hohne, M., Roske, I. and Schreier, E. (2005). Detection of enteric viruses and bacterial indicators in German environmental waters. *Arch. Virol.*

Pyle, B., Broadaway, S.C. and McFeters, G.A. (1999). Sensitive detection of *Escherichia coli O157:H7* in food and water by immunomagnetic separation and solid-phase laser cytometry. *Appl. Environ. Microbiol.*, **65**(5), 1966–1972.

Pyle, B., Broadaway, S.C., McFeters, G.A. (1995). A rapid, direct method for enumerating respiring enterohemorrhagic *Escherichia coli* O157:H7 in water. *Appl. Environ. Microbiol.*, **61**(7), 2614–9.

Pyo, D. and Hahn, J.H. (2009). Determination of trace amount of cyanobacterial toxin in water by microchip based enzyme-linked immunosorbent assay. *J. Immunoassay Immunochem.*, **30**(1), 97–105.

Qiao, C., Zhang, H., Lai P., Gao, Z., Wang, L., Pu, J., Wu, D., Bai, Y., Gu, Q., Zhang, W. and Duan, X. (2008). Development and application of real-time quantitative RT-PCR assay for the detection of hepatitis E virus *Sheng Wu Gong Cheng Xue Bao*, **24**(5), 892–7.

Queralt, N., Bartolome, R. and Araujo, R. (2005). Detection of *Helicobacter pylori* DNA in human faeces and water with different levels of faecal pollution in the north-east of Spain. *J. Appl. Microbiol.*, **98**(4), 889–95.

Quignon, F., Sardin, M., Kiene, L. and Schwartzbrod, L. (1997). Poliovirus-1 inactivation and interaction with biofilm: a pilot-scale study. *Appl. Environ. Microbiol.*, **63**(3), 978–82.

Quintero-Betancourt, W., Peele, E.R. and Rose, J.B. (2002). *Cryptosporidium parvum* and *Cyclospora cayetanensis*: a review of laboratory methods for detection of these waterborne parasites. *J. Microbiol. Methods*, **49**(3), 209–24.

Rabold, J., Hoge, C.W., Shlim, D.R., Kefford, C., Rajah, R. and Echeverria, P. (1994). *Cyclospora* outbreak associated with chlorinated drinking water. *Lancet*, **344**(8933), 1360–1.

Radford, C., Minassian, D.C. and Dart, J.K. (2002). *Acanthamoeba keratitis* in England and Wales: incidence, outcome, and risk factors. *Br. J. Ophthalmol.*, **86**(5), 536–42.

Radhakrishna, M., Haseena, M., Nisha, K.V. and Maliya, P.S. (2003). Bacteriological study of bottled drinking water marketed in Mangalore. *J. Commun. Dis.*, **35**(2), 123–8.

Rahim, Z., Khan, S.I. and Chopra, A.K. (2004). Biological characterization of *Aeromonas* spp. isolated from the environment. *Epidemiol. Infect.*, **132**(4), 627–36.

Rajkowski, K. and Rice, E.W. (2001). Growth and recovery of selected gram-negative bacteria in reconditioned wastewater. *J. Food Prot.*, **64**(11), 1761–7.

Ramirez, N.E. and Sreevatsan, S. (2006). Development of a sensitive detection system for *Cryptosporidium* in environmental samples. *Vet. Parasitol.*, **136**(3–4), 201–13.

Rantala, A., Rizzi, E., Castiglioni, B., de Bellis, G. and Sivonen, K. (2008). Identification of hepatotoxin-producing cyanobacteria by DNA-chip. *Environ. Microbiol.*, **10**(3), 653–64.

Rapala, J., Lahti, K., Rasanen, L.A., Esala, A.L., Niemela, S.I. and Sivonen, K. (2002). Endotoxins associated with cyanobacteria and their removal during drinking water treatment. *Water Res.*, **36**(10), 2627–35.

Rapala, J., Niemelä, M., Berg, K.A., Lepistö, L. and Lahti, K. (2006). Removal of cyanobacteria, cyanotoxins, heterotrophic bacteria and endotoxins at an operating surface water treatment plant. *Water Sci. Technol.*, **54**(3), 23–8.

Raphael, R.A., Sattar, S.A. and Springthorpe, V.S. (1985). Long-term survival of human rotavirus in raw and treated river water. *Can. J. Microbiol.*, **31**, 124–128.

Rasmussen, J.P., Giglio, S., Monis, P.T., Campbell, R.J. and Saint, C.P. (2008). Development and field testing of a real-time PCR assay for cylindrospermopsin-producing cyanobacteria. *J. Appl. Microbiol.*, **104**(5), 1503–15.

Rautelin, H., Koota, K., von Essen, R., Jahkola, M., Siitonen, A. and Kosunen, T.U. (1990). Waterborne *Campylobacter jejuni* epidemic in a Finnish hospital for rheumatic diseases. *Scand J. Infect. Dis.*, **22**(3), 321–6.

Razzolini, M.T., Di Bari, M., Sanchez, P.S. and Sato, M.I. (2008). *Aeromonas* detection and their toxins from drinking water from reservoirs and drinking fountains. *J. Water Health*, **6**(1), 117–23.

Reetz, J., Rinder, H., Thomschke, A., Manke, H., Schwebs, M. and Bruderek, A. (2002). First detection of the microsporidium *Enterocytozoon bieneusi* in non-mammalian hosts (chickens). *Int. J. Parasitol.*, **32**, 785–787.

Regli, S., Rose, J.B., Haas, C.N. and Gerba, C.P. (1991). Modeling the risk from Giardia and viruses in drinking water. *Journal of the American Water Works Association*, **83**, 76–84.

Renter, D., Sargeant, J.M., Oberst, R.D. and Samadpour, M. (2003). Diversity, frequency, and persistence of *Escherichia coli O157* strains from range cattle environments. *Appl. Environ. Microbiol.*, **69**(1), 542–7.

Reynolds, K. (2004). Integrated cell culture/PCR for detection of enteric viruses in environmental samples. *Methods Mol. Biol.*, **268**, 69–78.

Rice, E.W., Clark, R.M. and Johnson, C.H. (1999). Chlorine inactivation of *Escherichia coli O157:H7*. *Emerg. Infect. Dis.*, **5**, 461–463.

Richardson, G., Thomas, D.R., Smith, R.M., Nehaul, L., Ribeiro, C.D., Brown, A.G. and Salmon, R.L. (2007). A community outbreak of *Campylobacter jejuni* infection from a chlorinated public water supply. *Epidemiol. Infect.*, **135**(7), 1151–8.

Rigsbee, W., Simpson, L.M. and Oliver, J.D. (1997). Detection of the viable but nonculturable state in *Escherichia coli O157:H7*. *J. Food Safety*, **16**, 225–262.

Riley, L., Remis, R., Helgerson, S., McGee, H., Wells, J., Davis, B., Hebert, R., Olcott, E., Johnson, L., Hargrett, N., Blake, P. and Cohen, M. (1983). Hemorrhagic colitis associated with a rare *Escherichia coli* serotype. *N. Engl. J. Med.*, **308**, 681–685.

Rimhanen-Finne, R., Horman, A., Ronkainen, P. and Hanninen, M.L. (2002). An IC-PCR method for detection of *Cryptosporidium* and *Giardia* in natural surface waters in Finland. *J. Microbiol. Methods*, **50**(3), 299–303.

Rinder, H., Thomschke, A., Dengjel, B., Gothe, R., Loscher, T. and Zahler, M. (2000). Close genotypic relationship between *Enterocytozoon bieneusi* from humans and pigs and first detection in cattle. *J. Parasitol.*, **86**(1), 185–8.

Risebro, H., de Franca Doria, M., Yip, H. and Hunter, P. (2006). Intestinal illness through drinking water in Europe. MICRORISK (contract n°EVK1-CT-2002–00123). Final report. (http://www.microrisk.com).

Riviere, D., Szczebara, F.M., Berjeaud, J.M., Frere, J. and Hechard, Y. (2005). Development of a real-time PCR assay for quantification of *Acanthamoeba* trophozoites and cysts. *J. Microbiol. Methods*, May 26 in press.

Robertson, L., Campbell, A.T. and Smith, H.V. (1992). Survival of *Cryptosporidium parvum* oocysts under various environmental pressures. *Appl. Environ. Microbiol.*, **58**(11), 3494–500.

Rochelle, P., De Leon, R., Stewart, M.H. and Wolfe, R.L. (1997a). Comparison of primers and optimization of PCR conditions for detection of *Cryptosporidium parvum* and *Giardia lamblia* in water. *Appl. Environ. Microbiol.*, **63**(1), 106–14.

Rochelle, P., Ferguson, D.M., Handojo, T.J., De Leon, R., Stewart, M.H. and Wolfe, R.L. (1997b). An assay combining cell culture with reverse transcriptase PCR to detect and determine the infectivity of waterborne *Cryptosporidium parvum*. *Appl. Environ. Microbiol.*, **63**(5), 2029–37.

Rodriguez-Lazaro, D., D'Agostino, M., Herrewegh, A., Pla, M., Cook, N. and Ikonomopoulos, J. (2005). Real-time PCR-based methods for detection of *Mycobacterium avium* subsp. *paratuberculosis* in water and milk. *Int. J. Food Microbiol.*, **101**(1), 93–104.

Rogers, J. and Keevil, C.W. (1992). Immunogold and fluorescein immunolabelling of *Legionella pneumophila* within an aquatic biofilm visualized by using episcopic differential interference contrast microscopy. *Appl. Environ. Microbiol.*, **58**(7), 2326–2330.

Rolle-Kampczyk, U., Fritz, G.J., Diez, U., Lehmann, I., Richter, M. and Herbarth, O. (2004). Well water–one source of *Helicobacter pylori* colonization. *Int. J. Hyg. Environ. Health*, **207**(4), 363–8.

Rollins, D. and Colwell R. (1986). Viable but nonculturable stage of *Campylobacter jejuni* and its role in survival in the natural aquatic environment. *Appl. Environ. Microbiol.*, **52**(3), 531–8.

Romano, G., Stampi, S., Zanetti, F., De Luca, G. and Tonelli, E. (1997). Occurrence of gram-negative bacteria in drinking water undergoing softening treatment. *Zentralbl. Hyg. Umweltmed.*, 2–3, 152–62.

Rondini, S., Mensah-Quainoo, E., Troll, H., Bodmer, T. and Pluschke, G. (2003). Development and application of real-time PCR assay for quantification of *Mycobacterium ulcerans* DNA. *J. Clin. Microbiol.*, **41**(9), 4231–7.

Ronner, A., Engvall, E.O., Andersson, L. and Kaijser, B. (2004). Species identification by genotyping and determination of antibiotic resistance in Campylobacter jejuni and *Campylobacter coli* from humans and chickens in Sweden. *Int. J. Food Microbiol.*, **96**(2), 173–9.

Rose, J.B. and Gerba, C.P. *et al.* (1991a). Survey of potable water supplies for *Cryptosporidium* and *Giardia*. *Environment Science and Technology*, **25**, 1393–1400.

Rose, J.B., Haas, C.N. and Regli, S. (1991b) Risk assessment and control of waterborne giardiasis. *American Journal of Public Health*, **81**, 709–713.

Rositano, J., Newcombe, G., Nicholson, B. and Sztajnbok, P. (2001). Ozonation of NOM and algal toxins in four treated waters. *Water Res.*, **35**(1), 23–32.

Roszak, D., Grimes, D.J. and Colwell, R.R. (1984). Viable but nonrecoverable stage of *Salmonella enteritidis* in aquatic systems. *Can. J. Microbiol.*, **30**(3), 334–8.

Roszak, D.B. and Colwell R.R. (1987). Survival strategies of bacteria in the natural environment. *Microbiol. Rev.*, **51**, 365–379.

Roy, D., Engelbrecht, R.S. and Chain, E.S.K. (1982). Comparative inactivation of six enteroviruses by ozone. *JAWWA*, **74**, 660–664.

Ruscoe, Q., Hill, S., Blackmore, T. and McLean, M. (2006). An outbreak of *Legionella pneumophila* suspected to be associated with spa pools on display at a retail store in New Zealand. *N. Z. Med. J.*, **119**(1243), U2253.

Rutjes, S.A., Italiaander, R., van den Berg, H.H., Lodder, W.J. and de Roda Husman, A.M. (2005). Isolation and detection of enterovirus RNA from large-volume water samples by using the NucliSens miniMAG system and real-time nucleic acid sequence-based amplification. *Appl. Environ. Microbiol.*, **71**(7), 3734–40.

Saeed, A., Abd, H., Edvinsson, B. and Sandström, G. (2009). *Acanthamoeba castellanii* an environmental host for *Shigella dysenteriae* and *Shigella sonnei*. *Arch. Microbiol.*, **191**(1), 83–8.

Said, B., Wright, F., Nichols, G.L., Reacher, M. and Rutter, M. (2003). Outbreaks of infectious disease associated with private drinking water supplies in England and Wales 1970–2000. *Epidemiol. Infect.*, **130**(3), 469–79.

Sailaja, B., Murhekar, M.V., Hutin, Y.J., Kuruva, S., Murthy, S.P., Reddy, K.S., Rao, G.M. and Gupte, M.D. (2009). Outbreak of waterborne hepatitis E in Hyderabad, India, 2005. *Epidemiol. Infect.*, **137**(2), 234–40.

Sails, A., Bolton, F.J., Fox, A.J., Wareing, D.R. and Greenway, D.L. (2002). Detection of *Campylobacter jejuni* and *Campylobacter coli* in environmental waters by PCR enzyme-linked immunosorbent assay. *Appl. Environ. Microbiol.*, **68**(3), 1319–24.

Sails, A., Bolton, F.J., Fox, A.J., Wareing, D.R. and Greenway, D.L. (1998). A reverse transcriptase polymerase chain reaction assay for the detection of thermophilic *Campylobacter* spp. *Mol. Cell. Probes.*, **12**(5), 317–22.

Sails, A., Fox, A.J., Bolton, F.J., Wareing, D.R. and Greenway, D.L. (2003). A real-time PCR assay for the detection of *Campylobacter jejuni* in foods after enrichment culture. *Appl. Environ. Microbiol.*, **69**(3), 1383–90.

Saint, C.P. (1998). A colony based confirmation assay for *Legionella* and *Legionella pneumophila* employing the EnviroAmp Legionella system and seroagglutination . *Lett. Appl. Microbiol.*, **26**(5), 377–81.

Saint, C.P. and Ho, L. (1999). A PCR test for the identification and discrimination of *Legionella longbeachae* serogroups 1 and 2. *J. Microbiol. Methods*, **37**(3), 245–53.

Sakai, H., Oguma, K., Katayama, H. and Ohgaki, S. (2007). Effects of low- or medium-pressure ultraviolet lamp irradiation on *Microcystis aeruginosa* and *Anabaena variabilis*. *Water Res.*, **41**(1), 11–8.

Sala Ferré, M.R., Arias, C., Oliva, J.M., Pedrol, A., García, M., Pellicer, T., Roura, P. and Domínguez, A. (2008). A community outbreak of Legionnaires' disease associated with a cooling tower in Vic and Gurb, Catalonia (Spain) in 2005. *Eur J. Clin. Microbiol. Infect. Dis.*

Sanchez, G., Populaire, S., Butot, S., Putallaz, T. and Joosten, H. (2006). Detection and differentiation of human hepatitis A strains by commercial quantitative real-time RT-PCR tests. *J. Virol. Methods*, **132**(1–2), 160–5

Sanchez, G. and Villena, C. *et al.* (2004). Hepatitis a virus: molecular detection and typing. *Methods Mol. Biol.*, **268**, 103–14.

Sandery, M., Stinear, T. and Kaucner, C. (1996). Detection of pathogenic *Yersinia enterocolitica* in environmental waters by PCR. *J. Appl. Bacteriol.*, **80**(3), 327–32.

Santo Domingo, J., Harmon, S. and Bennett, J. (2000). Survival of *Salmonella* species in river water. *Curr. Microbiol.*, **40**(6), 409–17.

Sarguna, P., Rao, A., Sudha Ramana, K.N. (2007). Outbreak of acute viral hepatitis due to hepatitis E virus in Hyderabad. *Indian J. Med. Microbiol.*, **25**(4), 378–82.

Sartorius, B., Andersson, Y., Velicko, I., De Jong, B., Löfdahl, M., Hedlund, K.O., Allestam, G., Wångsell, C., Bergstedt, O., Horal, P., Ulleryd, P. and Soderstrom, A. (2007). Outbreak of norovirus in Västra Götaland associated with recreational activities at two lakes during August 2004. *Scand J. Infect. Dis.*, **39**(4), 323–31.

Sautour, M., Mary, P., Chihib, N.E. and Hornez, J.P. (2003). The effects of temperature, water activity and pH on the growth of Aeromonas hydrophila and on its subsequent survival in microcosm water. *J. Appl. Microbiol.*, **95**(4), 807–13.

Savill, M., Hudson, J.A., Ball, A., Klena, J.D., Scholes, P., Whyte, R.J., McCormick, R.E. and Jankovic, D. (2001). Enumeration of *Campylobacter* in New Zealand recreational and drinking waters. *J. Appl. Microbiol.*, **91**(1), 38–46.

Schieman, D. (1990). *Yersinia enterocolitica* in drinking water. *Drinking Water Microbiology*. G. McFeters. New York, Springer-Verlag, pp. 322–40.

Schmitt, M. and D. Bitar (2005). An outbreak of legionnaires' disease in Lyon, France, April-May 2005: preliminary results. *Eurosurveillance weekly release: 26 May 2005*, **10**(5).

Schonberg-Norio, D., Takkinen, J., Hanninen, M.L., Katila, M.L., Kaukoranta, S.S., Mattila, L. and Rautelin, H. (2004). Swimming and *Campylobacter* infections. *Emerg. Infect. Dis.*, **10**(8), 1474–7.

Schroeder, J.M., Booton, G.C., Hay, J., Niszl, I.A., Seal, D.V. Markus, M.B., Fuerst, P.A. and Byers. T.J. (2001). Use of subgenic 18S ribosomal DNA PCR and sequencing for genus and genotype identification of acanthamoebae from humans with keratitis and from sewage sludge. *J. Clin. Microbiol.*, **39**, 1903–1911.

Schubert, R. (1991). Aeromonads and their significance as potential pathogens in water. *Soc Appl Bacteriol Symp Ser.*, **20**, 131S-135S.

Schulze-Robbecke, R., Janning, B. and Fischeder, R. (1992). Occurrence of mycobacteria in biofilm samples. *Tuber Lung Dis.*, **73**(3), 141–4.

Schuster, F. (2002). Cultivation of Pathogenic and Opportunistic Free-Living Amebas. *Clinical Microbiology Reviews*, **15**(3), 342–354.

Schvoerer, E. and Bonnet, F. *et al.* (2000). PCR detection of human enteric viruses in bathing areas, wastewaters and human stools in Southwestern France. *Res. Microbiol.*, **151**(8), 693–701.

Schvoerer, E., Bonnet, F., Dubois, V., Rogues, A.M., Gachie, J.P., Lafon, M.E. and Fleury, H.J. (1999). A hospital outbreak of gastroenteritis possibly related to the contamination of tap water by a small round structured virus. *J. Hosp. Infect.*, **43**(2), 149–54.

Schwab, K. and De Leon, R. *et al.* (1995). Concentration and purification of beef extract mock eluates from water samples for the detection of enteroviruses, hepatitis A virus, and Norwalk virus by reverse transcription-PCR. *Appl. Environ. Microbiol.*, **61**(2), 531–537.

Schwab, K., De Leon, R. and Sobsey, M.D. (1996). Immunoaffinity concentration and purification of waterborne enteric viruses for detection by reverse transcriptase PCR. *Appl. Environ. Microbiol.*, **62**(6), 2086–94.

Sellwood, J.A. and Wyn-Jones, P. (1995). A novel method for the detection of infectious rotavirus from water. *Water Science and Technology*, **31**(5–6), 367–370.

Selma, M.V., Beltrán, D., Allende, A., Chacón-Vera, E. and Gil, M.I. (2007). Elimination by ozone of *Shigella sonnei* in shredded lettuce and water. *Food Microbiol.*, **24**(5), 492–9.

Semenza, J.C. and Nichols, G. (2007). Cryptosporidiosis surveillance and water-borne outbreaks in Europe. *Euro Surveill.*, **12**(5), E13–4.

Sen, K. and Rodgers, M. (2004). Distribution of six virulence factors in *Aeromonas* species isolated from US drinking water utilities: a PCR identification. *J. Appl. Microbiol.*, **97**(5), 1077–86.

Sen K. (2005). Development of a rapid identification method for *Aeromonas* species by multiplex-PCR. *Can. J. Microbiol.*, **51**(11), 957–66.

Sen, K., Schable, N.A. and Lye, D.J. (2007). Development of an internal control for evaluation and standardization of a quantitative PCR assay for detection of *Helicobacter pylori* in drinking water. *Appl. Environ. Microbiol.*, **73**(22), 7380–7.

Senogles-Derham, P., Seawright, A., Shaw, G., Wickramisingh, W. and Shahin, M. (2003). Toxicological aspects of treatment to remove cyanobacterial toxins from drinking water determined using the heterozygous P53 transgenic mouse model. *Toxicon.*, **41**(8), 979–88.

September, S., Brozel, V.S. and Venter, S.N. (2004). Diversity of nontuberculoid Mycobacterium species in biofilms of urban and semiurban drinking water distribution systems. *Appl. Environ. Microbiol.*, **70**(12), 7571–3.

Seymour, I. and Appleton, H. (2001). Foodborne viruses and fresh produce. *J. Appl. Microbiol.*, **91**, 759–73.

Shahamat, M., Alavi, M., Watts, J.E., Gonzalez, J.M., Sowers, K.R., Maeder, D.W. and Robb, F.T. (2004). Development of two PCR-based techniques for detecting helical and coccoid forms of *Helicobacter pylori*. *J. Clin. Microbiol.*, **42**(8), 3613–9.

Sharma, K. and Arya, S.C. (1995). Detection of *Salmonella typhi* by nested PCR based on the ViaB sequence. *J. Clin. Microbiol.*, **33**(12), 3361.

Shaw, N.J., Villegas, L.F., Eldred, B.J., Gaynor, D.H., Warden, P.S. and Pepich, B.V. (2008). Modification to EPA Method 1623 to address a unique seasonal matrix effect encountered in some U.S. source waters. *J. Microbiol. Methods*, **75**(3), 445–8.

Shelton, D., Higgins, J.A., Van Kessel, J.A., Pachepsky, Y.A., Belt, K. and Karns, J.S. (2004). Estimation of viable *Escherichia coli O157* in surface waters using enrichment in conjunction with immunological detection. *J. Microbiol. Methods*, **58**(2), 223–31.

Sheng, J.W., He, M., Shi, H.C. and Qian, Y. (2006). A comprehensive immunoassay for the detection of microcystins in waters based on polyclonal antibodies. *Anal. Chim. Acta.*, **572**(2), 309–15.

Shieh, Y.C., Wong, C.I., Krantz, J.A. and Hsu, F.C. (2008). Detection of naturally occurring enteroviruses in waters using direct RT-PCR and integrated cell culture-RT-PCR. *J. Virol. Methods*, **149**(1), 184–9.

Shields, J. and Olson, B.H. (2003a). *Cyclospora cayetanensis*: a review of an emerging parasitic coccidian. *Int. J. Parasitol.*, **33**(4), 371–91.

Shields, J. and Olson, B.H. (2003b). PCR-restriction fragment length polymorphism method for detection of *Cyclospora cayetanensis* in environmental waters without microscopic confirmation. *Appl. Environ. Microbiol.*, **69**(8), 4662–9.

Shields, J.M., Hill, V.R., Arrowood, M.J. and Beach, M.J. (2008). Inactivation of *Cryptosporidium parvum* under chlorinated recreational water conditions. *J. Water Health*, **6**(4), 513–20.

Shin, G.A. and Sobsey, M.D. (2003). Reduction of Norwalk virus, poliovirus 1, and bacteriophage MS2 by ozone disinfection of water. *Appl. Environ. Microbiol.*, **69**, 3975–78.

Shin, G.A., Lee, J.K., Freeman, R. and Cangelosi, G.A. (2008). Inactivation of *Mycobacterium avium* complex by UV irradiation. *Appl. Environ. Microbiol.*, **74**(22), 7067–9.

Shin, G.A. and Sobsey, M.D. (2008). Inactivation of norovirus by chlorine disinfection of water. *Water Res.*, **42**(17), 4562–8.

Shin, J.H., Lee, H.K., Cho, E.J., Yu, J.Y. and Kang, Y.H. (2008). Targeting the rpoB gene using nested PCR-restriction fragment length polymorphism for identification of nontuberculous mycobacteria in hospital tap water. *J. Microbiol.*, **46**(6), 608–14.

Shoff, M.E., Rogerson, A., Kessler, K., Schatz, S. and Seal, D.V. (2008). Prevalence of Acanthamoeba and other naked amoebae in south Florida domestic water. *Journal of Water and Health*, **6**(1), 99–104.

Silvestry-Rodriguez, N., Bright, K.R., Uhlmann, D.R., Slack, D.C. and Gerba, C.P. (2007). Inactivation of *Pseudomonas aeruginosa* and *Aeromonas hydrophila* by silver in tap water. *J. Environ. Sci. Health Part A.*, **42**(11), 1579–84.

Simmons, G., Jury, S., Thornley, C., Harte, D., Mohiuddin, J. and Taylor, M. (2008). A Legionnaires' disease outbreak: a water blaster and roof-collected rainwater systems. *Water Res.*, **42**(6–7), 1449–58.

Sirikanchana, K., Shisler, J.L. and Mariñas, B.J. (2008). Inactivation kinetics of adenovirus serotype 2 with monochloramine. *Water Res.*, **42**(6–7), 1467–74.

Skraber, S., Gassilloud, B., Schwartzbrod, L. and Gantzer, C. (2004). Survival of infectious Poliovirus-1 in river water compared to the persistence of somatic coliphages, thermotolerant coliforms and Poliovirus-1 genome. *Water Res.*, **38**(12), 2927–33.

Slifko, T., Friedman, D., Rose, J.B. and Jakubowski, W. (1997). An in vitro method for detecting infectious *Cryptosporidium* oocysts with cell culture. *Appl. Environ. Microbiol.*, **63**(9), 3669–75.

Slifko, T., Smith, H.V. and Rose, J.B. (2000). Emerging parasite zoonoses associated with water and food. *Int. J. Parasitol.*, **30**(12–13), 1379–93.

Smith, A., Skilling, D.E., Cherry, N., Mead, J.H. and Matson, D.O. (1998). Calicivirus emergence from ocean reservoirs: zoonotic and interspecies movements. *Emerg. Infect. Dis.*, **4**(1), 13–20.

Smith, E. and Gerba, C.P. (1982). Development of a method for detection of human rotavirus in water and sewage. *Appl. Environ. Microbiol.*, **43**(6), 1440–50.

Smith, H., Paton, C.A., Mitambo, M.M. and Girdwood, R.W. (1997). Sporulation of *Cyclospora* sp. oocysts. *Appl. Environ. Microbiol.*, **63**(4), 1631–2.

Smith, J.J., Gunasekera, T.S., Barard,i, C.R., Veal, D. and Vesey, G. (2004). Determination of *Cryptosporidium parvum* oocyst viability by fluorescence in situ hybridization using a ribosomal RNA-directed probe. *J. Appl. Microbiol.*, **96**(2), 409–17.

Sobsey, M. and Glass, J.S. (1980). Poliovirus concentration from tap water with electropositive adsorbent filters. *Appl. Environ. Microbiol.*, **40**(2), 201–10.

Sökmen, M., Değerli, S. and Aslan, A. (2008). Photocatalytic disinfection of Giardia intestinalis and Acanthamoeba castellani cysts in water. *Exp Parasitol.*, **119**(1), 44–8.

Soler, L., Yanez, M.A., Chacon, M.R., Aguilera-Arreola, M.G., Catalan, V., Figueras, M.J. and Martinez-Murcia, A.J. (2004). Phylogenetic analysis of the genus *Aeromonas* based on two housekeeping genes. *Int. J. Syst. Evol. Microbiol.*, **54**(5), 1511–9.

Sorel, N., Guillot, E., Thellier, M., Accoceberry, I., Datry, A., Mesnard-Rouiller, L. and Miegeville, M. (2003). Development of an immunomagnetic separation-polymerase chain reaction (IMS-PCR) assay specific for *Enterocytozoon bieneusi* in water samples. *J. Appl. Microbiol.*, **94**(2), 273–9.

Sparfel, J., Sarfati, C., Liguory, O., Caroff, B., Dumoutier, N., Gueglio, B., Billaud, E., Raffi, F., Molina, J.M., Miegeville, M. and Derouin, F. (1997). Detection of microsporidia and identification of *Enterocytozoon bieneusi* in surface water by filtration followed by specific PCR. *J. Eukaryot.Microbiol.*, **44**(6), 78S.

Staes, C. and Schlenker, T. *et al.* (2000). Sources of infection among persons with acute hepatitis A and no identified risk factors during a sustained community-wide outbreak. *Pediatrics.*, **106**(4), E54.

Stanley, K., Cunningham, R. and Jones, K. (1998). Isolation of *Campylobacter jejuni* from groundwater. *J. Appl. Microbiol.*, **85**(1), 187–91.

Steele, M., McNab, B., Fruhner, L., DeGrandis, S., Woodward, D. and Odumeru, J.A. (1998). Epidemiological typing of *Campylobacter* isolates from meat processing plants by pulsed-field gel electrophoresis, fatty acid profile typing, serotyping, and biotyping. *Appl. Environ. Microbiol.*, **64**(7), 2346–9.

Stehr-Green, J., Nicholls, C., McEwan, S., Payne, A. and Mitchell, P. (1991). Waterborne outbreak of *Campylobacter jejuni* in Christchurch: the importance of a combined epidemiologic and microbiologic investigation. *N. Z. Med. J.*, **104**(918), 356–8.

Steinert, M., Birkness, K., White, E., Fields, B. and Quinn, F. (1998). *Mycobacterium avium* bacilli grow saprozoically in coculture with Acanthamoeba polyphaga and survive within cyst walls. *Appl. Environ. Microbiol.*, **64**, 2256–2261.

Steinert, M., Emody, L., Amann, R. and Hacker, J. (1997). Resuscitation of viable but nonculturable *Legionella pneumophila* Philadelphia JR32 by *Acanthamoeba castellanii*. *Appl. Environ. Microbiol.*, **63**(5), 2047–53.

Stender, H., Lund, K., Petersen, K.H., Rasmussen, O.F., Hongmanee, P., Miorner, H. and Godtfredsen, S.E. (1999). Fluorescence In situ hybridization assay using peptide nucleic acid probes for differentiation between tuberculous and nontuberculous mycobacterium species in smears of mycobacterium cultures. *J. Clin. Microbiol.*, **37**(9), 2760–5.

Sterling, C. and Ortega, Y. (1999). *Cyclospora*: an enigma worth unraveling. *Emerg. Infect. Dis.*, **5**, 48–53.

Stine, S., Vladich, F.D., Pepper, I.L. and Gerba, C.P. (2005). Development of a method for the concentration and recovery of microsporidia from tap water. *J. Environ. Sci. Health A Tox Hazard Subst. Environ. Eng.*, **40**(5), 913–25.

Stinear, T., Davies, J.K., Jenkin, G.A., Hayman, J.A., Oppedisano, F. and Johnson, P.D. (2000). Identification of *Mycobacterium ulcerans* in the environment from regions in Southeast Australia in which it is endemic with sequence capture-PCR. *Appl. Environ. Microbiol.*, **66**(8), 3206–13.

Stinear, T., Matusan, A., Hines, K. and Sandery, M. (1996). Detection of a single viable *Cryptosporidium parvum* oocyst in environmental water concentrates by reverse transcription-PCR. *Appl. Environ. Microbiol.*, **62**(9), 3385–90.

Stone, M. (1999). Transmission of *Helicobacter pylori*. *Postgrad. Med.*, **75**, 198.

Storey, M., Ashbolt, J. and Stenstrom, T.A. (2004a). Biofilms, thermophilic amoebae and *Legionella pneumophila*–a quantitative risk assessment for distributed water. *Water Sci. Technol.*, **50**(1), 77–82.

Storey, M., Langmark, J., Ashbolt, N.J. and Stenstrom, T.A. (2004b). The fate of legionellae within distribution pipe biofilms: measurement of their persistence, inactivation and detachment. *Water Sci. Technol.*, **49**(11–12), 269–75.

Strahl, E., Gillaspy, G.E. and Falkinham, J.O. 3rd. (2001). Fluorescent acid-fast microscopy for measuring phagocytosis of *Mycobacterium avium*, *Mycobacterium intracellulare*, and *Mycobacterium scrofulaceum* by *Tetrahymena pyriformis* and their intracellular growth. *Appl. Environ. Microbiol.*, **67**(10), 4432–9.

Straub, T., Daly, D.S., Wunshel, S., Rochelle, P.A., DeLeon, R. and Chandler, D.P. (2002). Genotyping *Cryptosporidium parvum* with an hsp70 single-nucleotide polymorphism microarray. *Appl. Environ. Microbiol.*, **68**(4), 1817–26.

Stroffolini, T. and Biagini, W. *et al.* (1990). An outbreak of hepatitis A in young adults in central Italy. *Eur. J. Epidemiol.*, **6**(2), 156–9.

Sturbaum, G.D., Ortega,Y.R., Gilman, R.H., Sterling, C.R., Cabrera, L. and Klein, D. (1998). Detection of *Cyclospora cayetanensis* in wastewater. *Appl. Environ. Microbiol.*, **64**, 2284–2286.

Sulaiman, I., Fayer, R., Lal, A.A., Trout, J.M., Schaefer, F.W. 3rd. and Xiao, L. (2003). Molecular characterization of microsporidia indicates that wild mammals Harbor host-adapted *Enterocytozoon* spp. as well as human-pathogenic Enterocytozoon bieneusi. *Appl. Environ. Microbiol.*, **69**(8), 4495–501.

Sulaiman, I., Jiang, J., Singh, A. and Xiao, L. (2004). Distribution of *Giardia* duodenalis genotypes and subgenotypes in raw urban wastewater in Milwaukee, Wisconsin. *Appl. Environ. Microbiol.*, **70**(6), 3776–80.

Surdeanu, M., Ciudin, L., Pencu, E. and Straut, M. (2003). Comparative study of three different DNA fingerprint techniques for molecular typing of *Shigella flexneri* strains isolated in Romania. *Eur. J. Epidemiol.*, **18**(7), 703–10.

Sutmoller, F., Azeredo, R.S., Lacerda, M.D., Barth, O.M., Pereira, H.G., Hoffer, E. and Schatzmayr, H.G. (1982). An outbreak of gastroenteritis caused by both rotavirus and *Shigella sonnei* in a private school in Rio de Janeiro. *Hyg (Lond).*, **88**(2), 285–93.

Swaddiwudhipong, W. and Kanlayanaphotporn, J. (2001). A common-source water-borne outbreak of multidrug-resistant typhoid fever in a rural Thai community. *J. Med. Assoc. Thai.*, **84**(11), 1513–7.

Swerdlow, D., Woodruff, B.A., Brady, R.C., Griffin, P.M., Tippen, S., Donnell, H.D. Jr., Geldreich, E., Payne, B.J., Meyer, A. Jr. and Wells, J.G. *et al.* (1992). A waterborne outbreak in Missouri of *Escherichia coli O157:H7* associated with bloody diarrhea and death. *Ann. Intern. Med.*, **117**(10), 812–9.

Szakal, D., Gado, I. and Pal, T. (2001). A colony blot immunoassay to detect enteroinvasive *Escherichia coli* and *Shigella* in water samples. *J. Appl. Microbiol.*, **90**(2), 229–36.

Szczuka, E. and Kaznowski, A. (2004). Typing of clinical and environmental *Aeromonas* sp. strains by random amplified polymorphic DNA PCR, repetitive extragenic palindromic PCR, and enterobacterial repetitive intergenic consensus sequence PCR. *J. Clin. Microbiol.*, **42**(1), 220–8.

Ta, A., Stout, J.E., Yu, V.L. and Wagener, M.M. (1995). Comparison of culture methods for monitoring *Legionella* species in hospital potable water systems and recommendations for standardization of such methods. *J. Clin. Microbiol.*, **33**(8), 2118–23.

Tacão, M., Moura, A., Alves, A., Henriques, I., Saavedra, M.J. and Correia, A. (2005). Evaluation of 16S rDNA- and gyrB-DGGE for typing members of the genus *Aeromonas*. *FEMS Microbiol. Lett.*, **246**(1), 11–8.

Taghi-Kilani, R., Gyurek, L.L., Millard, P.J., Finch, G.R. and Belosevic, M. (1996). Nucleic acid stains as indicators of *Giardia muris* viability following cyst inactivation. *Int. J. Parasitol.*, **26**(6), 637–46.

Taguchi, T., Shinozaki, Y., Takeyama, H., Haraguchi, S., Yoshino, M., Kaneko, M., Ishimori, Y. and Matsunaga, T. (2006). Direct counting of *Cryptosporidium parvum* oocysts using fluorescence in situ hybridization on a membrane filter. *J. Microbiol. Methods*, **67**(2), 373–80.

Takagi, M., Toriumi, H., Endo, T., Yamamoto, N. and Kuroki, T. (2008). An outbreak of cryptosporidiosis associated with swimming pools. *Kansenshogaku Zasshi.*, **82**(1), 14–9.

Talibart, R., Denis, M., Castillo, A., Cappelier, J.M. and Ermel, G. (2000). Survival and recovery of viable but noncultivable forms of *Campylobacter* in aqueous microcosm. *Int. J. Food Microbiol.*, **55**(1–3), 263–7.

Tanriverdi, S., Grinberg, A., Chalmers, R.M., Hunter, P.R., Petrovic, Z., Akiyoshi, D.E., London, E., Zhang, L., Tzipori, S., Tumwine, J.K. and Widmer, G. (2008). Inferences about the global population structures of Cryptosporidium parvum and Cryptosporidium hominis. *Appl. Environ. Microbiol.*, **74**(23), 7227–34.

Tansuphasiri, U., Vathanophas, K., Pariyanonda, A., Kittigul, L., Utrarachkij, F., Diraphat, P., Siripanichgon, K., Punchitton, S., Chitpirom, K. and Cheaochantanakij, N. (2000). Rapid detection of polioviruses in environmental water samples by one-step duplex RT-PCR. *Southeast Asian J. Trop Med. Public Health*, **31**(1), 47–56.

Tauxe, R.V., Hargrett-Bean, N., Patton, C.M. and Wachsmuth, I.K. (1988). Campylobacter isolates in the United States, 1982–1986. *MMWR CDC Surveill Summ.*, **37**(2), 1–13.

Taylor, J. and Gary, G.J. *et al.* (1981). Norwalk-related viral gastroenteritis due to contaminated drinking water. *Am. J. Epidemiol.*, **114**(4), 584–92.

Taylor, M., Cox, N., Vrey, M.A. and Grabow, W.O. (2001). The occurrence of hepatitis A and astroviruses in selected river and dam waters in South Africa. *Water Res.*, **35**(11), 2653–60.

Taylor, R., Falkinham, J.O. 3rd., Norton, C.D. and LeChevallier, M.W. (2000). Chlorine, chloramine, chlorine dioxide, and ozone susceptibility of *Mycobacterium avium*. *Appl. Environ. Microbiol.*, **66**(4), 1702–5.

Teh, B., Lin, J.T., Pan, W.H., Lin, S.H., Wang, L.Y., Lee, T.K. and Chen, C.J. (1994). Seroprevalence and associated risk factors of *Helicobacter pylori* infection in Taiwan. *Anticancer Res.*, **14**(3B), 1389–92.

Terzieva, S. and McFeters G.A. (1991). Survival and injury of *Escherichia coli*, *Campylobacter jejuni*, and *Yersinia enterocolitica* in stream water. *Can. J. Microbiol.*, **37**(10), 785–90.

Teunis, P.F.M., Chappell, C.L. and Okhuysen, P.C. (2002a). *Cryptosporidium* dose response studies: variation between isolates. *Risk analysis*, **22**(1), 175–183.

Teunis, P.F.M., Chappell, C.L. and Okhuysen, P.C. (2002b). *Cryptosporidium* dose response studies: variation between hosts. *Risk analysis*, **22**(3), 475–485.

Teunis, P.F.M. and Havelaar, A. (2000). The Beta Poisson dose-response model is not a single-hit model. *Risk Analysis*, **20**, 513–520.

Teunis, P.F.M., Nagelkerke, N.J.D. and Haas, C.N. (1999). Dose response models for infectious gastroenteritis. *Risk Analysis*, **19**, 1251–1260.

Teunis, P., Takumi, K. and Shinagawa, K. (2004). Dose response for infection by *Escherichia coli O157:H7* from outbreak data. *Risk Anal.*, **24**, 401–407.

Teunis, P.F.M., van den Brandhof, W., Nauta, M., Wagenaar, J., van den Kerhof, H. and van Pelt, W. (2005). A reconsideration of *Campylobacter* dose-response relation. *Epidemiol. Infect.*, **133**, 583–592.

Teunis, P.F.M., Moe, C.L., Liu, P., Miller, S.E., Lindesmith, L., Baric, R.S., Le Pendu, J. and Calderon, R.L. (2008). Norwalk virus: how infectious is it? *Journal of Medical Virology*, **80**, 1468–76.

Theron, J. and Cloete, T.E. (2002). Emerging waterborne infections: contributing factors, agents, and detection tools. *Crit. Rev. Microbiol.*, **28**(1), 1–26.

Theron, J., Morar, D., Du Preez, M., Brozel, V.S. and Venter, S.N. (2001). A sensitive seminested PCR method for the detection of *Shigella* in spiked environmental water samples. *Water Res.*, **35**(4), 869–74.

Thiriat, L., Sidaner, F. and Schwartzbrod, J. (1998). Determination of *Giardia* cyst viability in environmental and faecal samples by immunofluorescence, fluorogenic dye staining and differential interference contrast microscopy. *Lett. Appl. Microbiol.*, **26**, 237–242.

Thomas, C., Hill, D.J. and Mabey, M. (1999). Evaluation of the effect of temperature and nutrients on the survival of *Campylobacter* spp. in water microcosms. *J. Appl. Microbiol.*, **86**(6), 1024–32.

Thomas, P.D.P., R.C.G and Gazzard, B.G. (1999). Enteric viral infections as a cause of diarrhoea in the acquired immunodeficiency syndrome. *HIV Medicine*, **1**, 19–24.

Thomas, V.B.T., Nicolas, V., Robert, S., Loret, J.F. and Levi, Y. (2004). Amoebae in domestic water systems: resistance to disinfection treatments and implication in *Legionella* persistence. *J. Appl. Microbiol.*, **97**(5), 950–63.

Thomas, V., Loret, J.F., Jousset, M. and Greub, G. (2008). Biodiversity of amoebae and amoebae-resisting bacteria in a drinking water treatment plant. *Environmental Microbiology*, **10**(10), 2728–45.

Thompson, D.E., Rajal, V.B., De Batz, S. and Wuertz, S. (2006). Detection of *Salmonella* spp. in water using magnetic capture hybridization combined with PCR or real-time PCR. *J. Water Health*, **4**(1), 67–75.

Thompson, J. and Gravel, M.J. (1986). Family outbreak of gastroenteritis due to *Yersinia enterocolitica* serotype 0:3 from well water. *Can. J. Microbiol.*, **32**(8), 700–1.

Thompson, R. (2000). Giardiasis as a re-emerging infectious disease and its zoonotic potential. *Int. J. Parasitol.*, **30**(12–13), 1259–67.

Thomson, R., Carter, R., Gilpin, C., Coulter, C. and Hargreaves, M. (2008). Comparison of methods for processing drinking water samples for the isolation of *Mycobacterium avium* and *Mycobacterium intracellulare*. *Appl. Environ. Microbiol.*, **74**(10), 3094–8.

Thompson, R., Hopkins, R.M. and Homan, W.L. (2000). Nomenclature and genetic groupings of *Giardia* infecting mammals. *Parasitol Today*, **16**(5), 210–3.

Thurston-Enriquez, J., Haas, C.N., Jacangelo, J. and Gerba, C.P. (2003). Chlorine inactivation of adenovirus type 40 and feline calicivirus. *Appl. Environ. Microbiol.*, **69**(7), 3979–85.

Thurston-Enriquez, J.A., Haas, C.N., Jacangelo, J. and Gerba, C.P. (2005a). Inactivation of enteric adenovirus and feline calicivirus by chlorine dioxide. *Appl. Environ. Microbiol.*, **71**(6), 3100–5.

Thurston-Enriquez, J.A., Haas, C.N., Jacangelo, J. and Gerba, C.P. (2005b). Inactivation of enteric adenovirus and feline calicivirus by ozone. *Water Res.*, **39**(15), 3650–6.

Thurston-Enriquez, J., Watt, P., Dowd, S.E., Enriquez, R., Pepper, I.L. and Gerba, C.P. (2002). Detection of protozoan parasites and microsporidia in irrigation waters used for crop production. *J. Food Prot.*, **65**(2), 378–82.

Tillett, D., Parker, D.L. and Neilan, B.A. (2001). Detection of toxigenicity by a probe for the microcystin synthetase A gene (mcyA) of the cyanobacterial genus Microcystis: comparison of toxicities with 16S rRNA and phycocyanin operon (Phycocyanin Intergenic Spacer phylogenies. *Appl. Environ. Microbiol.*, **67**(6), 2810–8.

Tims, T. and Lim, D.V. (2003). Confirmation of viable *E. coli O157:H7* by enrichment and PCR after rapid biosensor detection. *J. Microbiol. Methods*, **55**(1), 141–7.

Tindall, B.J., Grimont, P.A., Garrity, G.M. and Euzeby, J.P. (2005). Nomenclature and taxonomy of the genus *Salmonella*. *Int. J. Syst. Evol. Microbiol.*, **55**(Pt 1): 521–4.

Tomaso, H., Pitt, T.L., Landt, O., Al Dahouk, S., Scholz, H.C., Reisinger, E.C., Sprague, L.D., Rathmann, I. and Neubauer, H. (2005). Rapid presumptive identification of *Burkholderia pseudomallei* with real-time PCR assays using fluorescent hybridization probes. *Mol. Cell. Probes.*, **19**(1), 9–20.

Tong, S., Yang, S., Lu, Z. and He, W. (1996). Laboratory investigation of ecological factors influencing the environmental presence of *Burkholderia pseudomallei*. *Microbiol. Immunol.*, **40**(6), 451–3.

Törökné, A., Vasdinnyei, R. and Asztalos, B.M. (2007). A rapid microbiotest for the detection of cyanobacterial toxins. *Environ. Toxicol.*, **22**(1), 64–8.

Torvinen, E., Lehtola, M.J., Martikainen, P.J. and Miettinen, I.T. (2007). Survival of *Mycobacterium avium* in drinking water biofilms as affected by water flow velocity, availability of phosphorus, and temperature. *Appl. Environ. Microbiol.*, **73**(19), 6201–7.

Touron, A., Berthe, T., Pawlak, B. and Petit, F. (2005). Detection of *Salmonella* in environmental water and sediment by a nested-multiplex polymerase chain reaction assay. *Res. Microbiol.*, **156**(4), 541–53.

Trakulsomboon, S., Dance, D.A., Smith, M.D., White, N.J. and Pitt, T.L. (1997). Ribotype differences between clinical and environmental isolates of *Burkholderia pseudomallei*. *J. Med. Microbiol.*, **46**(7), 565–70.

Troesch, A. and Nguyen, H. *et al.* (1999). *Mycobacterium* species identification and rifampin resistance testing with high-density DNA probe arrays. *J. Clin. Microbiol.*, **37**(1), 49–55.

Trujillo, A.A., McCaustland, K.A., Zheng, D.P., Hadley, L.A., Vaughn, G., Adams, S.M., Ando, T., Glass, R.I. and Monroe, S.S. (2006). TaqMan real-time reverse transcription-PCR for rapid detection, quantification, and typing of norovirus. *J. Clin. Microbiol.*, **44**(4), 1405–12.

Tsai, Y. and Tran, B. *et al.* (1994). Detection of poliovirus, Hepatitis A, and rotavirus from sewage and ocean Water by triplex reverse transcriptase PCR. *Appl. Environ. Microbiol.*, **60**(7), 2400–2407.

Tsvetkova, N., Schild, M., Panaiotov, S., Kurdova-Mintcheva, R., Gottstein, B., Walochnik, J., Aspock, H., Lucas, M.S. and Muller, N. (2004). The identification of free-living environmental isolates of amoebae from Bulgaria. *Parasitol. Res.*, **92**(5), 405–13.

Uhnoo, I., Wadell, G., Svensson, L. and Johansson, M.E. (1984). Importance of Enteric Adenoviruses 40 and 41 in acute gastroenteritis in infants and young Children. *J. Clin. Microbiol.*, **20**(3), 365–372.

Urakami, H., Ikarashi, K., Okamoto, K., Abe, Y., Ikarashi, T., Kono, T., Konagaya, Y. and Tanaka, N. (2007). Chlorine sensitivity of feline calicivirus, a norovirus surrogate. *Appl. Environ. Microbiol.*, **73**(17), 5679–82.

USEPA. (1998). Announcement of the drinking water contaminant candidate list. *Fed. Regist.*, **63**, 10273–10287.

USEPA. (1999a). Standard Method 1622: *Cryptosporidium* in water by filtration/IMS/FA EPA 821-R-99–001, US Environmental Protection Agency, Washington, DC.

USEPA. (1999b). Standard Method 1623: *Cryptosporidium* and *Giardia* in water by filtration/IMS/FA EPA 821-R-99–006, US Environmental Protection Agency, Washington, DC.

USEPA. (2005). Drinking Water Contaminant Candidate List 2. *submitted to Fed. Regist.*

Vaira, D., Ferron, P., Negrini, R., Cavazzini, L., Holton, J., Ainley, C., Londei, M., Vergura, M., Dei, R. and Colecchia, A. *et al.* (1992). Detection of *Helicobacter pylori*-like organisms in the stomach of some food-source animals using a monoclonal antibody. *Ital J. Gastroenterol.*, **24**(4), 181–4.

Van Asperen, I., Stijnen, C., Mank, T., De Boer, A., Groot, J., Medema, G., Ten Ham, P., Sluiters, J. and Borgdorff, M. (1996). An outbreak investigation of cryptosporidiosis. Report number 215700001. National Institute of Public Health and the Environment, Bilthoven, The Netherlands.

Van Belkum, A., Maas, H., Verbrugh, H. and Van Leeuwen, N. (1996). Serotyping, ribotyping, PCR-mediated ribosomal 16S-23S spacer analysis and arbitrarily primed PCR for epidemiological studies on *Legionella pneumophila*. *Res. Microbiol.*, **147**(5), 405–13.

Van Cauteren, D., Campèse, C., Jarraud, S., Maine, C. and Che, D. (2008). Les légionelloses survenues en France en 2007. *Bulletin épidémiologique hebdomadaire*, **30–31**, 276–80.

Van der Poel, W., Vinje, J., van Der Heide, R., Herrera, M.I., Vivo, A. and Koopmans, M.P. (2000). Norwalk-like calicivirus genes in farm animals. *Emerg. Infect. Dis.*, **6**(1), 36–41.

Van der Poel, W.H., Van der Heide, R., Verschoor, F., Gelderblom, H., Vinje, J. and Koopmans, M.P. (2003). Epidemiology of Norwalk-like virus infections in cattle in The Netherlands. *Vet Microbiol.*, 92**(4), 297–30.**

Van Heerden, J. and Ehlers, M. *et al.* (2003). Incidence of adenoviruses in raw and treated water. *Water Res.*, **37**(15), 3704–8.

van Keulen, H., Macechko, P.T., Wade, S., Schaaf, S., Wallis, P.M. and Erlandsen, S.L. (2002). Presence of human *Giardia* in domestic, farm and wild animals, and environmental samples suggests a zoonotic potential for giardiasis. *Vet. Parasitol.*, **108**(2), 97–107.

van Zyl, W. and Williams, P. *et al.* (2004). Application of a molecular method for the detection of group A rotaviruses in raw and treated water. *Water. Sci. Technol.*, **50**(1), 223–8.

Varma, M., Hester, J.D., Schaefer, F.W. 3rd, Ware, M.W. and Lindquist, H.D. (2003). Detection of Cyclospora cayetanensis using a quantitative real-time PCR assay. *J. Microbiol. Methods*, **53**(1), 27–36.

Vazquez-Novelle, M., Pazos, A.J., Abad, M., Sanchez, J.L., Perez-Paralle, M.L. (2005). Eight-hour PCR-based procedure for the detection of *Salmonella* in raw oysters. *FEMS Microbiol. Lett.*, **243**(1), 279–83.

Vesey, G. and Ashbolt, N. *et al.* (1998). The use of a ribosomal RNA targeted oligonucleotide probe for fluorescent labelling of viable *Cryptosporidium parvum* oocysts. *J. Appl. Microbiol.*, **85**, 429–440.

Via-Ordorika, L., Fastner, J., Kurmayer, R., Hisbergues, M., Dittmann, E., Komarek, J., Erhard, M. and Chorus, I. (2004). Distribution of microcystin-producing and non-microcystin-producing Microcystis sp. in European freshwater bodies: detection of microcystins and microcystin genes in individual colonies. *Syst. Appl. Microbiol.*, **27**(5), 592–602.

Victoria, M., Guimarães, F., Fumian, T., Ferreira, F., Vieira, C., Leite, J.P. and Miagostovich, M. (2008). Evaluation of an adsorption-elution method for detection of astrovirus and norovirus in environmental waters. *J. Virol. Methods*, **16**.

Vicuña-Reyes, J.P., Luh, J. and Mariñas, B.J. (2008). Inactivation of Mycobacterium avium with chlorine dioxide. *Water Research,* **42**, 1531–38.

Vieira, J., Azevedo, M.T., Azevedo, S.M., Honda, R.Y. and Correa, B. (2005). Toxic cyanobacteria and microcystin concentrations in a public water supply reservoir in the Brazilian Amazonia region. *Toxicon.*, **45**(7), 901–909.

Villena, C., Gabrieli, R., Pinto, R.M., Guix, S., Donia, D., Buonom, E., Palombi, L., Cenko, F., Bino, S., Bosch, A. and Divizia, M. (2003). A large infantile gastroenteritis outbreak in Albania caused by multiple emerging rotavirus genotypes. *Epidemiol. Infect.*, **131**(3), 1105–10.

Villar, L.M., de Paula, V.S., Diniz-Mendes, L., Lampe, E. and Gaspar, A.M. (2006). Evaluation of methods used to concentrate and detect hepatitis A virus in water samples. *J. Virol. Methods*, **137**(2), 169–76.

Villari, P., Crispino, M., Montuori, P. and Boccia, S. (2003). Molecular typing of *Aeromonas* isolates in natural mineral waters. *Appl. Environ. Microbiol.*, **69**(1), 697–701.

Vincent, P. (1995). Transmission and acquisition of *Helicobacter pylori* infection: evidences and hypothesis. *Biomed. Pharmacother.*, **49**(1), 11–8.

Vinje, J. and Koopmans, M.P. (2000). Simultaneous detection and genotyping of Norwalk-like viruses by oligonucleotide array in a reverse line blot hybridization format. *J. Clin. Microbiol.*, **38**(7), 2595–601.

Vishwanathan, R. (1957). Infectious hepatitis in Delhi (1955–1956): a critical study; epidemiology. *Indian J. Med. Res.*, **45**, 1–30.

Vivier, J., Clay, C.G. and Grabow, W.O. (2001). Detection and rapid differentiation of human enteroviruses in water sources by restriction enzyme analysis. *Water. Sci. Technol.*, **43**(12), 209–12.

Vivier, J., Ehlers, M.M. and Grabow, W.O. (2004). Detection of enteroviruses in treated drinking water. *Water Res.*, **38**(11), 2699–705.

von Reyn, C., Maslow, J.N., Barber, T.W., Falkinham, J.O. 3rd and Arbeit, R.D. (1994). Persistent colonization of potable water as a source of *Mycobacterium avium* infection in AIDS. *Lancet.*, **343**(8906), 1137–41.

Waage, A., Vardund, T., Lund, V. and Kapperud, G. (1999a). Detection of low numbers of pathogenic *Yersinia enterocolitica* in environmental water and sewage samples by nested polymerase chain reaction. *J. Appl. Microbiol.*, **87**(6), 814–21.

Waage, A., Vardund, T., Lund, V. and Kapperud, G. (1999b). Detection of small numbers of Campylobacter jejuni and *Campylobacter coli* cells in environmental water, sewage, and food samples by a seminested PCR assay. *Appl. Environ. Microbiol.*, **65**(4), 1636–43.

Wadowsky, R., Butler, L.J., Cook, M.K., Verma, S.M., Paul, M.A., Fields, B.S., Keleti, G., Sykora, J.L. and Yee, R.B. (2004). Growth-supporting activity for *Legionella pneumophila* in tap water cultures and implication of hartmannellid amoebae as growth factors. *Appl. Environ. Microbiol.*, **70**(11), 6826–33.

Wagner-Wiening, C. and Kimmig, P. (1995). Detection of viable *Cryptosporidium parvum* oocysts by PCR. *Appl. Environ. Microbiol.*, **61**(12), 4514–6.

Wait, D. and Sobsey, M.D. (2001). Comparative survival of enteric viruses and bacteria in Atlantic Ocean seawater. *Water. Sci. Technol.*, **43**(12), 139–42.

Walker, J., Sonesson, A., Keevil, C.W. and White, D.C. (1993). Detection of *Legionella pneumophila* in biofilms containing a complex microbial consortium by gas chromatography-mass spectrometry analysis of genus-specific hydroxy fatty acids. *FEMS Microbiol. Lett.*, **113**(2), 139–44.

Wallis, P., Erlandsen, S.L., Isaac-Renton, J.L., Olson, M.E., Robertson, W.J. and van Keulen, H. (1996). Prevalence of Giardia cysts and *Cryptosporidium* oocysts and characterization of *Giardia* spp. isolated from drinking water in Canada. *Appl. Environ. Microbiol.*, **62**(8), 2789–97.

Wang, G. and Doyle, M.P. (1998). Survival of enterohemorrhagic *Escherichia coli O157:H7* in water. *J. Food Prot.*, **61**, 662–667.

Wang, X., Sturegard, E., Rupar, R., Nilsson, H.O., Aleljung, P.A., Carlen, B., Willen, R. and Wadstrom, T. (1997). Infection of BALB/c A mice by spiral and coccoid forms of *Helicobacter pylori*. *J. Med. Microbiol.*, **46**(8), 657–63.

Warburton, D., Austin, J.W., Harrison, B.H. and Sanders, G. (1998). Survival and recovery of *Escherichia coli O157:H7* in inoculated bottled water. *J. Food Prot.*, **61**(8), 948–52.

Ward, R., Bernstein, D.I., Young, E.C., Sherwood, J.R., Knowlton, D.R. and Schiff, G.M. (1986). Human rotavirus studies in volunteers: determination of infectious dose and serological response to infection. *J. Infect. Dis.*, **154**(5), 871–80.

Watson, C., Owen, R.J., Said, B., Lai, S., Lee, J.V., Surman-Lee, S. and Nichols, G. (2004). Detection of Helicobacter pylori by PCR but not culture in water and biofilm samples from drinking water distribution systems in England. *J. Appl. Microbiol.*, **97**(4), 690–8.

Way, J.S. and Josephson, K.L. *et al.* (1993). Specific detection of *Salmonella* spp. by multiplex polymerase chain reaction. *Appl. Environ. Microbiol.*, **59**, 1473–1479.

Weber, D. and Rutala, W.A. (2001). The emerging nosocomial pathogens *Cryptosporidium*, *Escherichia coli O157:H7*, *Helicobacter pylori*, and hepatitis C: epidemiology, environmental survival, efficacy of disinfection, and control measures. *Infect Control Hosp. Epidemiol.*, **22**(5), 306–15.

Weiss, L. (2001). Microsporidia: emerging pathogenic protist. *Acta Trop.*, **78**, 89–102.

Wellinghausen, N. and Frost, C. *et al.* (2001). Detection of Legionellae in hospital water samples by quantitative real-time lightcycler PCR. *Appl. Environ. Microbiol.*, **67**(9), 3985–3993.

Whan, L., Ball, H.J., Grant, I.R. and Rowe, M.T. (2005). Development of an IMS-PCR assay for the detection of *Mycobacterium avium* ssp. *paratuberculosis* in water. *Lett. Appl. Microbiol.*, **40**(4), 269–73.

WHO. (1993). *Guidelines for drinking water quality, 2nd ed.* Geneva.

WHO. (1998). *Guidelines for drinking-water quality, 2nd ed. Addendum to vol. 2. Health criteria and other supporting information.* Geneva, Switzerland, World Health Organization.

WHO. (1999). Toxic Cyanobacteria in Water: a guide to their public health consequences, monitoring and management, edited by J. Bartram and I. Chorus. Geneva.

WHO. (2000). Hepatitis A vaccines: WHO position paper. *Weekly Epidemiological Record*, **75**, 38–42.

WHO. (2002). Aeromonas. *Guidelines for Drinking Water Quality, 2nd ed. Addendum: Microbiological agents in drinking water.* Geneva, World Health Organization.

WHO. (2002). Enteric hepatitis viruses. *Guidelines for Drinking Water Quality, 2nd ed. Addendum: Microbiological agents in drinking water.* Geneva, World Health Organization.

WHO. (2002). Legionella. *Guidelines for Drinking Water Quality, 2nd ed. Addendum: Microbiological agents in drinking water.* Geneva, World Health Organization.

WHO. (2002). Protozoan parasites (*Cryptosporidium, Giardia, Cyclospora*). *Guidelines for Drinking Water Quality; Addendum to the second edition of the Guidelines for drinking water quality.* Geneva, World Health Organization.

WHO. (2003). *Emerging issues in water and infectious disease.* Geneva, World Health Organization.

WHO. (2004). Guidelines for Drinking-Water Quality; Third edition. Volume 1 Recommendations; Microbial fact sheets. Geneva.

WHO. (2004). *Pathogenic Mycobacteria in water: a Guide to public health consequences, Monitoring and Management.* Edited by Pedley, S., Bartram, J., Rees, G., Dufour, A. and Cotruvo, J. London, UK, IWA Publishing.

WHO. (2004). *Verocytotoxin-producing Escherichia coli and other diarrhoeagenic E coli. Waterborne Zoonoses: Identification, Causes and Control.* A.D. J.A. Cotruvo, G. Rees, J. Bartram, R. Carr, D.O. Cliver, G.F. Craun, R. Fayer and V.P.J. Gannon. London, IWA.

Wilson, K.M., Schembri, M.A., Baker, P.D. and Saint, C.P. (2000). Molecular characterization of the toxic cyanobacterium *Cylindrospermopsis raciborskii* and design of a species-specific PCR. *Appl. Environ. Microbiol.*, **66**(1), 332–8.

Winiecka-Krusnell, J., Wreiber, K., von Euler, A., Engstrand, L. and Linder, E. (2002). Free-living amoebae promote growth and survival of *Helicobacter pylori*. *Scand J. Infect. Dis.*, **34**(4), 253–6.

Wohlsen, T., Stewart, S., Aldridge, P., Bates, J., Gray, B. and Katouli, M. (2007). The efficiency of ozonated water from a water treatment plant to inactivate *Cryptosporidium* oocysts during two seasonal temperatures. *J. Water Health*, **5**(3), 433–40.

Wolf, S., Williamson, W.M., Hewitt, J., Rivera-Aban, M., Lin, S., Ball, A., Scholes, P. and Greening, G.E. (2007). Sensitive multiplex real-time reverse transcription-PCR assay for the detection of human and animal noroviruses in clinical and environmental samples. *Appl. Environ. Microbiol.*, **73**(17), 5464–70.

Wolter, A., Niessner, R. and Seidel, M. (2008). Detection of *Escherichia coli* O157:H7, *Salmonella typhimurium*, and *Legionella pneumophila* in water using a flow-through chemiluminescence microarray readout system. *Anal. Chem.*, **80**(15), 5854–63.

Wullings, B. and Van der Kooij, D. (2006). Occurence and genetic diversity of uncultured *Legionella* spp. in drinking water treated at temperatures below 15°C. *App. Environ. Microbiol.*, **72**(1), 157–66.

Xiao, L., Singh, A., Limor, J., Graczyk, T.K., Gradus, S. and Lal, A. (2001). Molecular characterization of *Cryptosporidium* oocysts in samples of raw surface water and wastewater. *Appl. Environ. Microbiol.*, **67**(3), 1097–101.

Xiao, L., Lal, A.A. and Jiang, J. (2004). Detection and differentiation of *Cryptosporidium* oocysts in water by PCR-RFLP. *Methods Mol. Biol.*, **268**, 163–76.

Yakoob, J., Hu, G.L., Fan, X.G., Yang, H.X., Liu, S.H., Tan, D.M., Li, T.G. and Zhang, Z. (2000). Diversity of *Helicobacter pylori* among Chinese persons with H. pylori infection. *APMIS*, **108**(7–8), 482–6.

Yamada S.O.K., Kato, R., Morimoto, K., Hayashi, Y., Ito, T., Matushita, S., Konishi, N., Kai, A. and Endoh, M. (1999). Outbreak case caused by different colicin type of *Shigella sonnei* in a day nursery in Tokyo (1998. *Kansenshogaku Zasshi.*, **73**(11), 1130–9.

Yamamoto, H., Hashimoto, Y. and Ezaki, T. (1993). Comparison of detection methods for Legionella species in environmental water by colony isolation, fluorescent antibody staining, and polymerase chain reaction. *Microbiol. Immunol.*, **37**(8), 617–22.

Yanez, M.A., Carrasco-Serrano, C., Barbera, V.M. and Catalan, V. (2005). Quantitative detection of *Legionella pneumophila* in water samples by immunomagnetic purification and real-time PCR amplification of the dotA gene. *Appl. Environ. Microbiol.*, **71**(7), 3433–41.

Yanez, M.A., Catalan, V., Apraiz, D., Figueras, M.J. and Martinez-Murcia, A.J. (2003). Phylogenetic analysis of members of the genus *Aeromonas* based on gyrB gene sequences. *Int. J. Syst. Evol. Microbiol.* **53**(Pt 3), 875–83.

Yang, N. and Yu, P. *et al.* (1988). Inapparent infection of hepatitis A virus. *Am. J. Epidemiol.*, **127**(3), 599–604.

Yap, E., Thong, T.W., Tan, A.L., Yeo, M., Tan, H.C., Loh, H., Teo, T.P., Thong, K.T., Singh, M. and Chan, Y.C. (1995). Comparison of *Pseudomonas pseudomallei* from humans, animals, soil and water by restriction endonuclease analysis. *Singapore Med. J.*, **36**(1), 60–2.

Yaradou, D.F., Hallier-Soulier, S., Moreau, S., Poty, F., Hillion, Y., Reyrolle, M., André, J., Festoc, G., Delabre, K., Vandenesch, F., Etienne, J. and Jarraud, S. (2007). Integrated real-time PCR for detection and monitoring of *Legionella pneumophila* in water systems. *Appl. Environ. Microbiol.*, **73**(5), 1452–6.

Yeats, J., Smuts, H., Serfontein, C.J. and Kannemeyer, J. (2005). Investigation into a school enterovirus outbreak using PCR detection and serotype identification based on the 5' non-coding region. *Epidemiol. Infect.*, **133**(6), 1123–30.

Yu, C.P., Farrell, S.K., Robinson, B. and Chu, K.H. (2008). Development and application of real-time PCR assays for quantifying total and aerolysin gene-containing aeromonas in source, intermediate, and finished drinking water. *Environ. Sci. Technol.*, **42**(4), 1191–200.

Yu, S.-Z. (1989). Drinking water and primary liver cancer. *Primary Liver Cancer*. Z.-Y. Tan, Wu, M.-C. and Xia, S.-S. Beijing, China Academics Publishers, pp. 30–37.

Yu, V. and Stout, J.E. (2000). Hospital characteristics associated with colonization of water systems by *Legionella* and risk of nosocomial legionnaires' disease: a cohort study of 15 hospitals. *Infect Control Hosp. Epidemiol.*, **21**(7), 434–5.

Zanetti, F., De Luca, G. and Stampi, S. (2000). Recovery of *Burkholderia pseudomallei* and *B. cepacia* from drinking water. *Int. J. Food Microbiol.*, **59**(1–2), 67–72.

Zhang, L.J., Wang, X.J., Bai, J.M., Fang, G., Liu, L.G., Zhang, Y. and Fontaine, R.E. (2008). An outbreak of hepatitis A in recently vaccinated students from ice snacks made from contaminated well water. *Epidemiol. Infect.*, **26**, 1–6.

Zoni, R., Zanelli, R., Riboldi, E., Bigliardi, L. and Sansebastiano, G. (2007). Investigation on virucidal activity of chlorine dioxide. experimental data on feline calicivirus, HAV and Coxsackie B5. *J. Prev. Med. Hyg.*, **48**(3), 91–5.

Zuckerman, A. and Thomas, H.C. (1993). *Viral Hepatitis: Scientific Basis and Clinical Management*. London, Churchill Livingstone.

Zuckerman, U. and Tzipori, S. (2006). Portable continuous flow centrifugation and method 1623 for monitoring of waterborne protozoa from large volumes of various water matrices. *J. Appl. Microbiol.*, **100**(6), 1220–7.

Zwart, G., Kamst-van Agterveld, M.P., van der Werff-Staverman, I., Hagen, F., Hoogveld, H.L. and Gons, H.J. (2005). Molecular characterization of cyanobacterial diversity in a shallow eutrophic lake. *Environ. Microbiol.*, **7**(3), 365–77.

Annex 1

Annex 1: Treatment data

Treatment data: clarification/rapid filtration

Pathogen	Treatment	Filter type	Water flow rate (gpm)	Filter loading rate[2] gpm/ft	Raw water quality		Effluent quality	PH	Scale	Reduction (log)	References
					Turbidity (NTU)	Concentration	Turbidity (NTU)				
Bacteria											
E. coli O157H7	coagulation + sedimentation + filtration	anth./sand	1	2–8	3.5–3.9	$6*10^7$ CFU/L	0.11–0.21	7.78 (average)	pilot	0.75 (average)	Harrington *et al.* (2003)
E. coli O157H7	coagulation + sedimentation + filtration	anth./sand/garnet	1	4	7.1–10.1	$6*10^7$ CFU/L	0.1–0.12	7.41 (average)	pilot	1.0 (average)	Harrington *et al.* (2003)
E. coli O157H7	coagulation + sedimentation + filtration	anth./sand	1	4	7.1–10.1	$6*10^7$ CFU/L	0.08–0.09	5.7 (average)	pilot	2.0 (average)	Harrington *et al.* (2003)
E. coli O157H7	coagulation + sedimentation + filtration	anth./sand	1	4		$6*10^7$ CFU/L			pilot	0.7 (average)	Xagoraraki *et al.* (2004)
E. coli ATCC25922	coagulation + sedimentation + filtration	GAC/sand/gravel	0.4–0.47	5–10		$1*10^8$ CFU/L	1		pilot	2.67 (optimal)	Gerba *et al.* (2003)
E. coli	coagulation + filtration	anth	20	5	0.3–4	$1*10^{7-8}$ CFU/L	<0.1		pilot	2.7–3.1	Hendricks *et al.* (2005)
Aeromonas hydrophila	coagulation + sedimentation + filtration	anth./sand	1	2–8	3.5–3.9	$2*10^7$ CFU/L	0.11–0.21	7.78 (average)	pilot	3.2 (average)	Harrington *et al.* (2003)
Aeromonas hydrophila	coagulation + sedimentation + filtration	anth./sand/garnet	1	4	7.1–10.1	$2*10^7$ CFU/L	0.1–0.12	7.41 (average)	pilot	3.5 (average)	Harrington *et al.* (2003)
Aeromonas hydrophila	coagulation + sedimentation + filtration	anth./sand	1	4	7.1–10.1	$2*10^7$ CFU/L	0.08–0.09	5.70 (average)	pilot	3.5 (average)	Harrington *et al.* (2003)
Aeromonas hydrophila	coagulation + sedimentation + filtration	anth./sand	1	4		$2*10^7$ CFU/L			pilot	2.5 (average)	Xagoraraki *et al.* (2004)
Viruses											
Adenovirus	coagulation + sedimentation + filtration	anth./sand	1	4		$1*10^8$/L	<0.2		pilot	2	Harrington *et al.* (2000)
Calicivirus	coagulation + sedimentation + filtration	GAC/sand/gravel	0.4–0.47	5–10		$1*10^6 TCID_{50}$/L	1		pilot	3.05 (optimal)	Gerba *et al.* (2003)
Enteroviruses	coagulation + sedimentation + filtration	varied			varied	$1*10^7$/L	0.4–3.5	6.8	full	1.85 (average)	Payment *et al.* (1985)
Enteroviruses	coagulation + sedimentation + filtration				varied	varied	0.4–2.0		full	0.85 (average)	Stetler *et al.* (1984)
Enteroviruses	coagulation + sedimentation				varied	$1*10^7$/L	0.9–6.8	6.8	full	0.65 (average)	Payment *et al.* (1985)
Enteroviruses	coagulation + sedimentation				varied	varied	0.4–2.0		full	0.3 (average)	Stetler *et al.* (1984)
Poliovirus	coagulation + sedimentation + filtration	anth./sand	1	4		$1*10^8$/L	<0.2		pilot	1.75	Harrington *et al.* (2000)
Polio 1 (vaccine)	coagulation + sedimentation				10	$6*10^5$/L	2.5	6.8–7.1	pilot	0.43	Wolf *et al.* (1974)
Polio 1 (vaccine)	coagulation + sedimentation				3.4	$1*10^5$/L	1.5	6.9–7.3	pilot	2.52	Wolf *et al.* (1974)
Polio (attenuated)	coagulation + filtration	anth	20	5	0.3–4	$1*10^{7-8}$/L	<0.1		pilot	2.2	Hendricks *et al.* (2005)
Echovirus 12	coagulation + filtration	anth	20	5	0.3–4	$1*10^{7-8}$/L	<0.1		pilot	2.4	Hendricks *et al.* (2005)
Coxsackievirus 2	coagulation + sedimentation				0.4ml SiO_2	$4.5*10^6$ PFU/L		6.2	Bench	1.4–2.0	Chang *et al.* (1958)
Rotavirus SA-11	coagulation + filtration	anth./sand		0.55		$1*10^6$ PFU/L			pilot	0.1	De Leon *et al.* (1986)
Rotavirus SA-11	coagulation + filtration	anth./sand/garnet		0.55		$1*10^6$ PFU/L			pilot	0.15	De Leon *et al.* (1986)

Pathogen	Treatment	Filter type	Water flow rate (gpm)	Filter loading rate^2 gpm/ft	Raw water quality		Effluent quality	PH	Scale	Reduction (log)	References
					Turbidity (NTU)	Concentration	Turbidity (NTU)				
Protozoa											
Giardia	coagulation + sedimentation + filtration	anth./sand	0.5	2.5	4–23	0.48–1.01/L	0.1–0.2	7.5 (average)	pilot	3.16 (average)	Nieminski (1994)
Giardia	coagulation + sedimentation + filtration	GAC/sand		0.75–1.2		3.78–5.80/L	0.19		pilot	>4.5 (optimal)	LeChevallier *et al.* (1991)
Giardia	coagulation + sedimentation + filtration	anth./sand	600		2.5–28	0.42–1/L	0.1–0.2	8.3 (average)	full	2.46 (average)	Nieminski (1994)
Giardia	coagulation + sedimentation + filtration	varied			<10	High	0.04–0.09		Pilot	3.4–5.1 (optimal)	Patania *et al.* (1995)
Giardia	coagulation + sedimentation + filtration	anth./sand	1	4	2–5	$3.9*10^2$/L	<0.1		pilot	1.5 (average)	Assavasilavasukul *et al.* (2008)
Giardia	coagulation + sedimentation + filtration	anth./sand	1	4	2–5	$6.1*10^4$/L	<0.1		pilot	3.8 (average)	Assavasilavasukul *et al.* (2008)
Giardia	coagulation + filtration	anth./sand	0.5	2.5	4–23	0.48–1.01/L	0.1–0.2	7.5 (average)	pilot	3.33 (average)	Nieminski (1994)
Giardia	coagulation + filtration	anth./sand/garnet	1	5	0.5–5	$5*10^3$/L	0.02–0.03	6.4–6.6	pilot	3.1–3.5 (optimal)	Ongerth *et al.* (1995)
Giardia	coagulation + filtration	anth./sand	600		2.5–28		0.1–0.2	8.3 (average)	full	3.52 (average)	Nieminski (1994)
Giardia	coagulation + filtration	anth.	20	5	0.3–4	$1*10^2$–$5*10^3$/L	<0.1		pilot	3.2–3.4	Hendricks *et al.* (2005)
Cryptosporidium	coagulation + sedimentation + filtration	anth./sand	0.5	2.5	4–23		0.1–0.2	7.5 (average)	pilot	2.82 (average)	Nieminski (1994)
Cryptosporidium	coagulation + sedimentation + filtration	GAC/sand		0.75–1.2		7.57–13	0.19		pilot	>5.3 (optimal)	LeChevallier *et al.* (1991)
Cryptosporidium	coagulation + sedimentation + filtration	GAC/sand		3–6	<10	high	0.04–0.09		pilot	0.2–3.1	Patania *et al.* (1995)
Cryptosporidium	coagulation + sedimentation + filtration	anth./sand		3–6	<10	high	0.04–0.09		pilot	0.4–1.1	Patania *et al.* (1995)
Cryptosporidium	coagulation + sedimentation + filtration	anth./sand	1	2–8	3.5–3.9	$1*10^5$/L	0.11–0.21	7.78 (average)	pilot	1.9 (average)	Harrington *et al.* (2003)
Cryptosporidium	coagulation + sedimentation + filtration	anth./sand/garnet	1	4	7.1–10.1	$1*10^5$/L	0.1–0.12	7.41 (average)	pilot	2.1 (average)	Harrington *et al.* (2003)
Cryptosporidium	coagulation + sedimentation + filtration	anth./sand	1	4	7.1–10.1	$1*10^5$/L	0.08–0.09	5.7 (average)	pilot	2.1 (average)	Harrington *et al.* (2003)
Cryptosporidium	coagulation + sedimentation + filtration	anth./sand	1	4		$1*10^5$/L			pilot	1.6 (average)	Xagoraraki *et al.* (2004)
Cryptosporidium	coagulation + sedimentation + filtration	anth./sand	600		2.5–28		0.1–0.2	8.3 (average)	full	2.1 (average)	Nieminski (1994)
Cryptosporidium	coagulation + sedimentation + filtration	anth./sand	1	4	2–5	$5.3*10^2$/L	<0.1		pilot	1.1 (average)	Assavasilavasukul *et al.* (2008)
Cryptosporidium	coagulation + sedimentation + filtration	anth./sand	1	4	2–5	$7*10^4$/L	<0.1		pilot	3.2 (average)	Assavasilavasukul *et al.* (2008)
Cryptosporidium	coagulation + filtration	anth./sand	0.5	2.5	4–23		0.1–0.2	7.5 (average)	pilot	2.9 (average)	Nieminski (1994)
Cryptosporidium	coagulation + filtration	varied			<10	high	0.04–0.09		pilot	2.7–5.9	Patania *et al.* (1995)
Cryptosporidium	coagulation + filtration	anth./sand/garnet	1	5	0.5–5	$5*10^3$	0.02–0.03	6.4–6.6	pilot	2.7–3.1 (optimal)	Ongerth *et al.* (1995)
Cryptosporidium	coagulation + filtration	anth./sand	600		2.5–28		0.1–0.2	8.3 (average)	full	2.74 (average)	Nieminski (1994)
Cryptosporidium	coagulation + filtration	anth.	20	5	0.3–4	$1*10^2$–$5*10^3$/L	<0.1		pilot	3.2–3.4	Hendricks *et al.* (2005)
Cryptosporidium	coagulation + filtration	sand (10cm depth)				$8*10^6$/L			pilot	0.5–1.6	Gitis (2008)
Cryptosporidium	coagulation + filtration	sand (40cm depth)				$8*10^6$/L			pilot	2.1–3.9	Gitis (2008)
Cryptosporidium	coagulation + filtration	sand (80cm depth)				$8*10^6$/L			pilot	3–>4	Gitis (2008)
Cryptosporidium	coagulation + sedimentation				<10	high	0.04–0.09		pilot	1.4–1.8	Patania *et al.* (1995)

Pathogen	Treatment	Filter type	Water flow rate (gpm)	Filter loading rate[2] gpm/ft	Raw water quality		Effluent quality	PH	Scale	Reduction (log)	References
					Turbidity (NTU)	Concentration	Turbidity (NTU)				
Protozoa											
Cryptosporidium	coagulation + sedimentation				1.2–1.4	$3*10^5$–$4*10^5$	<0.5	6	bench	0.81	Plummer *et al.* (1995)
Cryptosporidium	coagulation + DAF				1.2–1.4	$3*10^5$–$4*10^5$	0.33	4.3–6	bench	2–2.6 (optimal)	Plummer *et al.* (1995)
Cryptosporidium	filtration	sand			0.5–4	$3.4*10^7$–$7.9*10^7$		6.6–7.6	pilot	3.6–4.6 (average)	Yu *et al.* (2006)
Cryptosporidium	filtration	anth./sand			0.5–4	$3.4*10^7$–$7.9*10^7$		6.6–7.6	pilot	2.2–4.6 (average)	Yu *et al.* (2006)
Encephalitozoon intestinalis	coagulation + sedimentation + filtration	anth./sand	1	2–8	3.5–3.9	$2*10^5$/L	0.11–0.21	7.78 (average)	pilot	1.5 (average)	Harrington *et al.* (2003)
Encephalitozoon intestinalis	coagulation + sedimentation + filtration	anth./sand/garnet	1	4	7.1–10.1	$2*10^5$/L	0.1–0.12	7.41 (average)	pilot	1.3 (average)	Harrington *et al.* (2003)
Encephalitozoon intestinalis	coagulation + sedimentation + filtration	anth./sand	1	4	7.1–10.1	$2*10^5$/L	0.08–0.09	5.7 (average)	pilot	1.5 (average)	Harrington *et al.* (2003)
Encephalitozoon intestinalis	coagulation + sedimentation + filtration	anth./sand	1	4		$2*10^5$/L			pilot	1.3 (average)	Xagoraraki *et al.* (2004)
Encephalitozoon intestinalis	coagulation + sedimentation + filtration	GAC/sand/gravel	0.4–0.47	5–10		$1*10^6$/L	1		pilot	2.47 (optimal)	Gerba *et al.* (2003)
Acanthamoeba polyphaga	coagulation + sedimentation				1.2	$1*10^4$/L		7.8	bench	0.7–1.8	Loret *et al.* (2008)
Free-living amoebae	coagulation + sedimentation								Full (7 plants)	1.5 (average)	Loret *et al.* (2008)
Free-living amoebae	coagulation + sedimentation + filtration								Full (7 plants)	2.0 (average)	Loret *et al.* (2008)
Free-living amoebae	coagulation + sedimentation + filtration								Full (6 plants)	1–2	Hoffmann *et al.* (2001)

Treatment data: disinfection–chlorine

Pathogen	Temperature (°C)	PH	Inactivation (%)	CT (mg*min/L)	[Disinfectant] (mg/L)	Contact time (min)	Water source	Reference
Bacteria								
Salmonella typhi	25	7	99	0.4	1	0.4		King *et al.* (1988)
Salmonella typhi	25	7	99	0.5	0.5	1		King *et al.* (1988)
Salmonella typhi	25	7	99	0.5	0.25	2		King *et al.* (1988)
Salmonella typhi INTI			>99.999	12	2	6		Korol *et al.* (1995)
Shigella sonnei	25	7	99	0.5	1	0.5		King *et al.* (1988)
Shigella sonnei	25	7	99	0.5	0.5	1		King *et al.* (1988)
Shigella sonnei	25	7	99	0.4	0.25	1.6		King *et al.* (1988)
Yersinia enterocolitica	25	7	99	0.5	1	0.5		King *et al.* (1988)
Yersinia enterocolitica	25	7	99	0.5	0.5	1		King *et al.* (1988)
Yersinia enterocolitica	25	7	99	0.5	0.25	2		King *et al.* (1988)
Yersinia enterocolitica ATCC9610			>99.999	12	2	6		Korol *et al.* (1995)
Escherichia coli	5	6	99	0.04	0.1	0.4	CDF	LeChevallier *et al.* (1988)
Escherichia coli	5	10	99	0.92	1	0.92	CDF	LeChevallier *et al.* (1988)
Escherichia coli	25	6–7	99	0.03				Andriamirado *et al.* (2005)
Escherichia coli	room		99	0.119	0.119	1	CDF	Baker *et al.* (2002)
E. coli O157H7 ATCC35150	20–22	7	99	0.1	0.4–0.5		TAP	Chauret *et al.* (2003)
E. coli O157H7 FRIK2516	20–22	7	99	0.1	0.4–0.5		TAP	Chauret *et al.* (2003)
E. coli O157H7 FRIK2536	20–22	7	99	0.05	0.4–0.5		TAP	Chauret *et al.* (2003)
E. coli O157H7 FRIK2542	20–22	7	99	0.05–0.1	0.4–0.5		TAP	Chauret *et al.* (2003)
Aeromonas hydrophila (TW11)	21	6	99	1.4	0.2	7	CDF	Massa *et al.* (1999)
Aeromonas hydrophila (TW27)	21	6	99	0.2	0.2	1	CDF	Massa *et al.* (1999)
Aeromonas hydrophila (clinical)	5	7.8	90	1	0.05	20	TAP	Sisti *et al.* (1998)
Aeromonas hydrophila (clinical)	5	7.8	90	6	0.3	20	TAP	Sisti *et al.* (1998)
Aeromonas hydrophila I.N.M.			>99.999	12	2	6		Korol *et al.* (1995)
Legionella pneumophila	5	6	99	>50–>320	1–4		BDF	Jacangelo *et al.* (2002)
Legionella pneumophila	5	7	99	50–250	1–4		BDF	Jacangelo *et al.* (2002)
Legionella pneumophila	5	8	99	250–>1000	1–4		BDF	Jacangelo *et al.* (2002)
Legionella pneumophila	15	6	99	100–>320	1–4		BDF	Jacangelo *et al.* (2002)
Legionella pneumophila	15	7	99	60–>320	1–4		BDF	Jacangelo *et al.* (2002)
Legionella pneumophila	15	8	99	25–710	1–4		BDF	Jacangelo *et al.* (2002)
Legionella pneumophila	25	6	99	40–500	1–4		BDF	Jacangelo *et al.* (2002)
Legionella pneumophila	25	7	99	100–160	1–4		BDF	Jacangelo *et al.* (2002)
Legionella pneumophila	25	8	99	130–250	1–4		BDF	Jacangelo *et al.* (2002)
Helicobacter pylori	5	6	99	6.65	0.5	1.33	CDF	Johnson *et al.* (1997)

Pathogen	Temperature (°C)	PH	Inactivation (%)	CT (mg*min/L)	[Disinfectant] (mg/L)	Contact time (min)	Water source	Reference
Bacteria								
Helicobacter pylori ATCC 43504	room		99	0.299	0.299	1	CDF	Baker *et al.* (2002)
Mycobacterium avium strain A5	23	7	99.9	106	0.15		BDF	Taylor *et al.* (2000)
M. avium strain 1060	23	7	99.9	204	0.15		BDF	Taylor *et al.* (2000)
M. avium strain1508	23	7	99.9	164	0.15		BDF	Taylor *et al.* (2000)
M. avium strain 5002	23	7	99,9	126	0.15		BDF	Taylor *et al.* (2000)
M. avium strain 5502	23	7	99.9	51	0.15		BDF	Taylor *et al.* (2000)
Mycobacterium intra-cellulare		7	99	>480	0.15		BDF	Pelletier (1987)
Mycobacterium fortuitum	5	6	99	>320–1000			BDF	Jacangelo *et al.* (2002)
Mycobacterium fortuitum	5	7	99	320–630			BDF	Jacangelo *et al.* (2002)
Mycobacterium fortuitum	5	8	99	>320–>1000			BDF	Jacangelo *et al.* (2002)
Mycobacterium fortuitum	15	6	99	90–320			BDF	Jacangelo *et al.* (2002)
Mycobacterium fortuitum	15	7	99	500–>1000			BDF	Jacangelo *et al.* (2002)
Mycobacterium fortuitum	15	8	99	>320–1000			BDF	Jacangelo *et al.* (2002)
Mycobacterium fortuitum	25	6	99	50–>320			BDF	Jacangelo *et al.* (2002)
Mycobacterium fortuitum	25	7	99	130–>320			BDF	Jacangelo *et al.* (2002)
Mycobacterium fortuitum	25	8	99	130–>320			BDF	Jacangelo *et al.* (2002)
Burkholderia pseudomallei	22–25	6.25–7	99	30–60	1	30–60		Howard *et al.* (2005)
Burkholderia pseudomallei			99.99	30	1	30		Howard *et al.* (2005)
Viruses								
Adenovirus 3	2	7.8	99.99	2.4	0.5	4.8	RSW	Liu *et al.* (1971)
Adenovirus 5	25	7.5	99	6	0.2	30	TAP	Abad *et al.* (1994)
Adenovirus 7A	2	7.8	99.99	6.25	0.5	12.5	RSW	Liu *et al.* (1971)
Adenovirus 12	2	7.8	99.99	6.75	0.5	13.5	RSW	Liu *et al.* (1971)
Adenovirus 40	5	6	99	<0.4			CDF	Jacangelo *et al.* (2002)
Adenovirus 40	5	7	99	<0.2			CDF	Jacangelo *et al.* (2002)
Adenovirus 40	5	8	99	<0.2			CDF	Jacangelo *et al.* (2002)
Adenovirus 40	23	7	99	<0.2			CDF	Jacangelo *et al.* (2002)
Adenovirus 40	23	8	99	<0.2			CDF	Jacangelo *et al.* (2002)
Adenovirus 40	5	6	99	0.05			BDF	Thurston-Enriquez *et al.* (2003)
Adenovirus 40	5	6	99.9	0.11			BDF	Thurston-Enriquez *et al.* (2003)
Adenovirus 40	5	6	99.99	0.22			BDF	Thurston-Enriquez *et al.* (2003)
Adenovirus 40	5	7	99	0.15			BDF	Thurston-Enriquez *et al.* (2003)
Adenovirus 40	5	7	99.9	0.38			BDF	Thurston-Enriquez *et al.* (2003)
Adenovirus 40	5	7	99.99	0.75			BDF	Thurston-Enriquez *et al.* (2003)
Adenovirus 40	5	8	99	0.11			BDF	Thurston-Enriquez *et al.* (2003)

Pathogen	Temperature (°C)	PH	Inactivation (%)	CT (mg*min/L)	[Disinfectant] (mg/L)	Contact time (min)	Water source	Reference
Viruses								
Adenovirus 40	5	8	99.9	0.17			BDF	Thurston-Enriquez *et al.* (2003)
Adenovirus 40	5	8	99.99	0.24			BDF	Thurston-Enriquez *et al.* (2003)
Adenovirus 40	15	8–8.2	99	1.51			BDF	Thurston-Enriquez *et al.* (2003)
Adenovirus 40	15	8–8.2	99.9	9.69			BDF	Thurston-Enriquez *et al.* (2003)
Adenovirus 40	15	8–8.2	99.99	36.09			BDF	Thurston-Enriquez *et al.* (2003)
Feline calicivirus	5	6	99	<0.25			BDF	Jacangelo *et al.* (2002)
Feline calicivirus	5	7	99	<0.25			BDF	Jacangelo *et al.* (2002)
Feline calicivirus	5	8	99	<0.2			BDF	Jacangelo *et al.* (2002)
Feline calicivirus	15	7	99	<0.25			BDF	Jacangelo *et al.* (2002)
Feline calicivirus	15	8	99	<1			BDF	Jacangelo *et al.* (2002)
Feline calicivirus (dispersed)	5	6	99	0.02			BDF	Thurston-Enriquez *et al.* (2003)
Feline calicivirus (dispersed)	5	6	99.9	0.07			BDF	Thurston-Enriquez *et al.* (2003)
Feline calicivirus (dispersed)	5	6	99.99	0.19			BDF	Thurston-Enriquez *et al.* (2003)
Feline calicivirus (dispersed)	5	7	99	0.05			BDF	Thurston-Enriquez *et al.* (2003)
Feline calicivirus (dispersed)	5	7	99.9	0.06			BDF	Thurston-Enriquez *et al.* (2003)
Feline calicivirus (dispersed)	5	7	99.99	0.07			BDF	Thurston-Enriquez *et al.* (2003)
Feline calicivirus (dispersed)	5	8	99	0.18			BDF	Thurston-Enriquez *et al.* (2003)
Feline calicivirus (dispersed)	5	8	99.9	0.23			BDF	Thurston-Enriquez *et al.* (2003)
Feline calicivirus (dispersed)	5	8	99.99	0.27			BDF	Thurston-Enriquez *et al.* (2003)
Feline calicivirus (dispersed)	15	8–8.2	99	0.21			TAP	Thurston-Enriquez *et al.* (2003)
Feline calicivirus (dispersed)	15	8–8.2	99.9	0.56			TAP	Thurston-Enriquez *et al.* (2003)
Feline calicivirus (dispersed)	15	8–8.2	99.99	1.1			TAP	Thurston-Enriquez *et al.* (2003)
Norovirus Genogroup I	5	6	99	0.66	1	10	PBW	Shin *et al.* (2008)
Norovirus Genogroup I			99.9	0.99	1	10	PBW	Shin *et al.* (2008)
Norovirus Genogroup I			99.99	1.3	1	10	PBW	Shin *et al.* (2008)
Poliovirus 1	5	6	99	1.05	0.5	2.1		SDWC (1980)
Poliovirus 1	5	6	99	2.1	1	2.1		SDWC (1980)
Poliovirus 1	15	6	99	1	1	1		SDWC (1980)
Poliovirus 3	25–28	8	99	5.36	0.67	8		Crawford (1988); Malley (2000)
Poliovirus 1	5	10	99	10.5	0.5	21		SDWC (1980)
Poliovirus 1	15	10	99	3.5	1	3.5		SDWC (1980)
Poliovirus 1	5	6	99	0.6–1.8			BDF	Jacangelo *et al.* (2002)
Poliovirus 1 (aggregated)	5	6	99	1.5–3			BDF	Jacangelo *et al.* (2002)
Poliovirus 1 (aggregated)	5	7	99	<1–10			BDF	Jacangelo *et al.* (2002)
Poliovirus 1	5	6	99	0.93			BDF	Thurston-Enriquez *et al.* (2003)

Pathogen	Temperature (°C)	PH	Inactivation (%)	CT (mg*min/L)	[Disinfectant] (mg/L)	Contact time (min)	Water source	Reference
Viruses								
Poliovirus 1	5	6	99.9	2.87			BDF	Thurston-Enriquez *et al.* (2003)
Poliovirus 1	5	6	99.99	6.36			BDF	Thurston-Enriquez *et al.* (2003)
Coxsackie A2	0–5	7–7.5	99	12				White (1999)
Coxsackie A2	10	7–7.5	99	8				White (1999)
Coxsackie A2	0–5	7.5–8	99	20				White (1999)
Coxsackie A2	10	7.5–8	99	15				White (1999)
Coxsackie A2	0–5	8–8.5	99	30				White (1999)
Coxsackie A2	10	8–8.5	99	20				White (1999)
Coxsackie A2	0–5	8.5–9	99	35				White (1999)
Coxsackie A2	10	8.5–9	99	22				White (1999)
Coxsackie A5	2	7.8	99.99	16.8	0.5	33.5	RSW	Liu *et al.* (1971)
Coxsackie A9	5	6	99	0.14	0.46–0.49	0.3	CDF	Engelbrecht *et al.* (1980)
Coxsackie A9	5	10–10.01	99	0.72	0.48–0.5	1.5	CDF	Engelbrecht *et al.* (1980)
Coxsackie A9	2	7.8	99.99	3.4	0.5	6.8	RSW	Liu *et al.* (1971)
Coxsackie B1	2	7.8	99.99	4.3	0.5	8.5	RSW	Liu *et al.* (1971)
Coxsackie B3	2	7.8	99.99	8.1	0.5	16.2	RSW	Liu *et al.* (1971)
Coxsackie B5	5	6	99	1.751	0.515	3.4	CDF	Crawford (1988); Malley (2000)
Coxsackie B5	5	6	99	3.25	0.5	6.5	BDF	Sobsey *et al.* (1988)
Coxsackie B5	5	8	99	9.5	0.5	19	BDF	Sobsey *et al.* (1988)
Coxsackie B5	25	8.1	99	59	11.8	5	TAP	Grabow *et al.* (1984)
Coxsackie B5	25	8.5	99	108	27	4	HTE	Grabow *et al.* (1984)
Coxsackie B5	5	6–6.06	99	1.73	0.51–0.52	3.4	CDF	Engelbrecht *et al.* (1980)
Coxsackie B5	5	7.81–7.82	99	2.16	0.48–0.50	4.5	CDF	Engelbrecht *et al.* (1980)
Coxsackie B5	5	9.93–10.05	99	33	0.50–0.51	66	CDF	Engelbrecht *et al.* (1980)
Coxsackie B5	2	7.8	99.99	19.8	0.5	39.5	RSW	Liu *et al.* (1971)
Echovirus 1	4.8–5.2	6	99	0.24	0.48–0.49	0.5	CDF	Engelbrecht *et al.* (1980)
Echovirus 1	4.8–5.2	7.81–7.82	99	0.56	0.47–0.49	1.2	CDF	Engelbrecht *et al.* (1980)
Echovirus 1	4.8–5.2	10–10.4	99	47	0.49–0.51	96	CDF	Engelbrecht *et al.* (1980)
Echovirus 5	5	6	99	0.65	0.5	1.3	CDF	Engelbrecht *et al.* (1980)
Echovirus 5	5	10	99.99	13.5	0.5	27	CDF	Engelbrecht *et al.* (1980)
Echovirus 1	2	7.8	99.99	13.1	0.5	26.1	RSW	Liu *et al.* (1971)
Echovirus 7	2	7.8	99.99	3.6	0.5	7.1	RSW	Liu *et al.* (1971)
Echovirus 9	2	7.8	99.99	6.2	0.5	12.4	RSW	Liu *et al.* (1971)
Echovirus 11	2	7.8	99.99	6.7	0.5	13.4	RSW	Liu *et al.* (1971)
Echovirus 12	2	7.8	99.99	7.25	0.5	14.5	RSW	Liu *et al.* (1971)

Pathogen	Temperature (°C)	PH	Inactivation (%)	CT (mg*min/L)	[Disinfectant] (mg/L)	Contact time (min)	Water source	Reference
Viruses								
Echovirus 19	2	7.8	99.99	10	0.5	20	RSW	Liu *et al.* (1971)
Rotavirus	5	6–7	99	0.05				Andriamirado *et al.* (2005)
Rotavirus	25	6–7	99	0.01				Andriamirado *et al.* (2005)
Hepatitis A HM175 (dispersed)	5	6	99.99	2.3	0.36	6.5	BDF	Sobsey *et al.* (1990)
Hepatitis A HM175 (cell associated)	5	6	99.99	29	0.35	85	BDF	Sobsey *et al.* (1990)
Hepatitis A HM175 (dispersed)	5	8	99.99	2	0.37	5.5	BDF	Sobsey *et al.* (1990)
Hepatitis A HM175 (cell associated)	5	8	99.99	27	0.39	70	BDF	Sobsey *et al.* (1990)
Hepatitis A HM175 (dispersed)	5	10	99.99	19.3	0.39	50	BDF	Sobsey *et al.* (1990)
Hepatitis A HM175 (cell associated)	5	10	99.99	104	0.41	253	BDF	Sobsey *et al.* (1990)
Protozoa								
Cryptosporidium	25	7	90	4800			BDF	Korich *et al.* (1990)
Cryptosporidium	25	7	99	7200	80	90	BDF	Korich *et al.* (1990)
Cryptosporidium	25	7	99.9	8400			BDF	Korich *et al.* (1990)
Giardia lamblia	5	8	99	144			BDF	SWTR (1991)
Giardia lamblia	10	8	90	54				USEPA (1989)
Giardia lamblia	10	8	99.9	162				USEPA (1989)
Giardia lamblia	15	6	99	35			BDF	SWTR (1991)
Giardia lamblia	5	6	99	70			BDF	SWTR (1991)
Giardia lamblia	5	7	99	99			BDF	SWTR (1991)
Giardia lamblia	15	7	99	50			BDF	SWTR (1991)
Giardia lamblia	15	8	99	72			BDF	SWTR (1991)
Giardia lamblia	25	6	99	17			BDF	SWTR (1991)
Giardia lamblia	25	7	99	25			BDF	SWTR (1991)
Giardia lamblia	25	8	99	36			BDF	SWTR (1991)
Giardia muris	5	6	90	75	2.5	30	BDF	Rice *et al.* (1982)
Giardia muris	5	8	90	120	2.5	48	BDF	Rice *et al.* (1982)
Giardia muris	5	7	99	449–1012	23.8–78.5	5.7–42.6	BDF	Leahy *et al.* (1987)
Giardia muris	25	7	99	25.5–44.8	2.8–7.1	5.7–42.6	BDF	Leahy *et al.* (1987)
Giardia muris	25	9	99	177	11.1	15.6	BDF	Leahy *et al.* (1987)
Giardia muris	25	5	99	71	4.4	16.3	BDF	Leahy *et al.* (1987)
Encephalitozoon intestinalis	5	6	99	36–42	1		BDF	Jacangelo *et al.* (2002)
Encephalitozoon intestinalis	5	7	99	24–114	1		BDF	Jacangelo *et al.* (2002)
Encephalitozoon intestinalis	5	8	99	42–72	1		BDF	Jacangelo *et al.* (2002)
Encephalitozoon intestinalis	5	6	99	12.8	1		BDF	John *et al.* (2005)
Encephalitozoon intestinalis	5	7	99	35.6	1		BDF	John *et al.* (2005)

Pathogen	Temperature (°C)	PH	Inactivation (%)	CT (mg*min/L)	[Disinfectant] (mg/L)	Contact time (min)	Water source	Reference
Protozoa								
Encephalitozoon intestinalis	5	8	99	68.8	1		BDF	John *et al.* (2005)
Encephalitozoon intestinalis	25	7	99.9	32	2	16	BDF	Wolk *et al.* (2000)
Encephalitozoon intestinalis	5	6	99.9	57.2	1		BDF	John *et al.* (2005)
Encephalitozoon intestinalis	5	7	99.9	82.2	1		BDF	John *et al.* (2005)
Encephalitozoon intestinalis	5	8	99.9	128	1		BDF	John *et al.* (2005)
Encephalitozoon intestinalis	21–25	7	98.14	8	2	4	CDF	Johnson *et al.* (2003)
Encephalitozoon intestinalis	21–25	7	98.55	10	2	5	CDF	Johnson *et al.* (2003)
Encephalitozoon intestinalis	21–25	7	>99.99	12	2	6	CDF	Johnson *et al.* (2003)
Encephalitozoon cuniculi	21–25	7	99.95	8	2	4	CDF	Johnson *et al.* (2003)
Encephalitozoon cuniculi	21–25	7	99.97	10	2	5	CDF	Johnson *et al.* (2003)
Encephalitozoon cuniculi	21–25	7	99.98	12	2	6	CDF	Johnson *et al.* (2003)
Encephalitozoon cuniculi	21–25	7	99.99	16	2	8	CDF	Johnson *et al.* (2003)
Encephalitozoon hellem	21–25	7	80.05	8	2	4	CDF	Johnson *et al.* (2003)
Encephalitozoon hellem	21–25	7	92.05	10	2	5	CDF	Johnson *et al.* (2003)
Encephalitozoon hellem	21–25	7	99.79	12	2	6	CDF	Johnson *et al.* (2003)
Encephalitozoon hellem	21–25	7	99.99	16	2	8	CDF	Johnson *et al.* (2003)
Acanthamoeba polyphaga (cysts)	20–22	7.5–8	99	1300			CDF	Loret *et al.* (2008)
Acanthamoeba polyphaga (cysts)	20–22	7.5–8	99.9	2000			CDF	Loret *et al.* (2008)
Acanthamoeba polyphaga (cysts)	20–22	7.5–8	99.99	3500			CDF	Loret *et al.* (2008)
Acanthamoeba sp. 4A	25	7.3–7.4	99.99	960–7200	4–8	24h		DeJonckheere *et al.* (1976)
Acanthamoeba castellanii (1501)	25	7	99.9	24	0.8	30	BDF	Cursons *et al.* (1980)
Acanthamoeba castellanii (1501)	25	7	99.99	22.5	0.75	30	BDF	Cursons *et al.* (1980)
Acanthamoeba castellanii (1501)	25	7	99.91	18.75	0.625	30	BDF	Cursons *et al.* (1980)
Acanthamoeba culbertsoni A-1	25	7	99.9	28.5	0.95	30	BDF	Cursons *et al.* (1980)

Treatment data: disinfection–chloramines

Pathogen	Disinfectant	Temperature (°C)	pH	Inactivation (%)	CT (mg*min/L)	[Disinfectant] (mg/L)	Contact time (min)	Water source	Reference
Bacteria									
Salmonella typhi	chloramines	20–25	7	98.7	72	1.2	60		Butterfield *et al.* (1946)
Salmonella typhi	chloramines	25	8	99	14	1	14		Ward *et al.* (1982)
Shigella dysenteriae	chloramines	20–25	7	99.9	12	1.2	10		Butterfield *et al.* (1946)
Shigella dysenteriae	chloramines	20–25	8.5	99.3	48	1.2	40		Butterfield *et al.* (1946)
Shigella sonnei	chloramines	20–25	8.5	99.9	72	1.2	60		Butterfield *et al.* (1946)
Shigella paradysenteriae	chloramines	20–25	8.5	96.6	24	1.2	20		Butterfield *et al.* (1946)
Escherichia coli	monochloramine	5	9	99	175	1	175		SDWC (1980)
Escherichia coli	monochloramine	15	9	99	64	1	64		SDWC (1980)
Escherichia coli	monochloramine	25	9	99	40.2	1.2	33.5		SDWC (1980)
Escherichia coli	monochloramine	25	6	99	7	1	7		Crawford (1988)
Escherichia coli	monochloramine	25	8	99	38	1	38		Crawford (1988)
Escherichia coli	monochloramine	10	7	99	12.5	0.5	25	BDF	Knochel (1991)
Escherichia coli	monochloramine	room	7.5	99	11			CDF	Baker *et al.* (2002)
E. coli O157H7 ATCC35150	monochloramine	20–22	8	99	10	1.25–1.5			Chauret *et al.* (2003)
E. coli O157H7	monochloramine	20–22	8	99	20	1.25–1.5			Chauret *et al.* (2003)
Aeromonas hydrophila	monochloramine	20	8	99	5.6	5	1.12	BDF	Sirikanchana *et al.* (2002)
Aeromonas hydrophila	monochloramine	20	8	99	5.6	10.5	0.53	BDF	Sirikanchana *et al.* (2002)
Aeromonas hydrophila	monochloramine	20	8	99.9	7.1	5	1.42	BDF	Sirikanchana *et al.* (2002)
Aeromonas hydrophila	monochloramine	20	8	99.9	7.1	10.5	0.677	BDF	Sirikanchana *et al.* (2002)
Aeromonas hydrophila	monochloramine	20	8	99	8.9	1	8.9	BDF	Sirikanchana *et al.* (2002)
Aeromonas hydrophila	monochloramine	20	8	99.9	11.2	1	11.2	BDF	Sirikanchana *et al.* (2002)
A. hydrophila E9	monochloramine	10	7	99	18	0.5	36	BDF	Knochel (1991)
A. hydrophila 174	monochloramine	10	7	99	6.5	0.5	13	BDF	Knochel (1991)
A. sobria M4	monochloramine	10	7	99	1.5	0.5	3	BDF	Knochel (1991)
Legionella pneumophila	chloramines	5	7	99	1000–>1000	Cl2:NH4=3:1		BDF	Jacangelo *et al.* (2002)
Legionella pneumophila	chloramines	5	7	99	>320–>1000	Cl2:NH4=7:1		BDF	Jacangelo *et al.* (2002)
Legionella pneumophila	chloramines	25	7	99	>630–>1000	Cl2:NH4=3:1		BDF	Jacangelo *et al.* (2002)
Legionella pneumophila	chloramines	25	7	99	320–>1000	Cl2:NH4=7:1		BDF	Jacangelo *et al.* (2002)
Helicobacter pylori	monochloramine	room	7.5	99	9.5			CDF	Baker *et al.* (2002)
Mycobacterium avium strain A5	monochloramine	23	7	99.9	97			BDF	Taylor *et al.* (2000)
M. avium strain (1060)	monochloramine	23	7	99.9	458			BDF	Taylor *et al.* (2000)
M. avium strain (1508)	monochloramine	23	7	99.9	548			BDF	Taylor *et al.* (2000)
M. avium strain 5002	monochloramine	23	7	99.9	1710			BDF	Taylor *et al.* (2000)

Pathogen	Disinfectant	Temperature (°C)	pH	Inactivation (%)	CT (mg*min/L)	[Disinfectant] (mg/L)	Contact time (min)	Water source	Reference
Bacteria									
M. avium strain 5502	monochloramine	23	7	99.9	91			BDF	Taylor *et al.* (2000)
Mycobacterium fortuitum	chloramine	20	7	90	162.5	3.25	50	BDF	Engelbrecht *et al.* (1977)
Mycobacterium fortuitum	chloramine	5	7	99	>320–630	Cl2:NH3-N=3:1		BDF	Jacangelo *et al.* (2002)
Mycobacterium fortuitum	chloramine	5	7	99	100–250	Cl2:NH3-N=7:1		BDF	Jacangelo *et al.* (2002)
Mycobacterium fortuitum	chloramine	25	7	99	>320–>1000	Cl2:NH3-N=3:1		BDF	Jacangelo *et al.* (2002)
Mycobacterium fortuitum	chloramine	25	7	99	300–630	Cl2:NH3-N=7:1		BDF	Jacangelo *et al.* (2002)
Burkholderia pseudomallei	monochloramine	24	8	99.8	1440	NaOCl:NH4Cl=4:1		PBW	Howard *et al.* (2005)
Burkholderia pseudomallei	monochloramine	24	8	99.96	2880	NaOCl:NH4Cl=4:1		PBW	Howard *et al.* (2005)
Viruses									
Adenovirus 40	chloramines	5–6	7	99	>360	Cl2:NH4=3:1		BDF	Jacangelo *et al.* (2002)
Adenovirus 40	chloramines	5–6	7	99	70–360	Cl2:NH4=7:1		BDF	Jacangelo *et al.* (2002)
Adenovirus 40	chloramines	19–20	7	99	60–120	Cl2:NH4=3:1		BDF	Jacangelo *et al.* (2002)
Adenovirus 40	chloramines	19–20	7	99	20	Cl2:NH4=7:1		BDF	Jacangelo *et al.* (2002)
Feline calicivirus	chloramines	5.0–6.5	7	99	1–15	Cl2:NH4=3:1		BDF	Jacangelo *et al.* (2002)
Feline calicivirus	chloramines	6.5	7	99	5	Cl2:NH4=7:1		BDF	Jacangelo *et al.* (2002)
Feline calicivirus	chloramines	20–23	7	99	1–15	Cl2:NH4=3:1		BDF	Jacangelo *et al.* (2002)
Feline calicivirus	chloramines	20–24	7	99	1–3	Cl2:NH4=7:1		BDF	Jacangelo *et al.* (2002)
Poliovirus 1	chloramines	19	7	99	120	1, Cl2:NH3=3:1		BDF	Jacangelo *et al.* (2002)
Poliovirus 1	chloramines	19	7	99	180	1, Cl2:NH3=7:1		BDF	Jacangelo *et al.* (2002)
Poliovirus 1	monochloramine	15	10	99	900	10	90		SDWC (1980)
Poliovirus 1	monochloramine	25	10	99	320	10	32		SDWC (1980)
Coxsackie A9	dichloramine	15	4.5	99	300	10	30		Esposito (1974)
Coxsackie A9	dichloramine	15	4.5	99	300	1	300		Esposito (1974)
Coxsackie A9	dichloramine	15	4.5	99	300	0.5	600		Esposito (1974)
Rotavirus	chloramine	5	8–9	99	6480				Andriamirado *et al.* (2005)
Rotavirus	chloramine	25	8–9	99	3810				Andriamirado *et al.* (2005)
Rotavirus SA11	chloramine	5	8	99	4034	10	366–402		Berman *et al.* (1984)
Hepatitis A HM175 (dispersed)	monochloramine	5	8	99.99	1225	9.7	116	BDF	Sobsey *et al.* (1990)
Hepatitis A HM175 (cell associated)	monochloramine	5	8	99.99	1740	10.2	170	BDF	Sobsey *et al.* (1990)
Protozoa									
Cryptosporidium	monochloramine	25	7	90	5600			BDF	Korich *et al.* (1990)
Cryptosporidium	monochloramine	25	7	99	>9600			BDF	Korich *et al.* (1990)
Cryptosporidium	monochloramine	22	8	99	>7200	15	240		Finch *et al.* (1994)
Giardia lamblia	chloramines	10	8	90	615				USEPA (1989)

Pathogen	Disinfectant	Temperature (°C)	pH	Inactivation (%)	CT (mg*min/L)	[Disinfectant] (mg/L)	Contact time (min)	Water source	Reference
Protozoa									
Giardia lamblia	chloramines	5	7	99	1470	Cl2:NH4=3:1		BDF	SWTR (1991)
Giardia lamblia	chloramines	5	7	99	1470	Cl2:NH4=7:1		BDF	SWTR (1991)
Giardia lamblia	chloramines	25	7	99	500	Cl2:NH4=3:1		BDF	SWTR (1991)
Giardia lamblia	chloramines	25	7	99	500	Cl2:NH4=7:1		BDF	SWTR (1991)
Giardia lamblia	chloramines	10	8	99.9	1850				USEPA (1989)
Giardia muris	chloramines	5	7	99	1400	6.35	220	BDF	Rubin *et al.* (1983)
Giardia muris	chloramines	15	9	99	600	1.5–30		BDF	Rubin *et al.* (1983)
Giardia muris	chloramines	15	6	99	1000	1.5–30		BDF	Rubin *et al.* (1983)

Treatment data: disinfection–chlorine dioxide

Pathogen	Temperature (°C)	pH	Inactivation (%)	CT (mg*min/L)	[Disinfectant] (mg/L)	Contact time (min)	Water source	Reference
Bacteria								
Escherichia coli	20	6.5	99	0.18	0.5	0.35	RSW	LeChevallier *et al.* (1988)
Escherichia coli	10	6.5	99	0.24	0.5	0.47		Bernade *et al.* (1967)
Escherichia coli	5	6.5	99	0.38	0.75	0.5	PBW	LeChevallier *et al.* (1988)
Escherichia coli	25	7	99	0.28	0.5	0.55	PBW	LeChevallier *et al.* (1988)
Escherichia coli	15	7	99	0.38	0.8	0.47		Bernade *et al.* (1967)
Escherichia coli	5	7	99	0.49	0.5	0.98		Bernade *et al.* (1967)
Aeromonas hydrophila (M800)	0–4	8	99	0.11–0.14			BDF	Medema *et al.* (1991)
Aeromonas hydrophila (M800)	0–4	8	99	0.04–0.14			BDF	Medema *et al.* (1991)
Legionella pneumophila	5	6	99	> 320–> 1000	1		BDF	Jacangelo *et al.* (2002)
Legionella pneumophila	25	6	99	50–200	1		BDF	Jacangelo *et al.* (2002)
Legionella pneumophila	5	8	99	> 250–630	1		BDF	Jacangelo *et al.* (2002)
Legionella pneumophila	25	8	99	50–130	1		BDF	Jacangelo *et al.* (2002)
Mycobacterium avium strain 1060	23	7	99.9	8	0.1–1.2		BDF	Taylor *et al.* (2000)
Mycobacterium avium strain 5002	23	7	99.9	11	0.1–1.2		BDF	Taylor *et al.* (2000)
Mycobacterium avium strain 5502	23	7	99.9	2	0.1–1.2		BDF	Taylor *et al.* (2000)
Mycobacterium avium ATCC15769	5	6, 8, 10	99.9	36	0.5–5		PBW, BBW	Vicuña-Reyes *et al.* (2008)
Mycobacterium avium ATCC15769	10	6, 8, 10	99.9	21	0.5–5		PBW, BBW	Vicuña-Reyes *et al.* (2008)
Mycobacterium avium ATCC15769	20	6, 8, 10	99.9	7	0.5–5		PBW, BBW	Vicuña-Reyes *et al.* (2008)
Mycobacterium avium ATCC15769	30	6, 8, 10	99.9	3	0.5–5		PBW, BBW	Vicuña-Reyes *et al.* (2008)
Mycobacterium fortuitum	5	6	99	< 10–< 20			BDF	Jacangelo *et al.* (2002)
Mycobacterium fortuitum	25	6	99	< 10–< 20			BDF	Jacangelo *et al.* (2002)
Mycobacterium fortuitum	5	8	99	< 10–< 20			BDF	Jacangelo *et al.* (2002)
Mycobacterium fortuitum	25	8	99	< 10–< 20			BDF	Jacangelo *et al.* (2002)
Viruses								
Adenovirus 40	5	6	99	< 0.5			BDF	Jacangelo *et al.* (2002)
Adenovirus 40	15	6	99	< 0.5			BDF	Jacangelo *et al.* (2002)
Adenovirus 40	23	6	99	< 0.1			BDF	Jacangelo *et al.* (2002)
Adenovirus 40	5	7	99	< 0.3			BDF	Jacangelo *et al.* (2002)
Adenovirus 40	15	7	99	< 0.3			BDF	Jacangelo *et al.* (2002)
Adenovirus 40	23	7	99	< 0.1			BDF	Jacangelo *et al.* (2002)
Adenovirus 40	5	8	99	< 0.3			BDF	Jacangelo *et al.* (2002)
Adenovirus 40	5	6	99.99	> 0.77–< 1.53	0.5	3	BDF	Thurston-Enriquez *et al.* (2005b)
Adenovirus 40	15	6	99.99	> 0.49–< 0.74	0.5	2	BDF	Thurston-Enriquez *et al.* (2005b)
Adenovirus 40	5	8	99.99	> 0.8–< 1.59	0.5	3	BDF	Thurston-Enriquez *et al.* (2005b)

Pathogen	Temperature (°C)	pH	Inactivation (%)	CT (mg*min/L)	[Disinfectant] (mg/L)	Contact time (min)	Water source	Reference
Viruses								
Adenovirus 40	15	8	99.99	<0.12	0.5	2	BDF	Thurston-Enriquez *et al.* (2005b)
Feline calicivirus	5	6	99	0.6–15			BDF	Jacangelo *et al.* (2002)
Feline calicivirus	15	6	99	1–5			BDF	Jacangelo *et al.* (2002)
Feline calicivirus	25	6	99	2–15			BDF	Jacangelo *et al.* (2002)
Feline calicivirus	5	7	99	3–15			BDF	Jacangelo *et al.* (2002)
Feline calicivirus	15	7	99	2–5			BDF	Jacangelo *et al.* (2002)
Feline calicivirus	25	7	99	1–2			BDF	Jacangelo *et al.* (2002)
Feline calicivirus	5	8	99	0.3–1			BDF	Jacangelo *et al.* (2002)
Feline calicivirus	5	6	99.99	>20.2–<30.3	1	40	BDF	Thurston-Enriquez *et al.* (2005b)
Feline calicivirus	15	6	99.99	>4.2–<6.72	1	10	BDF	Thurston-Enriquez *et al.* (2005b)
Feline calicivirus	5	8	99.99	>0.68	1	40	BDF	Thurston-Enriquez *et al.* (2005b)
Feline calicivirus	15	8	99.99	<0.18	1	10	BDF	Thurston-Enriquez *et al.* (2005b)
Poliovirus 1	4	8	99	1	1		BDF	Jacangelo *et al.* (2002)
Poliovirus 1	5	7	99	5.4	0.8	6.8		Warriner (1967); Cronier (1977)
Poliovirus 1	15	7	99	1.3	0.3	4.2		Warriner (1967); Cronier (1977)
Poliovirus 1	25	7	99	1.08	0.3	3.6		Warriner (1967); Cronier (1977)
Coxsackie A9	15	7	99	0.4	0.3	1.2		Warriner (1967); Cronier (1977)
Coxsackie A9	15	7	99	0.34	0.5	0.05		Warriner (1967); Cronier (1977)
Coxsackie A9	15	7	99	0.2	0.8	0.25		Warriner (1967); Cronier (1977)
Rotavirus	5	6–7	99	2.1				Andriamirado *et al.* (2005)
Rotavirus	25	6–7	99	0.2				Andriamirado *et al.* (2005)
Rotavirus SA11	5	6	99	0.2–0.3	0.5–1.0	0.2–0.6	BDF	Berman *et al.* (1984)
Hepatitis A HM175 (dispersed)	5	6	99.99	12.6	0.34	37	BDF	Sobsey *et al.* (1990)
Hepatitis A HM175 (cell associated)	5	6	99.99	9.6	0.27	35	BDF	Sobsey *et al.* (1990)
Hepatitis A Nj-3	5	6	99.99	50	5	10	CDF	Li J.W. *et al.* (2004)
Hepatitis A Nj-3	5	6	99.9999	75	7.5	10	CDF	Li J.W. *et al.* (2004)
Protozoa								
Cryptosporidium	25	6–7	90	52			BDF	Korich *et al.* (1990)
Cryptosporidium	25	6–7	90	>78			BDF	Korich *et al.* (1990)
Cryptosporidium	10	7	90	>60				Ransome *et al.* (1993)
Giardia lamblia	10	8	90	7.7				USEPA (1989)
Giardia lamblia	5	6	99	17			BDF	SWTR (1991)
Giardia lamblia	25	6	99	7.3			BDF	SWTR (1991)
Giardia lamblia	5	8	99	17			BDF	SWTR (1991)

Pathogen	Temperature (°C)	pH	Inactivation (%)	CT (mg*min/L)	[Disinfectant] (mg/L)	Contact time (min)	Water source	Reference
Protozoa								
Giardia lamblia	25	8	99	7.3			BDF	SWTR (1991)
Giardia lamblia	10	8	99.9	23				USEPA (1989)
Giardia muris	5	7	99	10.7	0.1–5.55	1.3–168	BDF	Leahy *et al.* (1987); Leahy (1985)
Giardia muris	25	5	99	5.8	0.26–1.2	4.0–24	BDF	Leahy *et al.* (1987); Leahy (1985)
Giardia muris	25	7	99	5.1	0.21–1.12	3.3–28.8	BDF	Rubin (1988)
Giardia muris	25	9	99	2.7	0.15–0.81	2.1–19.2	BDF	Rubin (1988)
Acanthamoeba polyphaga (cysts)	20–22	7.5–8	99	20			CDF	Loret *et al.* (2008)
Acanthamoeba polyphaga (cysts)	20–22	7.5–8	99.9	50			CDF	Loret *et al.* (2008)
Acanthamoeba polyphaga (cysts)	20–22	7.5–8	99.99	80			CDF	Loret *et al.* (2008)
Acanthamoeba castellanii (1501)	25	7	99	19.5	2.5–0.6	30	BDF	Cursons *et al.* (1980)
Acanthamoeba culbertsoni A-1	25	7	99	18	1.775	30	BDF	Cursons *et al.* (1980)

Treatment data: disinfection–ozone

Pathogen	Temperature (°C)	pH	Inactivation (%)	CT (mg*min/L)	[Disinfectant] (mg/L)	Contact time (min)	Water source	Reference
Bacteria								
Salmonella typhimurium	25	7.2	99	3.3	0.48	6.88	DW	Lezcano *et al.* (1999)
Salmonella typhimurium	25	7.2	99	4.2	0.58	7.18	DW	Lezcano *et al.* (1999)
Salmonella typhimurium	25	7.2	99	3.5	1.04	3.38	DW	Lezcano *et al.* (1999)
Salmonella typhimurium	25	7.2	99	2.2	1.75	1.24	DW	Lezcano *et al.* (1999)
Salmonella typhi INTI			>99.999	2.1	0.35	6		Korol *et al.* (1995)
Shigella sonnei	25	7.2	99	1.8	0.5	3.5	DW	Lezcano *et al.* (1999)
Shigella sonnei	25	7.2	99	2.1	0.61	3.5	DW	Lezcano *et al.* (1999)
Shigella sonnei	25	7.2	99	2.4	0.88	2.7	DW	Lezcano *et al.* (1999)
Shigella sonnei	25	7.2	99	2.7	1.87	1.44	DW	Lezcano *et al.* (1999)
Shigella sonnei	25	7.2	99	2.8	2	1.4	DW	Lezcano *et al.* (1999)
Yersinia enterocolitica ATCC9610			>99.999	2.1	0.35	6		Korol *et al.* (1995)
Escherichia coli	1	7.2	99	0.022	0.065	0.33		SDWC (1980)
Escherichia coli	11	6	99	0.027	0.01	0.275		SDWC (1980)
Escherichia coli	11	6	99	0.035	0.01	0.35		SDWC (1980)
Escherichia coli	12	7	99	0.001	0.0006	1.7		SDWC (1980)
Escherichia coli	12	7	99	0.002	0.0023	1.03		SDWC (1980)
Escherichia coli	12	7	99	0.004	0.0125	0.33		SDWC (1980)
Escherichia coli	room		99	0.09				Baker *et al.* (2002)
E. coli ATCC8739			>99.999	2.1	0.35	6		Korol *et al.* (1995)
Aeromonas hydrophila I.N.M.			>99.999	2.1	0.35	6		Korol *et al.* (1995)
Legionella pneumophila	5	7	99	<4–<12			BDF	Jacangelo *et al.* (2002)
Legionella pneumophila	15	7	99	<1–<10			BDF	Jacangelo *et al.* (2002)
Legionella pneumophila	25	7	99	<1–<5			BDF	Jacangelo *et al.* (2002)
Helicobacter pylori	room		99	0.24			CDF	Baker *et al.* (2002)
Mycobacterium avium strain 1060	23		99.9	0.17			ODF	Taylor *et al.* (2000)
Mycobacterium avium strain 5002	23		99.9	0.12			ODF	Taylor *et al.* (2000)
Mycobacterium avium strain 5502	23		99.9	0.1			ODF	Taylor *et al.* (2000)
Mycobacterium fortuitum	24	7	99	0.53	0.8–1.08	0.58	ODF	Farooq *et al.* (1977)
Mycobacterium fortuitum	5	7	99	<1–320			BDF	Jacangelo *et al.* (2002)
Mycobacterium fortuitum	15	7	99	<1–4			BDF	Jacangelo *et al.* (2002)
Mycobacterium fortuitum	25	7	99	<1–13			BDF	Jacangelo *et al.* (2002)

Pathogen	Temperature (°C)	pH	Inactivation (%)	CT (mg*min/L)	[Disinfectant] (mg/L)	Contact time (min)	Water source	Reference
Viruses								
Adenovirus 40	5.0–7.0	7	99	< 0.1– < 0.2			ODF	Jacangelo *et al.* (2002)
Adenovirus 40	5	7	99.99	0.07–0.6	0.3–0.49		BDF	Thurston-Enriquez (2005a)
Feline calicivirus	5	7	99	< 0.01– < 0.14			BDF	Jacangelo *et al.* (2002)
Feline calicivirus	5	7	99.99	< 0.01–0.03	0.06–1.00		BDF	Thurston-Enriquez (2005a)
Norovirus	5	7	99.9	0.06	0.37	10s		Shin *et al.* (2003)
Poliovirus 1	5	7.2	99	< 0.04	< 0.3	0.13		SDWC (1980)
Poliovirus 1	20	7	99	< 0.005	< 0.03	0.16		SDWC (1980)
Poliovirus 1	24	7	99	0.12	0.245	0.5		SDWC (1980)
Poliovirus 1	25	7	99	0.42	0.042	10		SDWC (1980)
Poliovirus 1	5	7.2	99	0.2	0.15–0.2	0.4–1.5	BDF	Roy *et al.* (1982)
Poliovirus 2	25	7.2	99	0.72	0.15	4.83	BDF	Roy *et al.* (1982)
Coxsackie A9			99	0.0043	0.027	0.16	ODF	Emerson *et al.* (1982)
Coxsackie A9			99.2	0.0033	0.01	0.33	ODF	Emerson *et al.* (1982)
Coxsackie B5	20	7.2	99	1.3	0.25	5.5	ASE	Harakeh *et al.* (1985)
Coxsackie B5	20	7.2	99	0.64	0.32	2	ASE	Harakeh *et al.* (1985)
Coxsackie B3	25	7	99	0.095				SDWC (1980)
Coxsackie B3	25	7	99.9	0.6				SDWC (1980)
Coxsackie B5	25	7	99	0.053				SDWC (1980)
Coxsackie B5	25	7	99.9	0.076				SDWC (1980)
Echovirus 1	20	7.2	99	2.6	0.26	10	ASE	Harakeh *et al.* (1985)
Rotavirus SA11	4	6–8	99	0.019–0.064	0.1–0.3	0.12–0.19		Vaughn *et al.* (1987)
Hepatitis A	10	6	> 99.9999	< 0.098	1.43	5s	PBW	Hall *et al.* (1993)
Hepatitis A	6	8.1	> 99.99	< 0.16	1.28	5s	PBW	Hall *et al.* (1993)
Hepatitis A	5	9.9	> 99.99	< 0.115	1.32	5s	PBW	Hall *et al.* (1993)
Hepatitis A	3	8.3	> 99.99	< 0.027	0.3	5s	PBW	Hall *et al.* (1993)
Protozoa								
Cryptosporidium	25	7	90	3				Korich *et al.* (1990)
Cryptosporidium	25	7	99	3.75				Korich *et al.* (1990)
Cryptosporidium	7	6.9	99	7	0.16–1.3	5–15		Finch *et al.* (1993)
Cryptosporidium	22	6.9	99	3.5	1.07–1.9	5–15		Finch *et al.* (1993)
Cryptosporidium	22–25	8	99	5.5	2.4	2.3		Owens (1994)
Cryptosporidium	10	7	99	18				Ransome *et al.* (1993)
Cryptosporidium	25	7	99.9	4.5				Korich *et al.* (1990)
Cryptosporidium parvum	15	7	99	19.5	1–3	7.5	ODF	Li L. *et al.* (2004)

Pathogen	Temperature (°C)	pH	Inactivation (%)	CT (mg*min/L)	[Disinfectant] (mg/L)	Contact time (min)	Water source	Reference
Protozoa								
Cryptosporidium parvum	22	7	98.38	10	1	10		Keegan *et al.* (2003)
Cryptosporidium parvum	22	7	99.35	20	2	10		Keegan *et al.* (2003)
Cryptosporidium parvum	22	7	99.98	50	5	10		Keegan *et al.* (2003)
Giardia lamblia	10	8	90	0.48				USEPA (1989)
Giardia lamblia	5	7	99	1.3			ODF	SWTR (1991)
Giardia lamblia	15	7	99	0.63			ODF	SWTR (1991)
Giardia lamblia	25	7	99	0.32			ODF	SWTR (1991)
Giardia lamblia	10	8	99.9	1.5				USEPA (1989)
Giardia lamblia	5	7	99.9	0.53	0.11–0.48	0.94–5.0	PBW	Wickramanayake *et al.* (1984)
Giardia lamblia	22	6.85	>99.996	13.52	1.04	13	PBW	Finch *et al.* (1993)
Giardia lamblia	22	6.85	>99.95	1.52	0.76	2	PBW	Finch *et al.* (1993)
Giardia lamblia	22	6.85	>99.97	4.825	0.965	5	PBW	Finch *et al.* (1993)
Giardia lamblia	22	6.85	>99.995	4.275	0.855	5	PBW	Finch *et al.* (1993)
Giardia lamblia	22	6.85	>99.5	3.05	0.61	5	PBW	Finch *et al.* (1993)
Giardia muris	5	7	99	1.94	0.15–0.7	2.8–12.9	PBW	Wickramanayake *et al.* (1984)
Giardia muris	22	6.7	>99.994	1.825		5	PBW	Finch *et al.* (1993)
Giardia muris	22	6.7	>99.994	2.225		5	PBW	Finch *et al.* (1993)
Giardia muris	22	6.7	>99.92	0.65		5	PBW	Finch *et al.* (1993)
Giardia muris	22	6.7	>99.92	0.825		5	PBW	Finch *et al.* (1993)
Giardia muris	22	7.6	>99.9992	2.65		5	PBW	Finch *et al.* (1993)
Encephalitozoon intestinalis	5	7	99	0.3–0.4			BDF	Jacangelo *et al.* (2002)
Encephalitozoon intestinalis	5	7	99	0.841	1		DW	John *et al.* (2005)
Encephalitozoon intestinalis	5	7	99	0.591	1		DW	John *et al.* (2005)
Encephalitozoon intestinalis	5	7	99.9	1.4	0.5		DW	John *et al.* (2005)
Encephalitozoon intestinalis	5	7	99.9	1.08	0.5		DW	John *et al.* (2005)
Acanthamoeba polyphaga (cysts)	20–22	7.5–8	99	5			CDF	Loret *et al.* (2008)
Acanthamoeba polyphaga (cysts)	20–22	7.5–8	99.9	10			CDF	Loret *et al.* (2008)
Acanthamoeba polyphaga (cysts)	20–22	7.5–8	99.99	15			CDF	Loret *et al.* (2008)
Acanthamoeba spp	25	7	>98.9	1.6	0.4	4	DW	Langlais *et al.* (1986)
Acanthamoeba castellanii (1501)	25	7	99	2.34	6.05–0.078	30	ODF	Cursons *et al.* (1980)
Acanthamoeba culbertsoni A-1	25	7	99	2.4	6.05–0.08	30	ODF	Cursons *et al.* (1980)

Treatment data: disinfection–UV

Pathogen	Reduction (%)	UV Dose (mWs/cm^2)	Water source	Reference
Bacteria				
Salmonella typhi	90	2.5		Montiel *et al.* (1990)
Salmonella typhi ATCC19430	90	1.8		Wilson *et al.* (1992)
Salmonella typhi ATCC19430	99	4.8		Wilson *et al.* (1992)
Salmonella typhi ATCC19430	99.9	6.4		Wilson *et al.* (1992)
Salmonella typhi ATCC19430	99.99	8.2		Wilson *et al.* (1992)
Salmonella typhi ATCC6539	90	2.7		Chang *et al.* (1985)
Salmonella typhi ATCC6539	99	4.1		Chang *et al.* (1985)
Salmonella typhi ATCC6539	99.9	5.5		Chang *et al.* (1985)
Salmonella typhi ATCC6539	99.99	7.1		Chang *et al.* (1985)
Salmonella typhi (environmental spp.)	99.99	51		Hijnen *et al.* (2006)
Salmonella enteritis	90	4		Montiel *et al.* (1990)
Salmonella enteritis (human feces)	90	5		Tosa *et al.* (1998)
Salmonella enteritis (human feces)	99	7		Tosa *et al.* (1998)
Salmonella enteritis (human feces)	99.9	9		Tosa *et al.* (1998)
Salmonella enteritis (human feces)	99.99	10		Tosa *et al.* (1998)
Salmonella typhosa	90	2.2		Andriessen (1985)
Salmonella enteritis (human feces)	90	7.5		Tosa *et al.* (1998)
Salmonella enteritis (human feces)	99	12		Tosa *et al.* (1998)
Salmonella enteritis (human feces)	99.9	15		Tosa *et al.* (1998)
Salmonella enteritis (human feces)	90	2		Tosa *et al.* (1998)
Salmonella enteritis (human feces)	99	4		Tosa *et al.* (1998)
Salmonella enteritis (human feces)	99.9	6		Tosa *et al.* (1998)
Shigella dysenteriae	90	2.2		Montiel *et al.* (1990)
Shigella dysenteriae ATCC29027	90	0.5		Wilson *et al.* (1992)
Shigella dysenteriae ATCC29027	99	1.2		Wilson *et al.* (1992)
Shigella dysenteriae ATCC29027	99.9	2		Wilson *et al.* (1992)
Shigella dysenteriae ATCC29027	99.99	3		Wilson *et al.* (1992)
Shigella dysenteriae (environmental spp.)	99.99	11		Hijnen *et al.* (2006)
Shigella sonnei	90	3		Montiel *et al.* (1990)
Shigella sonnei ATCC9290	90	3.2		Chang *et al.* (1985)
Shigella sonnei ATCC9290	99	4.9		Chang *et al.* (1985)
Shigella sonnei ATCC9290	99.9	6.5		Chang *et al.* (1985)
Shigella sonnei ATCC9290	99.99	8.2		Chang *et al.* (1985)
Shigella sonnei (environmental spp.)	99.99	26		Hijnen *et al.* (2006)
Shigella paradysenteriae	90	1.7		Montiel *et al.* (1990)
Shigella flexneri	90	1.7		Montiel *et al.* (1990)
Yersinia enterocolitica ATCC27729	90	1.7		Wilson *et al.* (1992)
Yersinia enterocolitica ATCC27730	99	2.8		Wilson *et al.* (1992)
Yersinia enterocolitica ATCC27731	99.9	3.7		Wilson *et al.* (1992)
Yersinia enterocolitica ATCC27732	99.99	4.6		Wilson *et al.* (1992)
Yersinia enterocolitica (environmental spp.)	99.99	13		Hijnen *et al.* (2006)
Campylobacter jejuni ATCC43429	90	1.6		Wilson *et al.* (1992)
Campylobacter jejuni ATCC43429	99	3.4		Wilson *et al.* (1992)
Campylobacter jejuni ATCC43429	99.9	4		Wilson *et al.* (1992)
Campylobacter jejuni ATCC43429	99.99	4.6		Wilson *et al.* (1992)
Campylobacter jejuni (environmental spp.)	99.99	14		Hijnen *et al.* (2006)

Pathogen	Reduction (%)	UV Dose (mWs/cm^2)	Water source	Reference
Bacteria				
Escherichia coli	90	3		Hoff *et al.* (1981)
Escherichia coli	99	6		Hoff *et al.* (1981)
Escherichia coli	99.9	9		Hoff *et al.* (1981)
Escherichia coli	99.99	12		Hoff *et al.* (1981)
Escherichia coli	99.999	15		Hoff *et al.* (1981)
E. coli O157:H7 ATCC43894	90	1.5		Wilson *et al.* (1992)
E. coli O157:H7 ATCC43894	99	2.8		Wilson *et al.* (1992)
E. coli O157:H7 ATCC43894	99.9	4.1		Wilson *et al.* (1992)
E. coli O157:H7 ATCC43894	99.99	5.6		Wilson *et al.* (1992)
E. coli O157 (environmental spp.)	99.99	19		Hijnen *et al.* (2006)
Aeromonas hydrophila ATCC7966	90	1.1	PBW	Wilson *et al.* (1992)
Aeromonas hydrophila ATCC7966	99	2.6	PBW	Wilson *et al.* (1992)
Aeromonas hydrophila ATCC7966	99.9	3.9	PBW	Wilson *et al.* (1992)
Aeromonas hydrophila ATCC7966	99.99	5	PBW	Wilson *et al.* (1992)
Aeromonas hydrophila	90	3		USEPA (2002)
Aeromonas hydrophila	99	8		USEPA (2002)
Legionella pneumophila	90	0.38		Montiel *et al.* (1990)
Legionella pneumophila 43660	90	3.1		Wilson *et al.* (1992)
Legionella pneumophila 43660	99	5		Wilson *et al.* (1992)
Legionella pneumophila 43660	99.9	6.9		Wilson *et al.* (1992)
Legionella pneumophila 43660	99.99	9.4		Wilson *et al.* (1992)
Legionella pneumophila (environmental spp.)	99.99	11–30		Hijnen *et al.* (2006)
Helicobacter pylori (Texas isolate)	99.99	<8	PBW	Hayes *et al.* (2006)
Helicobacter pylori 43504	99.99	<8	PBW	Hayes *et al.* (2006)
Helicobacter pylori 49503	99.99	<8	PBW	Hayes *et al.* (2006)
Mycobacterium avium–intra-cellulare	99	14	GM	David (1973)
Mycobacterium avium	90	6 (low pressure)	PBS	Shin *et al.* (2008)
Mycobacterium avium	90	6 (medium pressure)	PBS	Shin *et al.* (2008)
Mycobacterium avium	99.99	24 (low pressure)	PBS	Shin *et al.* (2008)
Mycobacterium avium	99.99	22 (medium pressure)	PBS	Shin *et al.* (2008)
Mycobacterium fortuitum	90	66	PBW	USEPA (2002)
Mycobacterium fortuitum	99	103	PBW	USEPA (2002)
Mycobacterium intra-cellulare	90	7	PBW	USEPA (2002)
Mycobacterium intra-cellulare	99	25	PBW	USEPA (2002)
Burkholderia pseudomallei	98	90		Howard *et al.* (2005)
Viruses				
Adenovirus 40 ATCC Dugan	90	29.5	SDW	Meng *et al.* (1996)
Adenovirus 40 ATCC Dugan	99	59.4	SDW	Meng *et al.* (1996)
Adenovirus 40 ATCC Dugan	99.9	89.8	SDW	Meng *et al.* (1996)
Adenovirus 40 ATCC Dugan	99.99	120.9	SDW	Meng *et al.* (1996)
Adenovirus 41 ATCC TAK	90	23.6	SDW	Meng *et al.* (1996)
Adenovirus 41 ATCC TAK	99	50	SDW	Meng *et al.* (1996)
Adenovirus 41 ATCC TAK	99.9	88	SDW	Meng *et al.* (1996)
Adenovirus 41 ATCC TAK	99.99	111.8	SDW	Meng *et al.* (1996)
Adenovirus 40	90	58	PBW	Jacangelo *et al.* (2002)
Adenovirus 40	99	111	PBW	Jacangelo *et al.* (2002)
Adenovirus 40	99.9	170	PBW	Jacangelo *et al.* (2002)

Pathogen	Reduction (%)	UV Dose (mWs/cm^2)	Water source	Reference
Viruses				
Adenovirus 2	90	78	PBW	USEPA (2002)
Adenovirus 2	99.9	119	PBW	USEPA (2002)
Adenovirus	99.99	186	TAP	Malley (2000)
Adenovirus 2, 15, 40, 41	90	42 (low pressure)		Hijnen *et al.* (2006)
Adenovirus 2, 41	90	25 (medium pressure)		Hijnen *et al.* (2006)
Feline calicivirus	90	9	BDF	Jacangelo *et al.* (2002)
Feline calicivirus	90	16.8	PBW	USEPA (2002)
Feline calicivirus	99	17	BDF	Jacangelo *et al.* (2002)
Feline calicivirus	99	22		Duizer *et al.* (2004)
Feline calicivirus	99.9	25.2	PBW	USEPA (2002)
Feline calicivirus	99.9	34		Duizer *et al.* (2004)
Feline calicivirus	99.9	26	BDF	Jacangelo *et al.* (2002)
Feline calicivirus	99.9	12		De Roda Husman *et al.* (2004)
Feline calicivirus	99.99	38		Hijnen *et al.* (2006)
Bovine calicivirus	99.99	21		Hijnen *et al.* (2006)
Canine calicivirus	99.99	41		Hijnen *et al.* (2006)
Canine calicivirus	90	21		Duizer *et al.* (2004)
Canine calicivirus	99.9	34		Duizer *et al.* (2004)
Canine calicivirus	99.9	20		De Roda Husman *et al.* (2004)
Murine norovirus	90	10	PBS	Lee *et al.* (2008)
Murine norovirus	99.84	20	PBS	Lee *et al.* (2008)
Murine norovirus	99.95	25	PBS	Lee *et al.* (2008)
Murine norovirus	99.999	30	PBS	Lee *et al.* (2008)
Poliovirus 1 LSc2ab	90	4	SDW	Meng *et al.* (1996)
Poliovirus 1 LSc2ab	99	9	SDW	Meng *et al.* (1996)
Poliovirus 1 LSc2ab	99.9	12	SDW	Meng *et al.* (1996)
Poliovirus 1 LSc2ab	99.99	21	SDW	Meng *et al.* (1996)
Poliovirus 1 ATCC Mahoney	90	6		Harris *et al.* (1987)
Poliovirus 1 ATCC Mahoney	99	14		Harris *et al.* (1987)
Poliovirus 1 ATCC Mahoney	99.9	23		Harris *et al.* (1987)
Poliovirus 1 ATCC Mahoney	99.99	30		Harris *et al.* (1987)
Poliovirus 1 Lsc2ab	90	5.6		Chang *et al.* (1985)
Poliovirus 1 Lsc2ab	99	11		Chang *et al.* (1985)
Poliovirus 1 Lsc2ab	99.9	16		Chang *et al.* (1985)
Poliovirus 1 Lsc2ab	99.99	22		Chang *et al.* (1985)
Poliovirus 1	99.99	30		Hijnen *et al.* (2006)
Coxsackie A	99.99	30		Malley (2000)
Coxsackie B5	90	6.9	PBW	Battigelli *et al.* (1993)
Coxsackie B5	99	14	PBW	Battigelli *et al.* (1993)
Coxsackie B5	99.9	21	PBW	Battigelli *et al.* (1993)
Coxsackie B5	90	18	PBW	USEPA (2002)
Coxsackie B5	99.9	27	PBW	USEPA (2002)
Coxsackie B5	99.99	34		Hijnen *et al.* (2006)
Coxsackie B3	90	16	PBW	USEPA (2002)
Coxsackie B3	99.9	24.5	PBW	USEPA (2002)
Coxsackie A9	99	1.66	RSW	Hill *et al.* (1970)
Echovirus 1	99	1.25	RSW	Hill *et al.* (1970)

Pathogen	Reduction (%)	UV Dose (mWs/cm^2)	Water source	Reference
Viruses				
Echovirus 11	99	1.41	RSW	Hill *et al.* (1970)
Rotavirus WA	99.99	50	TAP	Malley (2000)
Rotavirus SA2	99.99	40	TAP	Malley (2000)
Rotavirus SA11	90	8		Montiel *et al.* (1990)
Rotavirus SA11	90	7.6		Battigelli *et al.* (1993)
Rotavirus SA11	99	15		Battigelli *et al.* (1993)
Rotavirus SA11	99.9	23		Battigelli *et al.* (1993)
Rotavirus SA11	90	7.1	PBS	Chang *et al.* (1985)
Rotavirus SA11	99	15	PBS	Chang *et al.* (1985)
Rotavirus SA11	99.9	23	PBS	Chang *et al.* (1985)
Rotavirus SA11	90	9.1		Wilson *et al.* (1992)
Rotavirus SA11	99	19		Wilson *et al.* (1992)
Rotavirus SA11	99.9	26		Wilson *et al.* (1992)
Rotavirus SA11	99.99	36		Wilson *et al.* (1992)
Rotavirus SA11	99.99	39		Hijnen *et al.* (2006)
Hepatitis A (with safety factor)	99	21		SWTR (1991)
Hepatitis A (with safety factor)	99.9	36		SWTR (1991)
Hepatitis A (without safety factor)	99	7		SWTR (1991)
Hepatitis A (without safety factor)	99.9	12		SWTR (1991)
Hepatitis A (HM175)	90	4		Battigelli *et al.* (1993)
Hepatitis A (HM175)	99	8		Battigelli *et al.* (1993)
Hepatitis A (HM175)	99.9	12		Battigelli *et al.* (1993)
Hepatitis A (HM175)	99.99	16		Battigelli *et al.* (1993)
Hepatitis A (HM175)	90	5.1		Wilson *et al.* (1992)
Hepatitis A (HM175)	99	14		Wilson *et al.* (1992)
Hepatitis A (HM175)	99.9	22		Wilson *et al.* (1992)
Hepatitis A (HM175)	99.99	30		Wilson *et al.* (1992)
Hepatitis A	90	5.5		Wiedenmann *et al.* (1993)
Hepatitis A	99	9.8		Wiedenmann *et al.* (1993)
Hepatitis A	99.9	15		Wiedenmann *et al.* (1993)
Hepatitis A	99.99	21		Wiedenmann *et al.* (1993)
Hepatitis A	99.99	6–15	GW	Malley (2000).
Hepatitis A	99.99	16	TAP	Malley (2000)
Hepatitis A	99.99	22		Hijnen *et al.* (2006)
Protozoa				
Cyptosporidium	99.99	123		Clancy *et al.* (1998)
Cryptosporidium	99.9	12		Hijnen *et al.* (2006)
Cryptosporidium parvum	97	3	PBW	Keegan *et al.* (2003)
Cryptosporidium parvum	98.42	6	PBW	Keegan *et al.* (2003)
Cryptosporidium parvum	98.3	12	PBW	Keegan *et al.* (2003)
Cryptosporidium parvum	99.3	20	PBW	Keegan *et al.* (2003)
Cryptosporidium parvum	99.34	30	PBW	Keegan *et al.* (2003)
Cryptosporidium parvum	99.98	500	PBW	Keegan *et al.* (2003)
Cryptosporidium parvum	99.99	1000	PBW	Keegan *et al.* (2003)
Cryptosporidium hominis	90	3 (medium pressure)		Johnson *et al.* (2005)
Giardia lambia	99	<1		Linden *et al.* (2001)
Giardia muris	99	<6		Craick *et al.* (2000)

Pathogen	Reduction (%)	UV Dose (mWs/cm^2)	Water source	Reference
Protozoa				
Giardia	99.9	11		Hijnen *et al.* (2006)
Encephalitozoon intestinalis	90	2	BDF	Jacangelo *et al.* (2002)
Encephalitozoon intestinalis	99	5	BDF	Jacangelo *et al.* (2002)
Encephalitozoon intestinalis	99.9	7	BDF	Jacangelo *et al.* (2002)
Encephalitozoon intestinalis	99.99	9	BDF	Jacangelo *et al.* (2002)
Encephalitozoon intestinalis	90	2.2	PBW	USEPA (2002)
Encephalitozoon intestinalis	99.9	8.4	PBW	USEPA (2002)
Encephalitozoon intestinalis	99.9	8.43	PBW	John *et al.* (2003)
Encephalitozoon intestinalis	96–99	3		Huffman *et al.* (2001)
Encephalitozoon intestinalis	>99.99	9		Huffman *et al.* (2001)
Acanthamoeba castellanii	90	35		Montiel *et al.* (1990)
Acanthamoeba castellanii	99	70	SBW	Chang *et al.* (1985)
Acanthamoeba spp.	90	40		Hijnen *et al.* (2006)

ASE: Activated Sludge Effluent
BBW: Borate Buffered Water
BDF: Buffer Demand Free
CDF: Chlorine Demand Free
DW: Distilled Water
GM: Growth Medium
GW: Ground Water
HTE: Human Tank Effluent
ODF: Ozone Demand Free
PBS: Phosphate Buffered Saline
PBW: Phosphate Buffered Water
RSW: Raw Surface Water
SBF: Sterile Buffered Water
SDW: Sterile Distilled Water
TAP: Tap Water

Treatment data: References

Abad, F.X. and Pinto, R.M. *et al.* (1994). Disinfection of Human Enteric Viruses in Water by Copper and Silver in Combination with Low Level Chlorine. *Appl. Environ. Microbiol.*, **60**(7), 2377–2383.

Andriamirado, L. and Julve, J.M. *et al.* (2005). *MEMENTO TECHNIQUE DE L'EAU 10EME EDITON.* paris.

Andrissen, H. (1985). *Le Traitement des Eaux de Recyclage industrielles par Rayonnement Ultraviolet.* 38ème journée Internationale de CEBEDEAU, Bruxelles.

Assavasilavasukul, P., Lau, B.L., Harrington, G.W., Hoffman, R.M. and Borchardt, M.A. (2008). Effect of pathogen concentrations on removal of *Cryptosporidium* and *Giardia* by conventional drinking water treatment. *Water Res.*, **42**(10–11), 2678–90.

Baker, K.H. and Hegarty, J.P. *et al.* (2002). Effect of Oxiding Disinfectants(Chlorine, Monochloramine, and Ozone) on Helicobacter Pylori. *Appl. Environ. Microbiol.*, **68**(2), 981–984.

Battigelli, D.A. and Sobsey, M.D. *et al.* (1993). The Inactivation of Hepatitis A Virus and Other Model Viruses by UV Irradiation. *Wat. Sci. Tech.*, **27**(3–4), 339–342.

Berman, D. and Hoff, J.C. (1984). Inactivation of Simian Rotavirus SA11 by Chlorine, Chlorine dioxide and Chloramine. *Appl. Environ. Microbiol.*, **48**, 317–323.

Bernade, M.A. and Snow, W.B. *et al.* (1967). Chlorine Dioxide-Disinfection Temperature Effects. *J. Appl. Bacteriol.*, **30**, 159–169.

Butterfield, C.T. and Wattie, E. (1946). Influence of PH and temperature on the survival of coliforms and enteric pathogens when exposed to chloramines. *Publ. Health Repts.*, **61**, 157–192.

Chang, J.C. and Osoff, S.F. *et al.* (1985). UV Inactivation of Pathogenic and Indicator Microorganisms. *Appl. Environ. Microbiol.*, **49**(6), 1361–1365.

Chang, S.L. and Stevenson, R.E. *et al.* (1958). Removal of Coxsackie and Bacterial Viruses in Water of known Chemical Content by Flocculation with Aluminum Sulfate or Ferric Chloride under Various Testing Conditions. *Am. J. Public Health*, **48**(1), 51–61.

Chauret, C. and Miller, T. *et al.* (2003). *Inactivation by Chlorine and Monochloramine of Environmental and Culture Collection Strains of Escherichia coli O157:H7 and selected Heterotrophic Bacteria.* Water Quality Technology Conference, Philadelphia, PA.

Clancy, J.L. and Bolton, J.B. *et al.* (1998). *Inactivation of cryptosporidium Oocysts Using Medium-Pressure Ultraviolet Light in Bench and pilot study scale Studies.* Water Quality and Technology Conference, San Diego, USA.

Clancy, J.L. and Hargy, T.M. *et al.* (1998). UV Light Inactivation of Cryptosporidium Oocysts. *J. Am. Water Works Assoc.*, **90**(9), 92–102.

Craick, S.A., Finch, G.R., Bolton, J.R. and Belosevic, M. (2000). Inactivation of Giardia muris cysts using medium pressure ultraviolet radiation in filtered drinling water. *Water Research*, **34**(18), 4325

Crawford, R.A. (1988). Halamine water disinfection. *Critical Review Environ. Controle.*, **18**(2), 133–175.

Cronier, S.D. (1977). Destruction by Chloride Dioxide of Viruses and Bacteria in Water. *Thesis.* Cincinnati, University of Cincinnati.

Cursons, R.T.M. and Brown, T.J. *et al.* (1980). Effect of Disinfectants on Pathogenic Free-Living Amoebae: in Axenic Conditions. *Appl. Environ. Microbiol.*, **40**(1), 62–66.

David, H.L. (1973). Response of Mycobacteria to Ultraviolet Light Reduction. *Am. Review of Respiratory Disease*, **108**, 1175–1185.

DeJonckheere, J. and Van DeVoorde, H. (1976). Differences in Destruction of Cysts of Pathogenic and nonpathogenic Naegleria and Acanthamoeba by Chlorine. *Appl. Environ. Microbiol.*, **31**, 294–297.

DeLeon, R. and Singh, S.N. *et al.* (1986). Microorganism Removal from Wastewater by Rapid mixed Media Filtration. *Wat. Res.*, **20**(5), 583–587.

DeRoda Husman, A.N. and Bijkerk, P. *et al.* (2004). Calicivirus Inactivation by Nonionizing (253.7-Nanometer-Wavelength [UV] and Ionizing (Gamma) Radiation. *Appl. Environ. Microbiol.*, **70**(9), 5089–5093.

Duizer, E. and Bijkerk, P. *et al.* (2004). Inactivation of Caliciviruses. *Appl. Environ. Microbiol.*, **70**(8), 4538–4543.

Emerson, M.A. and O.J. Sproul, *et al.* (1982). Ozone inactivation of cell-associated viruses. *Appl. Environ. Microbiol.*, **43**, 603–608.

Engelbrecht, R.S. and Severin, B.F. *et al.* (1977). New Microbial Indicators of Disinfection Efficiency. *EPA-600/2–052.* Cincinnati, O.H., U.S. Environmental Protection Agency.

Engelbrecht, R.S. and Weber, M.J. *et al.* (1980). Comparative inactivation of viruses by chlorine. *Appl. Environ. Microbiol.*, **40**, 249–256.

Esposito, P.P. (1974). The inactivation of virus in water by dichloramines. Ohio, University Cincinnati. M.S.c, **121**.

Farooq, S. and Chain, E.S.K. *et al.* (1977). Basic Concepts in Disinfection with Ozone. *J. Water Pollut. Control Fed.*, **49**, 1818–1831.

Finch, G.R. and Black, E.K. *et al.* (1993). Ozone inactivation of Cryptosporidium parvum in demand-free phosphate buffer determined by in vitro excystation and animal infectivity. *Appl. Environ. Microbiol.*, **59**(12), 4203–4210.

Finch, G.R. and Kathleen, B. *et al.* (1994). *Ozone and Chlorine Inactivation of Cryptosporidium.* Water Quality Technology Conference, San Francisco, CA, USA, American Water Works Association.

Gerba, C.P. and Riley, K.R. *et al.* (2003). Removal of Encephalitozoon Intestinalis, Calicivirus, and Coliphases by Conventional Drinking Water Treatment. *J. Environ Sci Health A Tox Hazaed Subst Environ Eng.*, **38**(7), 1259–1268.

Gitis, V. (2008) Rapid sand filtration of Cryptosporidium parvum: effects of media depth and coagulation. *Water Science and Technology: Water Supply*, **8**(2), 129–34.

Grabow, W.O. and Coubrough, K.P. *et al.* (1984). Inactivation of hepatitis A virus, other enteric viruses and indicator organisms in water by chlorination. *Wat. Sci. Tech.*, **17**, 657–664.

Hall, R.M. and Sobsey, M.D. (1993). Inactivation of Hepatitis A Virus and MS2 by Ozone and Ozone-Hydrogen peroxide in Buffered Water. *Wat. Sci. Tech.*, **27**(3/4), 371–378.

Harakeh, M.S. and Butler, M. (1985). Factors influenting the ozone inactivation of enteric viruses in effluent. *Ozone Sci. & Eng.*, **6**, 235–243.

Harrington, G.W. and Xagoraraki, I. *et al.* (2000). *Removal of Emerging Waterborne Pathogens by Pilot-Scale Conventional Treatment.* Water Quality Technology Conference, Salt Lake city, USA.

Harrington, G.W. and Xagoraraki, I. *et al.* (2003). Effect of Filtration Conditions on Removal of Emerging Waterborne Pathogens. *JAWWA*, **95**(12), 95–104.

Harris, G.D. and Adams, V.D. *et al.* (1987). Ultraviolet inactivation of selected bacteria and virus with photoreactivation of the bacteria. *Wat. Res.*, **21**(6), 687–692.

Hayes, S.L., White, K.M. and Rodgers, M.R. (2006) Assessment of the effectiveness of low-pressure UV light for inactivation of Helicobacter pylori. *Applied and Environmental Microbiology*, **72**(5), 3763–65.

Hendricks, D.W., Clunie, W.F., Sturbaum, G.D., Klein, D.A., Champlin, T.L., Kugrens, P., Hirsch, J., McCourt, B., Nordby, G.R., Sobsey, M.D., Hunt, D.J. and Allen, M.J. (2005) Filtration removals of microorganisms and particles. *Journal of Environmental Engineering*, **131**(12), 1621–32.

Hijnen, W.A.M., Beerendonk, E.F. and Medema, G.J. (2006) Inactivation credit of UV radiation for viruses, bacteria and protozoan (oo)cysts in water: a review. *Water Research*, **40**, 3–22.

Hill, W.F. and Hamblet, F.E. *et al.* (1970). Ultraviolet Devitalization of Eight Selected Enteric Viruses in Estuarine Water. *Appl. Microbiol.*, **19**, 805–812.

Hoff, J.C. and Geldreich, E.E. (1981). Comparison of the Biocidal Efficiency of Alternative Disinfectants. *JAWWA*, **73**(1), 40–44.

Hoffmann, R. and Michel, R. (2001). Distribution of free-living amoeba (FLA) during preparation and supply of drinking water. *Int. J. Hyg. Environ. Health*, **203**, 215–219.

Howard, K. and Inglis, T.J.J. (2005). Disinfection of burkholderia pseudomallei in potable water. *Wat. Res.*, **39**, 1085–1092.

Huffman, D.E. and Gennaccaro, A. *et al.* (2001). Low- and Medium-Pressure UV Inactivation of Microsporidia Encephalitozoon Intestinalis. *Wat. Res.*, **36**(12), 3161–3164.

Jacangelo, J.G. and Patania, N.L. *et al.* (2002). Inactivation of waterborne Emerging Pathogens by selected Disinfectant. 90886, AWWARF.

John, D.E. and Haas, C.N. *et al.* (2005). Chlorine and Ozone Disinfection of Encephalitozoon Intestinalis Spores. *Wat. Res.*, **39**, 2369–2375.

John, D.E. and Nwachuku, N. *et al.* (2003). Development and Optimization of a Quantitative Cell Culture Infectivity assay for the Microsporidium Encephalitozoon intestinalis and Application to Ultraviolet Light Inactivation. *J. Microbiol Methods.*, **52**(2), 183–196.

Johnson, A.M. and Linden, K. *et al.* (2005). UV Inactivation of Cryptosporidium Homonis as Measured in Cell Culture. *Appl. Environ. Microbiol.*, **71**(5), 2800–2802.

Johnson, C.H. and Marshall, M.M. *et al.* (2003). Chlorine Inactivation of Spores of Encephalitozoon spp. *Appl. Environ. Microbiol.*, **69**(2), 1325–1326.

Johnson, C.H. and Rice, E.W. *et al.* (1997). Inactivation of Helicobacter pylori by chlorination. *Appl. Environ. Microbiol.*, **63**, 4969–4970.

Keegan, A.R. and Fanok, S. *et al.* (2003). Cell Culture-Taqman PCR Assay for Evaluation of Cryptosporidium parvum Disinfection. *Appl. Environ. Microbiol.*, **69**(5), 2505–2511.

King, C.H. and Shorts, E.B. *et al.* (1988). Survival of Coliforms and Bacterial Pathogens within Protozoa during Chlorination. *Appl. Environ. Microbiol.*, **54**(12), 3023–3033.

Knochel, S. (1991). Chlorine Resistance of Motile Aeromonas Spp. *Wat. Sci. Tech.*, **24**(2), 327–330.

Korich, D.G. and Mead, J.R. *et al.* (1990). Effects of ozone, chlorine dioxide, chlorine and monochloramine on Cryptosporidium parvum oocyst viability. *Appl. Environ. Microbiol.*, **56**(5), 1423–1428.

Korol, S., Fortunato, M.S. *et al.* (1995). Desinfeccion de Agua: Accion Comparativa del Ozono y Cloro sobre un Amplio Espectro Bacteriano. *Revista Argentina de Microbiologia*, **27**, 175–183.

Langlais, B. and Pierre, D. (1986). Action of Ozone on Trophozoites and Free Amoeba Cysts, Whether Pathogenic or not. *Ozone Sci. Eng.*, **8**, 187–198.

Leahy, J.G. (1985). Inactivation of Giardia Muris cysts by Chlorine and Chlorine dioxide. *Department of Civil Engineering*. Columbus, Ohio State University. M.Sc. Thesis, **115**.

Leahy, J.G. and Rubin, A.J. *et al.* (1987). Inactivation of Giardia muris cysts by free chlorine. *Appl. Environ. Microbiol.*, **53**, 1448–1453.

LeChevallier, M.W. and Cawthon, C.D. *et al.* (1988). Factors promoting survival of bacteria in chlorinated water supplies. *Appl. Environ. Microbiol.*, **54**(3), 649–654.

LeChevallier, M.W. and Norton, W.D. *et al.* (1991). Giardia and Cryptosporidium in Water Supplies. 90583, AWWARF.

Lee, J.E., Zoh, K.D. and Ko, G.P. (2008) Inactivation and UV disinfection of murine norovirus with TiO2 under various environmental conditions. *Applied and Environmental Microbiology*, **74**(7), 2111–17.

Lezcano, I., Rey, R.P. *et al.* (1999). Ozone Inactivation of Pseudomonas aeruginosa, Escherichia coli, Shigella sonnei and Salmonella typhimurium in water. *Ozone Sci. & Eng.*, **21**, 293–300.

Li, J.W. and Xin, Z.T. *et al.* (2004). Mechanisms of Inactivation of Hepatitis A Virus in Water by chlorine dioxide. *Wat. Res.*, **38**, 1514–1519.

Li, L. and Haas, C.N. (2004). Inactivation of Cryptosporidium parvum with ozone in treated drinking water. *Journal of Water Supply: Research and Technology-AQUA*, **53**(5), 287–297

Linden, K.G., Shin, A., Faubert, G., Cairns, W.L. and Sobsey, M.D. (2001). UV disinfection of Giardia lamblia cysts in water. Environmental Science and Technology, **36**(11), 2519–2522.

Liu, O.C. and H.R. Seraichekas, *et al.* (1971). *Relative Resistance of 20 Humain Enteric Viruses to Free Chlorine in Potomac River*. 13th Water Quality Conference, Univ. of Illinois Bulletin.

Loret, J.F., Jousset, M., Robert, S., Anselme, C., Saucedo, G., Ribas, F., Martinez, L. and Catalan, V. (2008). Elimination of free-living amoebae by drinking water treatment processes. *European Journal of Water Quality*, **39**(1), 37–50.

Malley, J.J. (2000). Ultraviolet disinfection in control of Microbes in Drinking water. *American Society of Civil Engineering*.

Massa, S. and Armuzzi, R. *et al.* (1999). Susceptibility to chlorine of Aeromonas hydrophila strains. *J. Appl. Microbiol.*, **86**, 169–173.

Medema, G.J. and Wondergem, E. *et al.* (1991). Effectivity of Chlorine Dioxide to Control Aeromonas in Drinking Water Distribution Systems. *Wat. Sci. Tech.*, **24**(2), 325–326.

Meng, Q.S. and Gerba, C.P. (1996). Comparative Inactivation of Enteric Adenovirus, Poliovirus and Coliphages by Ultraviolet Irradiation. *Wat. Res.*, **30**(11), 2665–2668.

Montiel, A. and Welte, B. (1990). *LA DESINFECTION*. Paris, sociéte Anonyme de Gestion des Eaux de Paris.

Nieminski, E.C. (1994). *Giardia and Cryptospordium Cysts Removal Through Direct Filtration and Conventional Treatment*. AWWA Annual Conference, New York City, NY.

Ongerth, J.E. and Pecoraro, J.P. (1995). Removing Cryptosporidium using multimedia filters. *JAWWA*, **87**(12), 83–89.

Owens, J.H. (1994). *Pilot-Scale Ozone Inactivation of Cryptosporidium and Giardia*. Water Quality Technology Conference, San Francisco, CA, USA, American Water Works Association.

Patania, N.I. and Jacangelo, J.G. *et al.* (1995). Optimization of filtration for cyst removal. 90699, AWWARF.

Payment, P. and Trudel, M. *et al.* (1985). Elimination of Viruses and Indicator Bacteria at Each Step of Treatment during Preparation of Drinking Water at Seven Water Treatment Plant. *Appl. Environ. Microbiol.*, **49**(6), 1418–1428.

Pelletier, P.A. and Dumoulin, G.C. (1987). *Mycobacteria in Public Water Supplies: Comparative Resistance to Chlorine*. Emerging Issue in Effluent Disinfection Workshop, Water Pollution Control Federation, Washington, DC.

Plummer, J.D. and Edzwald, J.K. *et al.* (1995). Removing Cryptosporidium by dissolved-air flotation. *JAWWA*, **87**(9), 85–95.

Ransome, M.E. and Whitmore, T.N. *et al.* (1993). Effect of disinfectants on the viability of Cryptosporidium parvum. *Water Supply*, **11**, 75–89.

Rice, E.W. and Hoff, J.C. *et al.* (1982). Inactivation of Giardia cysts by Chlorine. *Appl. Environ. Microbiol.*, **43**, 250–251.

Roy, D., Engelbrecht, R.S. *et al.* (1982). Comparative Inactivation of Six Enteroviruses by Ozone. *JAWWA*, **74**, 660–664.

Rubin, A.J. (1988). *Factors Affecting the Inactivation of Giardia Cysts by Monochloramine and Comparison with other Disinfectants.* Conference on Current Research in Drinking Water Treatment, Cincinnati, OH, USA.

Rubin, A.J. and Engel, J.P. *et al.* (1983). Disinfection of Amoebic Cysts in Water with Free Chlorine. *J. Water Pollut. Control Fed.*, **55**, 1174–1182.

SDWC–Safe Drinking Water Committee (1980). drinking water and Health. *National Acad. Sci.*, **2** (the disinfection of drinking water), 5–137.

Shin, G.A., Lee, J.K., Freeman, R. and Cangelosi, G.A. (2008). Inactivation of Mycobacterium avium complex by UV irradiation. *Applied and Environmental Microbiology*, **74**(22), 7067–69.

Shin, G.A. and Sobsey, M.D. (2003). Reduction of Norwalk virus, poliovirus 1, and bacteriophage MS2 by ozone disinfection of water. *Applied and Environmental Microbiology*, **69**, 3975–78.

Shin, G.A. and Sobsey, M.D. (2008). Inactivation of norovirus by chlorine disinfection of water. *Water Research*, **42**, 4562–68.

Sirikanchana, K. and Raskin, L. *et al.* (2002). *Inactivation Kinetics of Aeromonas Hydrophila with Monochloramine.* Water Quality Technology Conference, Seattle, USA, American Water Works Association.

Sisti, M. and Albano, A. *et al.* (1998). Bactericidal Effect of Chlorine on motile Aeromonas spp. in Drinking Water Supplies and Influence of Temperature on disinfection efficacy. *Letters Appl. Microbiol.*, **26**, 347–351.

Sobsey, M.D. and Fuji, T. *et al.* (1988). Inactivation of hepatitis A virus and model viruses in water by free chlorine and monochloramine. *Wat. Sci. Tech.*, **20**, 385–391.

Sobsey, M.D. and Takashi, F. *et al.* (1990). *Inactivation of Cell-associated and Dispersed Hepatitis A Virus by Free and Combined Chlorine and Chlorine Dioxide.* Water Quality Technology Conference, Philadelphia, PA, USA, AWWA.

Stetler, R. and Ward, R.L. *et al.* (1984). Enteric Virus and Indicator Levels in a Water Treatment System modified to Reduce Trihalomethane Production. *Appl. Environ. Microbiol.*, **47**(2), 319–324.

SWTR. (1991). Surface Water Treatment Rule. Guidance manual for compliance with the filtration and disinfection requirements for public water systems using surface water sources. USA–AWWA. Malcolm Pirnie and HDR Engineering, in: Jacangelo *et al.* (2002). Inactivation of waterborne emerging pathogens by selected disinfectants. AWWARF report n°90886.

Taylor, R.H. and Falkinham, J.O. III, *et al.* (2000). Chlorine, Chloramine, Chlorine Dioxide, and Ozone Susceptibility of Mycobacterium avium. *Appl. Environ. Microbiol.*, **66**(4), 1702–1705.

Thurston-Enriquez, J.A. and Haas, C.N. *et al.* (2003). Chlorine Inactivation of Adenovirus Type 40 and Feline Calicivirus. *Appl. Environ. Microbiol.*, **69**(7), 3979–3985.

Thurston-Enriquez, J.A., Haas, C.N., Jacangelo, J. and Gerba, C.P. (2005a) Inactivation of enteric adenovirus and feline calicivirus by ozone. *Water Research*, **39**, 3650–56.

Thurston-Enriquez, J.A., Haas, C.N., Jacangelo, J. and Gerba, C.P. (2005b) Inactivation of enteric adenovirus and feline calicivirus by chlorine dioxide. *Applied and Environmental Microbiology*, **71**(6), 3100–05.

Tosa, K. and Hirata, T. (1998). Photoreactivation of Salmonella following UV disinfection. *Health-Related Water Microbiology*, **10**, 168–174.

USEPA. (1989). Drinking water; national primary drinking water regulation; filtration, disinfection; turbidity, Giardia lamblia, viruses, Legionella, and heterotrophic bacterial; final rule. *Fed. Regist.*, **54**, 27486–27541.

USEPA. (2002). Treatability Screening for CCL Microbial Contaminants. Washington, D.C., United States Environmental Protection Agency.

Vaughn, J.M. and Chen, Y.S. *et al.* (1987). Inactivation of Human and Siman Rotaviruses by Ozone. *Appl. Environ. Microbiol.*, **53**, 2218–2221.

Vicuña-Reyes, J.P., Luh, J. and Mariñas, B.J. (2008) Inactivation of Mycobacterium avium with chlorine dioxide. *Water Research*, **42**, 1531–38.

Ward, N.R. and Wolte, R.L. *et al.* (1982). *The Inactivation of Total Count and Selected Gram Negative Bacteria by Inorganic Monochloramines and Dichloramines.* Water Quality Technology Conference, Nashville, Tenn.

Warriner, T.R. (1967). Inaktwering av poliovirus med klordioxid. *publikation A67*, Institute for vattenforsornging och avlopp.: 284–290.

White, G.C. (1999). *Handbook of Chlorination and Alternative Disinfectants*. New York, John Wiley and Sons, Inc.

Wickramanayake, G.B. and Rubin, A.J. *et al.* (1984). Inactivation of Naegleria and Giardia cysts in Water by Ozonation. *J. Water Pollut. Control Fed.*, **56**, 983.

Wiedenmann, A. and Fischer, B. *et al.* (1993). Disinfection of Hepatitis A virus and MS-2 coliphage in water by ultraviolet irradiation: Comparison of UV-susceptibility. *Wat. Sci. Tech.*, **27**(3–4), 335–338.

Wilson, B.R. and Roessler, P.F. *et al.* (1992). *Coliphage MS-2 as a UV Water Disinfection Efficacy Test Surrogate for Bacterial and Viral Pathogens*. Water Quality Technology Conference, Toronto.

Wolf, H.W. and Safferman, R.S. *et al.* (1974). Virus Inactivation During Tertiary Treatment. *JAWWA*, **66**(9), 526–531.

Wolk, D.W. and Johnson, C.H. *et al.* (2000). A Spore Counting Method and Cell Culture Model for Chlorine Disinfection Studies of Encephalitozoon syn. Septata intestinalis. *Appl. Environ. Microbiol.*, **66**(4), 1266–1273.

Xagoraraki, I. and Harrington, G.W. *et al.* (2004). Removal of Emerging Waterborne Pathogens and Pathogen Indicators by Pilot-scale Conventional Treatment. *JAWWA*, **96**(5), 102–113.

Yu, M.J., Ahn, S.K., Chung, S.H., Noh, S., Park, J.A., Rhim, Y.T., Jheong, W.H. and Chung, H.M. (2006). Evaluation of the rapid filtration system with particle size distribution and Cryptosporidium in different operating conditions. *Water Science and Technology: Water Supply*, **6**(1), 129–39.

IWA Publishing's authorised EU representative for General Product Safety Regulations is Diane D'Arras, 15 rue Duret, 75116 Paris, France, e-mail: safety@iwap.co.uk.

Printed and bound by CPI Group (UK) Ltd, Croydon, CR0 4YY
13/05/2026
02109604-0003